113 Advances in Polymer Science

Membrane-Mimetic Approach to Advanced Materials

By J.H. Fendler

With 134 Figures and 11 Tables

Springer-Verlag
Berlin Heidelberg GmbH

ISBN 978-3-662-14907-2   ISBN 978-3-540-47992-5 (eBook)
DOI 10.1007/978-3-540-47992-5

© Springer-Verlag Berlin Heidelberg 1994
Originally published by Springer-Verlag Berlin Heidelberg New York in 1994
Softcover reprint of the hardcover 1st edition 1994
Library of Congress Catalog Card Number 61-642

Typesetting: Macmillan India Ltd., Bangalore-25

SPIN: 10126167   02/3020 - 5 4 3 2 1 0 - Printed on acid-free paper

Editors

Preface

All creativity is an escape from the restraints of the conscious mind.
Arthur Koestler

The demand for advanced materials with superior mechanical, thermal, electrical, optical, magnetic, electro-optical, and electromagnetic properties is ever increasing. Most advanced materials, until recently, have been formed empirically by solid state methods. Further progress in the generation of advanced materials with preselected properties demands innovative chemical tailoring and, thus, a fundamental understanding of interactions and reactions at atomic, molecular, and supramolecular levels.

This contribution advocates a „wet" colloid-chemical approach, based on membrane-mimetic chemistry, to the preparation of advanced materials. Only the essential functions of the biological membrane - molecular organization, compartmentalization and discrimination - are imitated in membrane-mimetic chemistry. Membrane-like compartments are constructed and are employed in the in situ generation and stabilization of advanced materials. The membrane-mimetic approach is analogous to and inspired by biomineralization - the in vivo formation of inorganic crystals and/or amorphous particles in biological systems.

The importance of advanced materials and the rationale and philosophy of a membrane-mimetic approach to advanced materials are detailed in the Introduction. The Introduction also contains definitions of the terms used and the scope of the monograph. Preparation an characterization of the different membrane-mimetic compartments are described in the second section of the treatment. Merits and disadvantages of monolayers (Langmuir films), Langmuir-Blodgett (LB) films, self-assembled (SA) films, aqueous micelles, reversed micelles, micro-emulsions, surfactant vesicles, polymerized vesicles, polymeric vesicles, tubules, rods and related self-assembled structures, bilayer lipid membranes (BLMs), cast bilayers, polymers and polymeric membranes, zeolites, clays, porous glasses, and proteins are discussed in this section. The exploitation of these systems in the preparation of metallic catalytic particles, semiconductor particles and particulate films, conductors and superconductors, magnetic particles and particulate films, and advanced ceramic materials and their subsequent characterization and utilization are discussed in Sec. 3, 4, 5, 6, and 7, respectively. Ample references have been provided throughout this review to primary and secondary publications.

Decidated to my guiding spirit whose shadow I am.

Janos H. Fendler

Table of Contents

1 Introduction and Scope

1.1 *What Are Advanced Materials? – Definitions*

Advanced materials are vital to daily life in our world of high technology [1, 2]. The notebook computer I am drafting this very sentence on illustrates the marvel of advanced materials engineering. It is smaller than a typical issue of the *Journal of Chemical Physics* and weighs less than five pounds; yet it outperforms the very much bulkier personal computers used only a few years ago, let alone the "old" IBM electric typewriters.

The dramatic increase in the number of advanced materials described in recent materials science literature has been accompanied by the generation of a plethora of terminology and jargon. Advanced, smart, adaptive, intelligent, nanosized, nanostructured, nanofabricated, hierarchically organized, artificially structured, genetically engineered, self-organized, and supramolecular materials are all terms which are in common usage. Unfortunately, definitions are relative. We are all slaves of our time, scientific training, and experience. Physicists, chemists, biologists, and engineers often use their own language to describe the same phenomenon in materials science. Consequently, every effort will be made to utilize consistent and comprehensible definitions in the following text.

Advanced materials are the subject of the present monograph. The desired properties of *modern and advanced materials* are vastly superior to those which were available some 60 to 80 years ago [1]. The following examples serve to illustrate this point:

- the strength-to-density ratio of carbon fibers is some fifty-fold better than that of a cast iron rod;
- the efficiency of heat to mechanical energy conversion has been increased from 60% to 80% through the use of high-temperature ceramic turbo engines in place of traditional steam engines;
- the operating temperature of superconductors has increased from 23 K to 125 K and there is a every hope for the appearance of a material which will be superconductive at room temperature; and
- most significantly, the number of components per chip has increased exponentially to millions from that of hundreds in the 1960s.

Advanced materials have, of course, many other beneficial properties. Importantly, new advanced materials with unique properties are continuously being developed for novel, and as yet unknown, applications.

Smart and adaptive or intelligent materials adjust their mechanical properties upon the receipt of an external stimulus. In engineering language, they act as integrated sensors, processors, and actuators. For example, some polymers can be considered to be smart since they change their shapes with a change of temperature [3]; this property has been exploited in the development of all-weather lubricants in which the retention of constant viscosity is essential. Are these polymers adaptive? Yes, they can be considered to be if they can be made to remember and to respond to new stimuli. Are they intelligent? It is difficult to know. This example is meant to illustrate the subtle difference between adaptive and intelligent materials. It should be pointed out, however, that there is a healthy ongoing discussion about the meaning of adaptive, smart, and intelligent materials [4]. In the broader sense, these terms indicate man-made materials which can, to different degrees, self assemble, self diagnose, self repair, and recognize and discriminate physical and/or chemical stimuli, and, at the extreme, have the capability of learning and self replicating.

Nanosized materials or nanostructures have dimensions, as their name implies, in the 1–100 nanometer range [5, 6]. It is in this size region that the interaction between biology, chemistry, and physics is the most synergistic. Consequently, it is also an area which may yield truly advanced materials.

Nanostructures are, literally, facts of life in biology. Proteins, viruses, and bacteria are nanosized, three-dimensional structures which have been self assembled from smaller subunits. Although individual atoms in the subunits (polypeptides, for example) are covalently linked, assembly of the subunits is maintained by non-covalent (van der Waals, hydrogen-bonding, electrostatic, and hydrophobic) interactions.

As far as the chemist is concerned, nanosized materials are huge macromolecules (with molecular weights of the order of 10^6 to 10^{10}) constructed from millions of atoms. Atom-by-atom synthesis of nanostructures, via covalent bond formation, is a formidable task which has not as yet been achieved by synthetic chemists. Covalent polymerization is the best that chemists have done thus far [3]. Chemists have made spectacular progress, however, in forming *self-organized and supramolecular materials* in the size domain of nanostructures by the non-covalent bond assembly of molecules [7].

From the physicist's point of view, nanosized materials are size quantized; that is to say that their dimensions are comparable to the length of the de Broglie electron, the wavelengths of phonons, and the mean free paths of excitons [8–11]. Electron-hole confinement in nanosized spherical particles results in three-dimensional, quantum-size effects, i.e. in the formation of "quantum dots", "quantum crystallites", or "zero-dimensional excitons". In one-dimensional size quantization, the exciton is free to move only in two dimensions with the resultant formation of "quantum wells" or "two-dimensional excitons". In quantum wells, size quantization manifests in the growth direction, while bulk properties prevail in the other two dimensions. Finally, two-dimensional confinement of charge carriers (i.e. providing the exciton with only one-dimensional mobility) results in "quantum well wires". The importance of size and dimensionality quantizations is that they result in altered mechanical, chemical, electrical, optical, magnetic, electro-optical, and magneto-optical properties [8–11].

One kind of *nanostructured or nanofabricated material* has been generated by the introduction of high-density defects into a nanosized region of a perfect

Fig. 1a. Atomic structure of a two-dimensional nano-structured material. For the sake of clarity, the atoms in the centers of the "crystals" are indicated in *black*. The ones in the boundary core regions are represented by *open circles*. Both types of atoms are assumed to be chemically identical. **b** Atomic arrangement in a two-dimensional glass (hard sphere model). **c** Atomic structure of a two-dimensional nanostructured material consisting of elastically distorted crystallites. The distortion results from the incorporation of large solute atoms. In the vicinity of the large solute atoms, the lattice planes are curved as indicated in the crystallite on the *lower left side*. This is not so if all atoms have the same size as indicated in Fig. 1a [13]

crystal [12–14]. This results in atomic-scale disorders and lattice defects (Fig. 1), and manifests in grossly altered physical properties (density, ductility, and magnetism, to name a few). Nanostructured materials prepared by doping crystals with chemically different compounds led to the formation of crystalline, quasi-crystalline, and glassy alloys. Some of these materials have been described as "superstrong" and "supermagnetic" [12, 13].

It is interesting to note that nanostructured materials are synthesized in nature by a process known as biomineralization [15–18]. Most naturally occurring nanostructured materials are hierarchical composites and are often referred to as *hierarchically organized materials* [19–23]. This term implies an organization of materials in discrete steps, ranging from the atomic to the macroscopic scale. At each step, the components are held together by specific interactions and are organized for optimal overall performance. Tendons, which are responsible for connecting muscles to bones, provide a good illustration of hierarchical construction. A second example is the abalone shell, whose "bricks-and-mortar" hierarchial architecture is illustrated in Fig. 2 [23]. Calcium

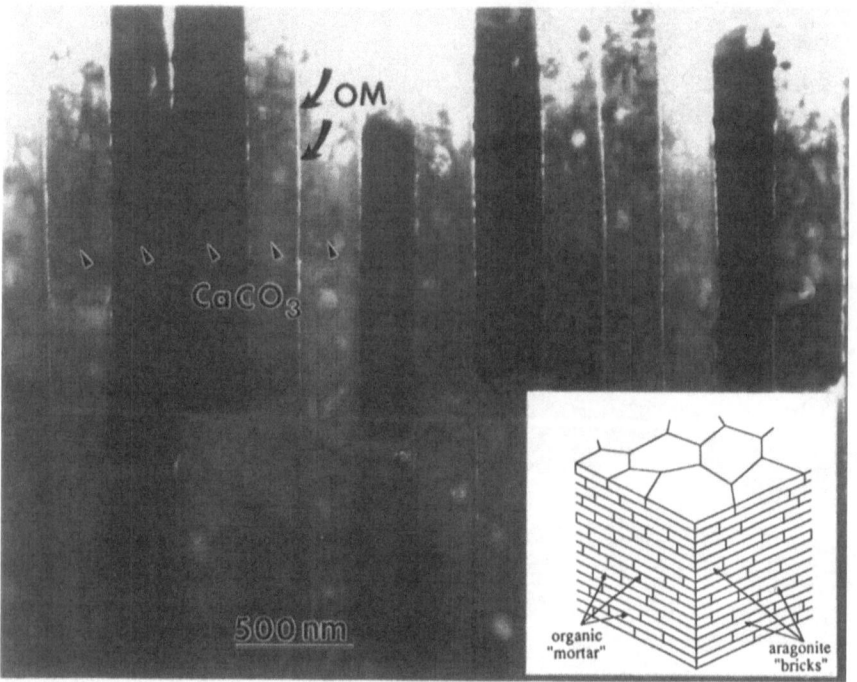

Fig. 2. Nacre of red abalone shell (*Halitotis refescens*), imaged here by transmission electron microscopy (*left*), has a bricks-and-mortar structure. The bricks are $CaCO_3$ (aragonite) platelets, and the mortar is a composite of macromolecules, including structural proteins and polysaccharides, that form a thin film around the platelets. The three-dimensional structure is depicted on the *right* [23]

carbonate (aragonite) bricks are bonded by soluble acidic proteins and poly-saccharide which are present in the abalone wall. This design imparts a great deal of structural strength to the shell and reduces the risk of cracking. Architectural lessons to be learned from nature, illustrated by these two examples, will be an important theme in the present monograph.

Nanostructured materials have also been formed by scanning tunneling microscopy (STM) [24], scanning electrochemical microscopy (SECM) [25], and atomic force microscopy (AFM) [26]. Recent reports on the modification of atomic sites at bare surfaces by STM [27] and the formation of nanometer-scale defects by STM [28] and AFM [29] illustrate the power of these techniques.

Differences between nanophysics and nanochemistry have been lucidly delineated in a recent review on nanochemistry [6]. The aim in nanophysics is to generate small particles by reducing the size of bulk materials by using engineering-type manipulations. Conversely, nanochemists synthesize their materials by chemical reactions from "atoms-up". A synergetic combination of these two approaches should not be overlooked, of course.

The term *artificially structured materials* implies a construction similar to nanostructured materials on a somewhat larger, on the whole unspecified, scale. The terms *biomaterials* and *genetically engineered materials* are self explanatory.

1.2 The Importance of Advanced Materials

The supreme importance of materials science and engineering has been recognized by national governments and international Agencies. In the United States, the National Research Council, the principal operating agency of the National Academy of Sciences and the National Academy of Engineering, issued a major report on "Materials Science and Engineering for the 1990s – Maintaining Competitiveness in the Age of Materials" in 1989 [1]. The material needs of eight U.S. industries (aerospace, automotive, biomaterials, chemical, electronics, energy, metals, and telecommunications), collectively employing over seven million people and generating sales in excess of $ 1.4 trillion, were critically examined in this report. It was considered that advanced materials with improved structural, optical, electronic, and magnetic peoperties, at a competitive cost, were essential for the future economic well-being of all of the cited industries. The future automobile should be made, for example, from materials that are strong, light-weight, reliable under adverse conditions, durable, and reusable and should be economically priced. It should be propelled by a light-weight, long-lasting, and efficient battery. Construction of this automobile will require well-coordinated and sustained advances in materials science and engineering.

Advanced materials are of paramount importance for the future of the electronics industry. Remarkable progress has been made in fitting increasingly greater numbers of components into an integrated circuit. This is illustrated by the often-shown electronic device density versus time semilogarithmic plot (Fig. 3) [30], as well as by the notebook computer, alluded to earlier. The

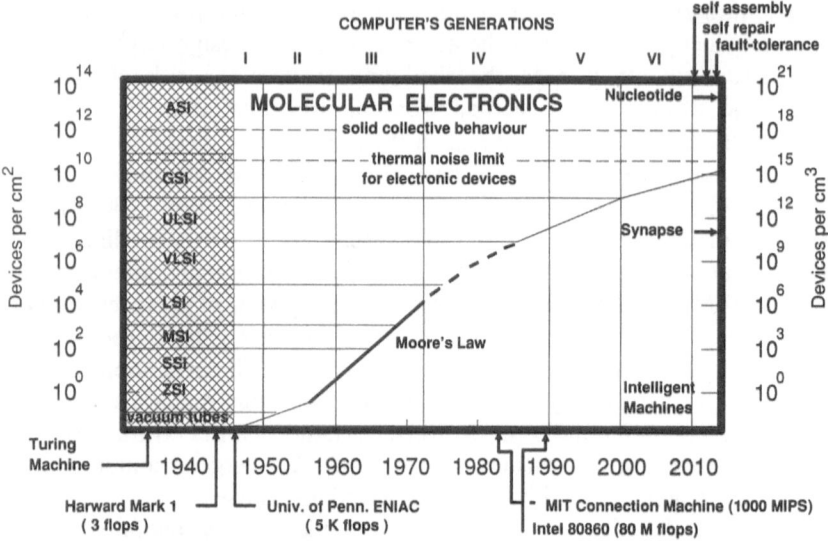

Fig. 3. Moore's law. The semilogarithmic plot of the device density in a computer as a function of the passage of time [30]

challenge is to increase still further the degree of miniaturization and the speed of operation at reduced energy consumption. This challenge cannot be met with electronic components whose function is based on the bulk properties of inorganic materials. In the vicinity of the nanometer regime, operation of these devices becomes unreliable due to quantum and thermal fluctuations. Have we reached the end of the road?

The time has come to make a quantum leap, both figuratively and literally. Short-term solutions may exploit organic semiconductors [31–33] and semiconductor quantum heterostructures [34, 35]. A long-term solution will be provided by molecular electronics [36–40] in which single molecules are employed as integrated circuits [41, 42].

Development of advanced materials, both for immediate and longer range uses, is imperative.

1.3 Rationale for a Colloid Chemical Approach

Most modern materials are formed empirically by solid-state methods. These methods generally involve more processing activity than chemical synthesis (for example, sintering of ceramic powders, modifying concrete by polymers, thermomechanical processing of alloys, layering polymeric membranes for special applications, implanting ions into semiconductors, and growing crystals). The industrial production of ferrofluids (liquid magnets) serves to illustrate the crudest end of materials processing. Commercial ferrofluids are prepared by the "brute-force" method of grinding iron oxide (Fe_2O_3) into relatively uniform, 30- to 50-nm-sized particles, followed by their dispersion in an organic solvent in the presence of a surfactant [43]. Band-gap engineering, a deliberate variation of the composition and doping of semiconductors in the submicron to subnanometer regime, is at the other end of the spectrum. Molecular beam epitaxy (MBE), liquid phase epitaxy (LPE), chemical vapor deposition (CVD), chemical beam epitaxy (CBE), metallo-organic molecular epitaxy (MOMBE), low-pressure chemical vapor deposition (LPCVD), and sputter and vapor deposition have been developed for band-gap engineering [44, 45]. State-of-the-art technology has reached atomic, layer-by-layer, molecular-beam epitaxy [46]. Nevertheless, it is fair to say that there is, at present, very little understanding of the mechanisms of these processes at the molecular level. Improvements have been incremental and have resulted, by and large, from clever engineering.

The development of a new generation of advanced materials demands innovative chemical tailoring, a task firmly based upon a fundamental under-standing of the interactions and reactions involved at the atomic level. Chemists have already risen to the occasion and have contributed significantly to materials science. Indeed, their accomplishments are documented in such recently estab-lised scientific journals as *Advanced Materials, Chemistry of Materials*, and the *Journal of Materials Chemistry*. Colloid chemistry is particularly well suited for advancing materials science since

- it has matured [47] and has become quantitative and predictive thanks to both the vast number of new techniques being utilized and theories being developed;
- the sizes and behaviors of many advanced materials enable them to be colloidal;
- materials processing often involves reactions at solid-solid, solid-liquid, or solid-gaseous interfaces (of which the manufacturing of ceramic membranes by the sol-gel process is an obvious example [48]);
- colloidal aggregates have served as containers and/or templates for the generation of advanced materials (the use of monolayers, multilayers, and other aggregates will be, in fact, the subject of discussion in the next section of this monograph); and
- most significantly, many biominerals, which constitute Mother Nature's response to advanced materials, can be considered to be colloidal systems; indeed, the mimetic approach to biomineralization, an important theme of the present text, is clearly centered upon colloid chemistry.

Colloid chemistry is, thus, pervasive to materials science and continues to make an ever-increasing contribution to the field.

1.4 Rationale and Philosophy of the Mimetic Approach to Advanced Materials

The membrane-mimetic approach to advanced materials, advocated here, is inspired by biomineralization [15–18]. Precipitation and/or cluster formation, occurring in biological cells, are nature's way of constructing teeth, bones, shells, and direction-finding equipment. Intense effort has been exerted for the rationalization of biomineralization in terms of known colloidal, crystallographic, and solid-state principles [15–18]. The availability of appropriate precursors in appropriate concentrations and the presence of a matrix was suggested to determine the outcome of nucleation and subsequent particle growth [49, 50]. Compartmentalization provided by membranes and cells was considered to be responsible for the inflow of ionic precursors and for the imposition of shape and size control over the incipient crystal growth. Proteins have been recognized as playing a paramount role in directing biomineralization [18].

Biological membranes are intimately involved in biosyntheses, energy transduction, information transmission, and cell recognition [51–53]. They define the very existence of the cell. The ardent desire to understand the structure and properties of membranes is not surprising. Insight has been obtained from studies of biomembranes and their constituents, in vivo and in vitro, as well as from the examination of a variety of membrane models [54]. Investigation of models has led to the rationalization of many properties, including osmotic activity, phase transition, phase separation, dissymmetry, fluidity, permeability, fusion and substrate mobilities into and out of membranes. There are, of course, no perfect models, and no single model can faithfully reproduce all of the complexities of the membrane ensemble in vivo.

We have been interested for some time in membrane mimicking [55–57]. A subtle, but important, difference exists between modeling and mimicking. Modeling (diminutive of Latin *modus* = form) implies a more or less faithful duplication of the original in a scaled-down version; whereas mimicking (Greek *mimetikos* = imitative; *mimeitahai* = to imitate) suggests only the imitation of the essential features. Modeling is Rembrandt's self-portrait, mimicking is the cubist rendering of the Nude Descending a Staircase by Marcel Duchamp. No attention is paid in mimicking to the faithful duplication or reconstitution of cell membranes. Only the essential components of membranes are placed on the chemist's canvas. Our often-quoted motto has been a caution against the slavish and needless copying of Mother Nature [55–57]. Substantial alteration of reaction products, rates, and stereochemistries has been observed in relatively simple surfactant assemblies which mimicked aspects of the biological membrane [55, 58].

The philosophy of membrane-mimetic chemistry may be illustrated by a comparison of plant photosynthesis with sacrificial water photoreduction in artificial systems; the process has been mediated by metal-catalyst-coated semiconductor colloids supported on polymerized vesicles (Fig. 4a) [59–64].

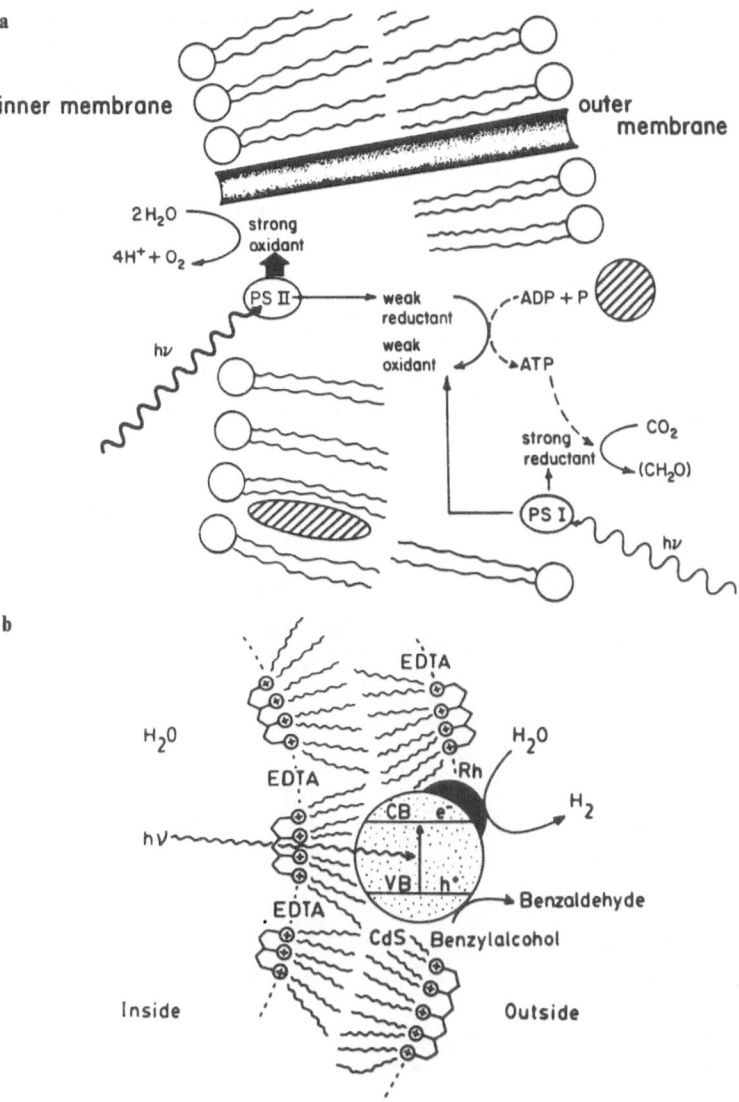

Fig. 4a. Schematic representation of a portion of the thylakoid membrane, thought to be involved in photosynthesis. Photosystem I (PSI), upon light (hv) absorption, produces a strong reductant and a weak oxidant. Photosystem II (PSII), upon light (hv) absorption, produces a strong oxidant and a weak reductant. Electron flow from the weak reductant to the weak oxidant is coupled to phosphorylation, which converts adenosine diphosphate (ADP) and inorganic phosphate (P) to adenosine triphosphate (ATP). With the aid of ATP, the strong reductant produced by PSI reduces carbon dioxide to carbohydrate. The strong oxidant, produced by PSII, oxidizes water molecules to molecular oxygen and protons, released in the inner membrane. Protons can be transmitted through channels going through the bilayers (indicated by the *dotted area*). Cholesterol (indicated by *shaded ovals*) adds to the rigidity of the membrane and proteins (indicated by *shaded circles*) may be bound to the surface [55]. **b** In the mimetic system, water (rather than CO_2) is reduced in the reduction half cycle to hydrogen at the expense of benzyl alcohol. The location of the CdS/Rh particle is based on chemical experiments [63]

Natural photosynthesis is best understood in terms of the zig-zag, or Z, scheme. Briefly, light is harvested by photosystem I (PSI) and photosystem II (PSII). These two systems operate in series; two photons are absorbed for every electron liberated from water. Light-induced charge separation in PSII leads to the formation of a strong oxidant, Z^+ ($E_0 = +0.8$ V), and a weak reductant, Q^- ($E_0 = 0$ V). Although the reduction potential of Z^+ is sufficient for water oxidation, the process of molecular oxygen evolution demands the accumulation of four positive charges. Electrons flow from Q^-, via a pool of plastoquinone (pQ) and other complex carriers, to a weak oxidant, Y ($E_0 = +0.4$ V), which is generated along with a strong reductant, X^- ($E_0 = -0.6$ V), in PSI. This electron flow is coupled to the $NADH^+ \leftrightarrows NADH$ cycle (via ferrodoxin). Protons enter the ATPase to react with ADP to form ATP (not shown in Fig. 4). With the aid of ATP, X^- reduces CO_2 to carbohydrates.

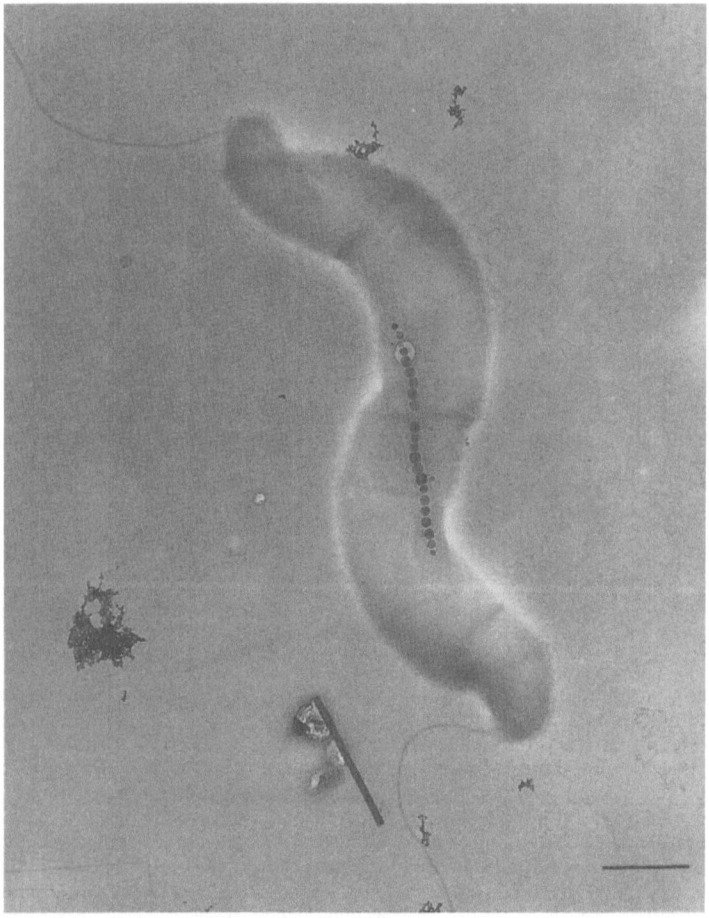

Fig. 5. Electron micrograph of particle chain in magnetotactic bacteria. The *bar* = 500 nm [66] (photo credit: Gorby Y, Blakemore R, Microbiology Department, University of New Hampshire)

In the artificial system Figure 4b, a polymerized surfactant vesicle is substituted for the thylakoid membrane. Energy is harvested by semiconductors, rather than by PSI and PSII. Electron transfer is rather simple. Water (rather than CO_2) is reduced in the reduction half cycle to hydrogen, at the expense of benzyl alcohol. In spite of these differences, the basic principles in plant and mimetic photosyntheses are similar. Components of *both* are compartmentalized. The sequence of events is identical in both systems: energy harvesting, vectorial charge separation, and reduction.

We have not as yet been able to mimic one of nature's most-accomplished materials scientists – the magnetotactic bacterium (*Aquaspirillum magnetotacticum*). Each of these bacteria is capable of producing 20 to 25, 45 ± 8-nm-diameter, spherical, single-domain Fe_3O_4 (magnetite) particles in the cytoplasmic membrane (Fig. 5) [65–67]. Formation of correctly sized and precisely oriented magnetite imparts to the bacterium a magnetic moment parallel to its axis of motility and assures that it swims downwards in both hemispheres [66]. Understanding the chemistry which the bacterium performs and reproducing it by using a membrane-mimetic approach is a worthy challenge to any reader of this monograph.

1.5 Scope

Accomplishments and potential of the mimetic approach to advanced materials is surveyed in the present monograph. Emphasis will be placed primarily on aqueous, "wet" colloidal methodologies since water is the milieu for the biochemical processes to be mimicked. The interpretation of the mimetic approach will be somewhat broad. It will allow a discussion of advanced materials prepared by molecular organization and compartmentalization, as well as of those generated in the different mimetic systems.

Description of the different mimetic systems will be the starting point of the presentation (Sect. 2). Preparation and characterization of monolayers (Langmuir films), Langmuir–Blodgett (LB) films, self-assembled (SA) monolayers and multilayers, aqueous micelles, reversed micelles, microemulsions, surfactant vesicles, polymerized vesicles, polymeric vesicles, tubules, rods and related SA structures, bilayer lipid membranes (BLMs), cast multibilayers, polymers, polymeric membranes, and other systems will be delineated in sufficient detail to enable the neophyte to utilize these systems. Ample references will be provided to primary and secondary sources.

Available results on the preparation, characterization, and utilization of metallic and catalytic particles (Sect. 3), semiconductor particles and particulate films (Sect. 4), conductors and superconductors (Sect. 5), magnetism and magnetic particles and particulate films (Sect. 6), and advanced ceramic materials (Sect. 7) will constitute the main body of the monograph. An attempt will be made to cover these materials exhaustively.

2 Preparation and Characterization of Compartments

Since its inception more than a decade ago [55], membrane-mimetic chemistry has blossomed into a mature discipline. Many new mimetic systems have been self assembled and constructed from a large variety of naturally occurring and synthetic amphiphatic and polymeric molecules [7, 68–75]. Several truly innovative techniques, including the surface forces apparatus [76–79], scanning tunneling microscopy [24, 80–89], scanning electrochemical microscopy [25], atomic force microscopy [26], fluorescence microscopy [90–93], Brewster angle microscopy [94, 95], and X-ray-diffraction using synchrotron sources [96–100] have been added to the armory of membrane-mimetic chemists. These, along with established methodologies, have permitted the characterization of self-assembled (SA) systems and their constituents at the atomic and molecular levels. A vastly improved understanding of the forces maintaining SA and man-made mimetic systems [101] and the burgeoning of molecular modeling of monolayers [68, 102, 103] were witnessed during the past decade. Many of these activities have been fuelled by the recognized and expected utilization of membrane-mimetic approaches in a variety of fields, including materials science.

Clarification of the nomenclature used in this monograph is, once again, required. Use of the term "monolayer" will be restricted to implying a mono-molecular, two-dimensional film floating on an aqueous solution (the subphase) in a Langmuir film balance (the trough). Monolayers will also, on occasion, be referred to as Langmuir films. A monolayer or several monolayers, *transferred to solid substrates*, will be denoted as "Langmuir-Blodgett (LB) films". "Self-assembled (SA) monolayers" will be used to designate monomolecularly thick films which are *spontaneously* formed upon the immersion of a solid substrate into a solution of appropriate surface-active molecules. "SA multilayers" will denote systems formed upon the covalent or non-covalent attachment of subsequent layer(s) of surfactants to the SA monolayer. The term "bilayer (black) lipid membranes" (BLMs) will be used to designate bimolecularly thick films spanning across a pinhole (1–2 mm) which separates an aqueous solution into two compartments. Aqueous micelles are spherical aggregates dynamically and cooperatively formed in water above a surfactant concentration known as the critical micelle concentration (CMC). Reversed micelles are surfactant aggregates in non-polar solvents. Their formation requires water; thus, they can be considered to be surfactant-entrapped water pools in non-polar solvents. Increasing the surfactant concentration and/or adding a third component (alcohol, for example) leads to the formation of larger aggregates: oil-in-water (o/w) or water-in-oil (w/o) microemulsions. "Small unilamellar vesicles" (SUVs) will refer to closed, spherical, single-bilayer surfactant and/or phospholipid aggregates dispersed in aqueous solutions. There is an important difference between "polymerized vesicles" and "polymeric vesicles". Use of the former term will be limited to vesicles constructed from monomeric surfactants and subsequently polymerized. "Polymeric vesicles" will be used to designate aggregates formed from polyelectrolytes. The same meanings will be attributed, of course, to "polymerized monolayers" and "polymeric monolayers". "Tubules", "rods", and "ribbons" will be used to indicate the shapes of relatively stable aggregates formed from surfactants or lipids in aqueous solutions. "Polymeric membranes" will be used to designate industrial membranes fabricated from synthetic polymers and used for separation or dialysis (for example).

Most of the compartments discussed here are constituted from naturally occurring phospholipids or synthetic surfactants. A common feature of both types of molecules is that they are amphiphatic (having distinct polar and apolar regions) [55]. Depending on their chemical structures, surfactants can be neutral, anionic, cationic, and zwitterionic. The hydrophobic part of the surfactants can be of various lengths (typically, $C_8n–C_{20}$) and may have one or several double bonds or consist of two or more chains. Surfactants have also been functionalized to contain desired reactive and/or reporter groups. Aggregation behavior depends on the nature and concentration of the surfactants, the nature of the solvent, and the method of preparation. The thermodynamically preferred structure formed from a given surfactant has been rationalized by considering its headgroup and tail volume ratio [101]. Structures of the most frequently used phospholipids and surfactants and their common names are collected in Table 1. The **bold numerals** written under the structures will be used consistently throughout the present monograph to refer to them.

Only a brief description of the different mimetic systems will be provided in this section. Selected examples of recent research will be given to enliven the presentation; references citing the original work should be consulted for greater details.

Table 1. List of structures

bis [2-(*n*-Hexadecanoyloxy)ethyl]methyl(4-vinylbenzyl)ammonium chloride

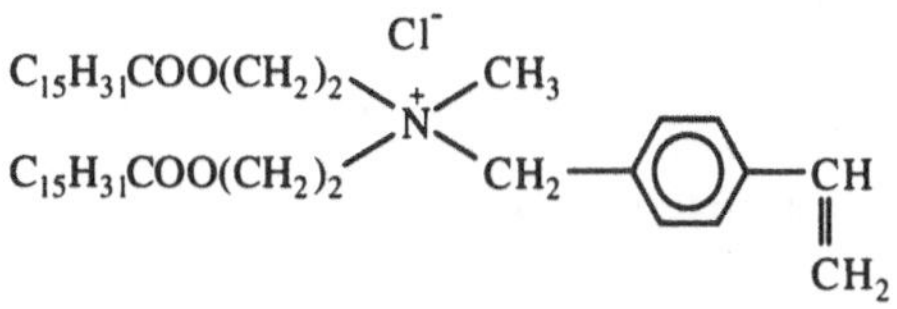

1

Dimethyl-*n*-hexadecyl[10-(*p*-vinylcarboxanilido)decyl]ammonium chloride

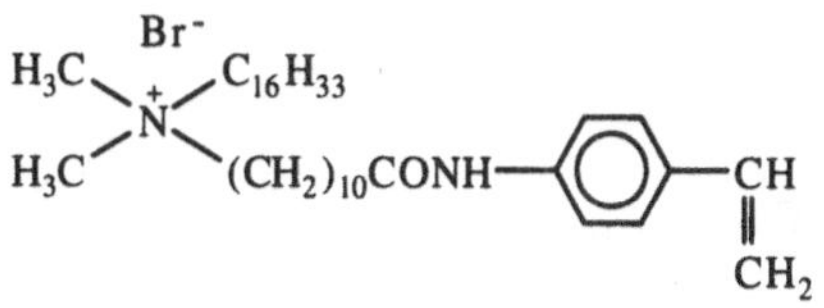

2

Table 1. Contd.

Dioctadecylmethyl-[2-[[4-vinylphenyl)oxy]carbonyl]ethyl]ammonium chloride

$$C_{18}H_{37}\diagdown \overset{Cl^-}{\underset{|}{\overset{+}{N}}}\diagup CH_3$$
$$C_{18}H_{37}\diagup \qquad \diagdown CH_2CH_2COO-\langle\bigcirc\rangle-CH{=}CH_2$$

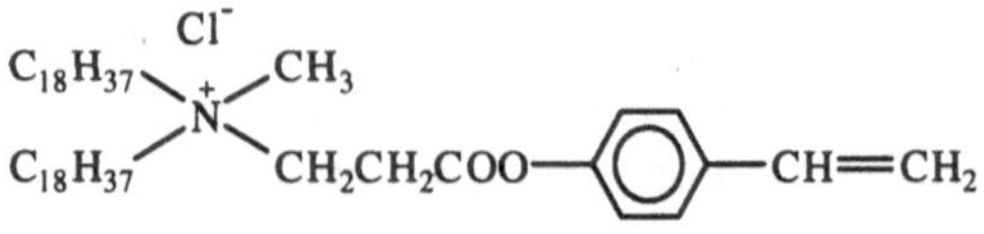

3

n-Hexadecyl[11-(4-vinylbenzylamido)undecyl]phosphate

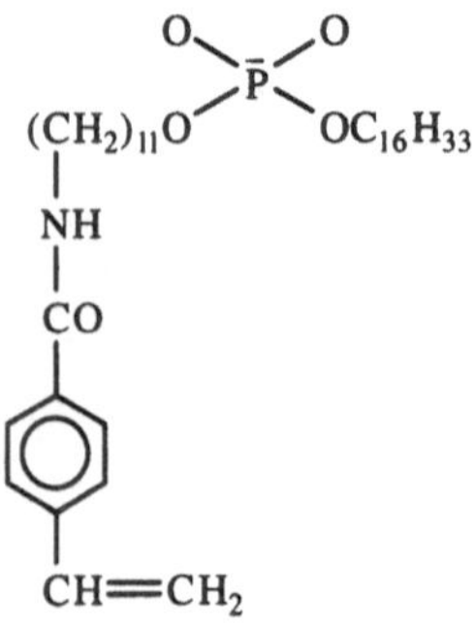

$$(CH_2)_{11}O\diagdown \overset{O}{\underset{}{\overset{\|}{P}}}\diagup OC_{16}H_{33}$$
$$|$$
$$NH$$
$$|$$
$$CO$$
$$\langle\bigcirc\rangle$$
$$CH{=}CH_2$$

4

Dioctadecyldimethylammonium bromide (DODAB)

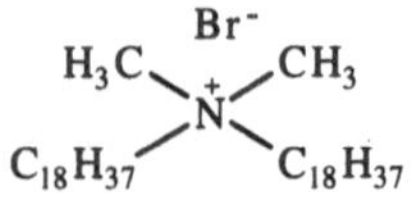

$$H_3C\diagdown \overset{Br^-}{\underset{|}{\overset{+}{N}}}\diagup CH_3$$
$$C_{18}H_{37}\diagup \qquad \diagdown C_{18}H_{37}$$

5

Dioctadecylmethylammonium chloride (DODAC)

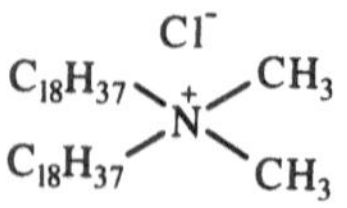

$$C_{18}H_{37}\diagdown \overset{Cl^-}{\underset{|}{\overset{+}{N}}}\diagup CH_3$$
$$C_{18}H_{37}\diagup \qquad \diagdown CH_3$$

6

Tetradecanoic acid

$$CH_3(CH_2)_{12}COOH$$

7

Table 1. Contd.

Octadecanoic acid

$$CH_3(CH_2)_{16}COOH$$

8

Eicosanoic acid

$$CH_3(CH_2)_{18}COOH$$

9

Docosanoic acid

$$CH_3(CH_2)_{20}COOH$$

10

Tetracosanoic acid

$$CH_3(CH_2)_{22}COOH$$

11

1-Tetradecanol

$$CH_3(CH_2)_{12}CH_2OH$$

12

9-Octadecenoic acid

$$CH_3(CH_2)_7CH\!=\!CH(CH_2)_7CO_2H$$

13

Phosphatidyl choline

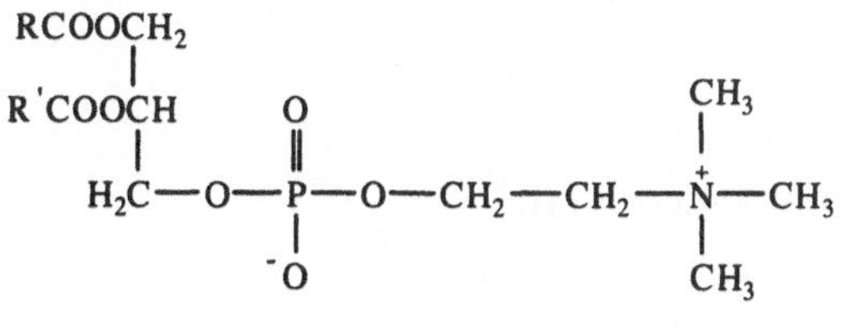

14

Phosphatidylethanolamine

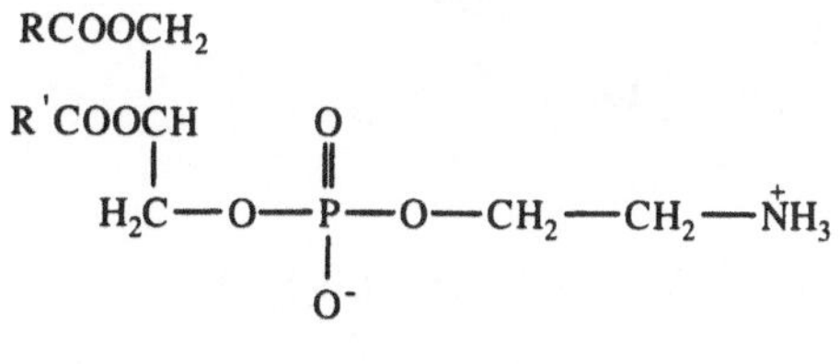

15

Table 1. Contd.

Phosphatidyl serine

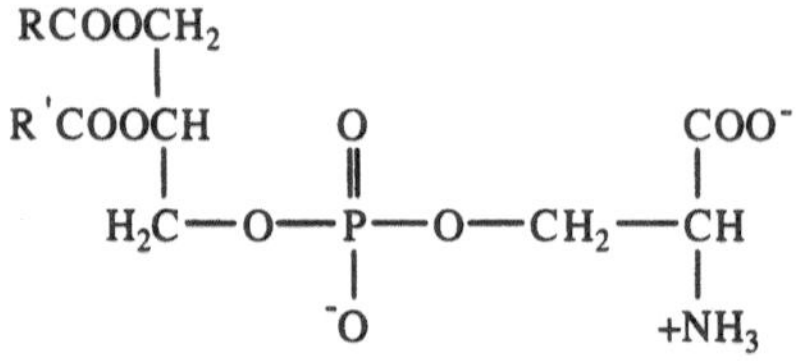

$$\begin{array}{c}
\text{RCOOCH}_2 \\
|\\
\text{R'COOCH} \qquad \text{O} \qquad\qquad\qquad \text{COO}^- \\
|\qquad\qquad\quad \| \qquad\qquad\qquad\quad | \\
\text{H}_2\text{C}-\text{O}-\text{P}-\text{O}-\text{CH}_2-\text{CH} \\
|\qquad\qquad\qquad\qquad\quad | \\
\text{}^-\text{O} \qquad\qquad\qquad\quad {}^+\text{NH}_3
\end{array}$$

16

Phosphatidic acid

$$\begin{array}{c}
\text{CH}_2\text{O}-\text{CO}-\text{R}' \\
\vdots \\
\text{H}_2\text{O}_3\text{P}-\text{O}\blacktriangleright\text{C}\blacktriangleleft\text{H} \\
\vdots \\
\text{CH}_2\text{O}-\text{CO}-\text{R}''
\end{array}$$

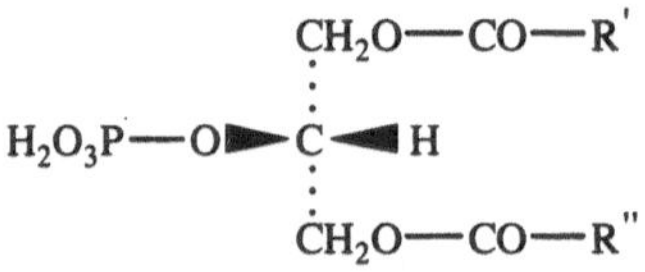

17

Dipalmitoyl phosphatidyl choline (DPPC)

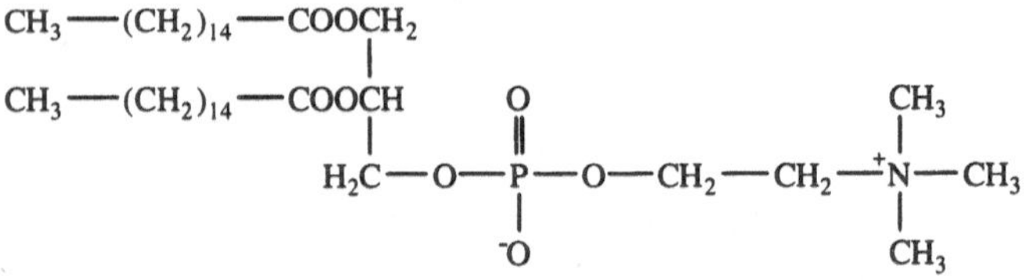

$$\begin{array}{c}
\text{CH}_3-(\text{CH}_2)_{14}-\text{COOCH}_2 \\
|\\
\text{CH}_3-(\text{CH}_2)_{14}-\text{COOCH} \qquad \text{O} \qquad\qquad\qquad\qquad \text{CH}_3 \\
|\qquad\qquad\quad \| \qquad\qquad\qquad\qquad\quad | \\
\text{H}_2\text{C}-\text{O}-\text{P}-\text{O}-\text{CH}_2-\text{CH}_2-{}^+\text{N}-\text{CH}_3 \\
|\qquad\qquad\qquad\qquad\qquad\qquad\quad | \\
\text{}^-\text{O} \qquad\qquad\qquad\qquad\qquad\quad \text{CH}_3
\end{array}$$

18

Dimyristoyl phosphatidic acid

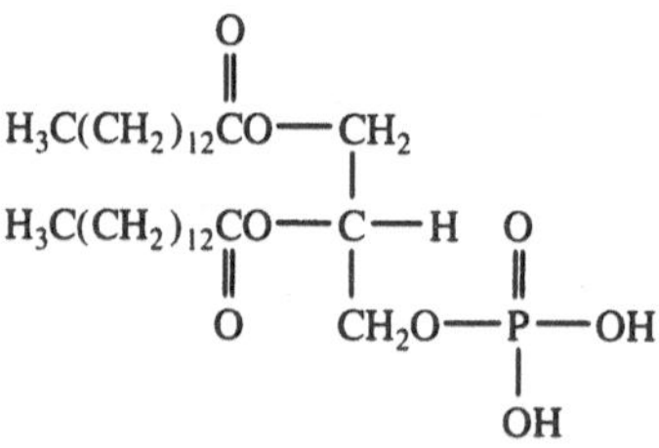

$$\begin{array}{c}
\qquad\qquad\qquad \text{O} \\
\qquad\qquad\qquad \| \\
\text{H}_3\text{C}(\text{CH}_2)_{12}\text{CO}-\text{CH}_2 \\
|\\
\text{H}_3\text{C}(\text{CH}_2)_{12}\text{CO}-\text{C}-\text{H} \quad \text{O} \\
\|\qquad\qquad | \qquad\quad \| \\
\text{O} \qquad\quad \text{CH}_2\text{O}-\text{P}-\text{OH} \\
|\\
\text{OH}
\end{array}$$

19

A diacetylene carboxylic acid

$$\text{CH}_3(\text{CH}_2)_{11}\text{C}\equiv\text{C}-\text{C}\equiv\text{C}-(\text{CH}_2)_8\text{CO}_2\text{H}$$

20

Table 1. Contd.

Polymerizable phosphatidylcholine

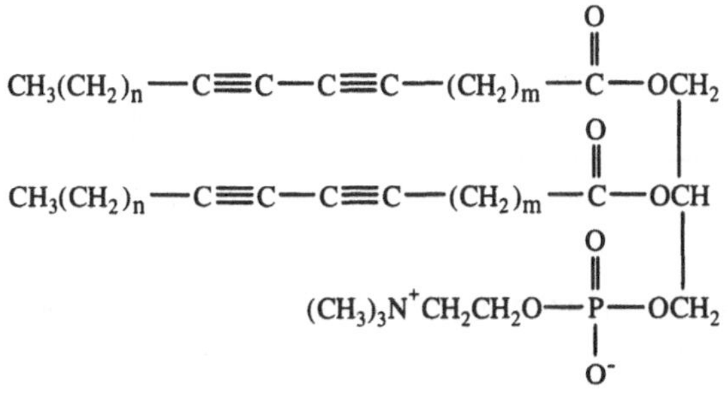

$$CH_3(CH_2)_n-C\equiv C-C\equiv C-(CH_2)_m-\overset{\overset{\displaystyle O}{\|}}{C}-OCH_2$$

$$CH_3(CH_2)_n-C\equiv C-C\equiv C-(CH_2)_m-\overset{\overset{\displaystyle O}{\|}}{C}-OCH$$

$$(CH_3)_3N^+CH_2CH_2O-\overset{\overset{\displaystyle O}{\|}}{\underset{\underset{\displaystyle O^-}{|}}{P}}-OCH_2$$

21

n = 7 − 16
m = 5 − 11

Glyceryl monooleate (GMO)

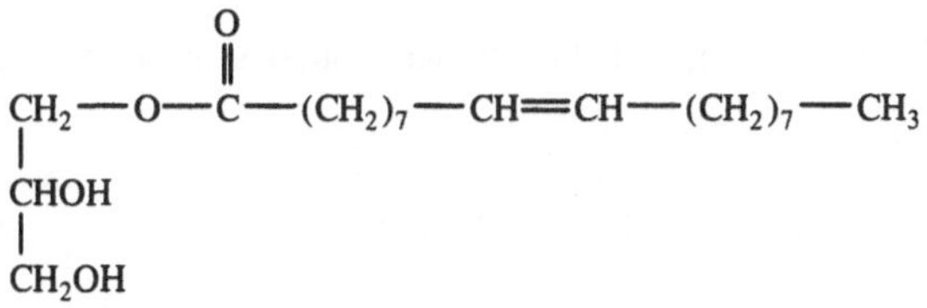

$$CH_2-O-\overset{\overset{\displaystyle O}{\|}}{C}-(CH_2)_7-CH=CH-(CH_2)_7-CH_3$$
$$|$$
$$CHOH$$
$$|$$
$$CH_2OH$$

22

2,3-Bis(hexadecanoyloxy)propyl-P-methacroyl-3,6,9-trioxanoyl dimethylammonium iodide

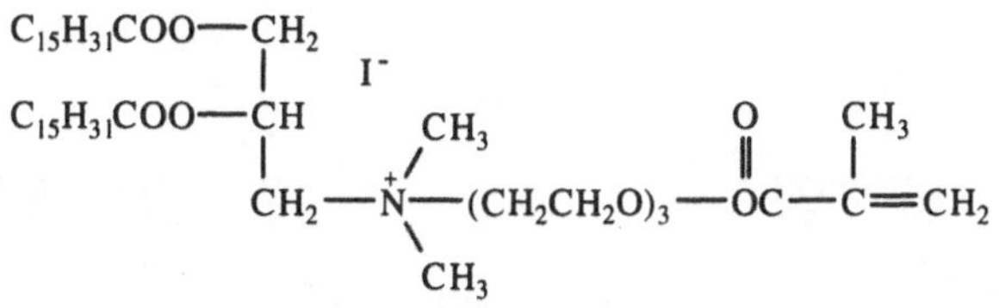

$$C_{15}H_{31}COO-CH_2$$
$$C_{15}H_{31}COO-CH \quad I^-$$
$$CH_2-\overset{+}{\underset{\underset{\displaystyle CH_3}{|}}{N}}-(CH_2CH_2O)_3-OC-\overset{\overset{\displaystyle CH_3}{|}}{C}=CH_2$$

23

12-Methacryloyl-3,6,9,12-tetraoxadodecyl 3-(N-dioctadecylcarbamoyl)propionate

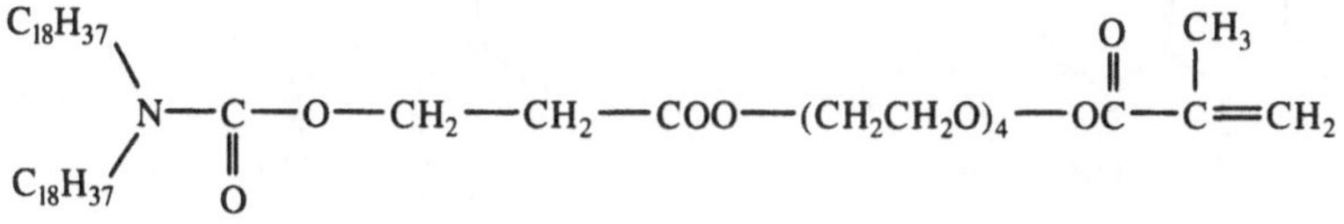

$$\overset{\displaystyle C_{18}H_{37}}{\underset{\displaystyle C_{18}H_{37}}{>}}N-\overset{\overset{\displaystyle }{|}}{\underset{\underset{\displaystyle O}{\|}}{C}}-O-CH_2-CH_2-COO-(CH_2CH_2O)_4-OC-\overset{\overset{\displaystyle CH_3}{|}}{C}=CH_2$$

24

Table 1. Contd.

2,3-Bis(hexadecyloxy)propyl 12-methacrylovl-3,6,9,12-tetraoxadodecyl succinate

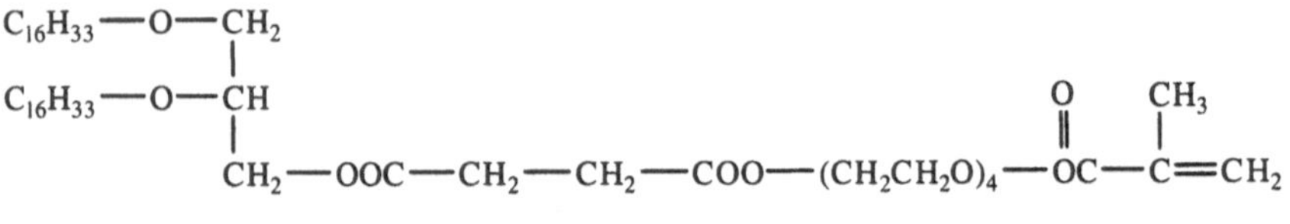

$$C_{16}H_{33}-O-CH_2$$
$$C_{16}H_{33}-O-CH$$

$CH_2-OOC-CH_2-CH_2-COO-(CH_2CH_2O)_4-OC-C=CH_2$, with $\overset{O}{\overset{\|}{}}$ and CH_3

25

Sodium 2,3-bis(hexadecyloxy)propyl-12-methacryloyl-3,6,9,12-tetraoxadodecylphosphate

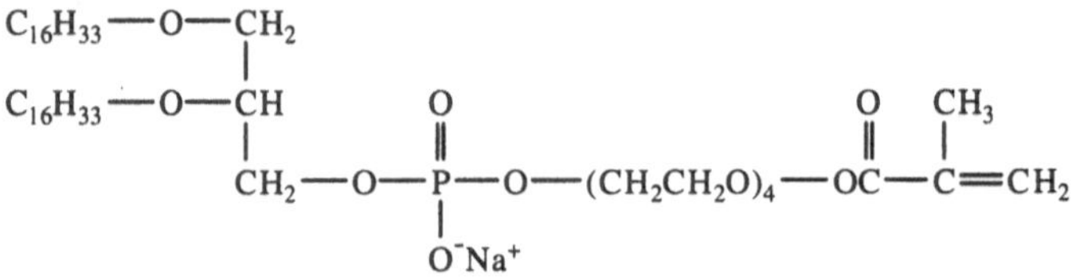

$$C_{16}H_{33}-O-CH_2$$
$$C_{16}H_{33}-O-CH$$

$CH_2-O-P-O-(CH_2CH_2O)_4-OC-C=CH_2$, with phosphate O^-Na^+ and CH_3

26

N',N''-Bis[11-(sorboyloxy)undecyl](pyridinium-N-ylpropionyl)-L-glutamide bromide

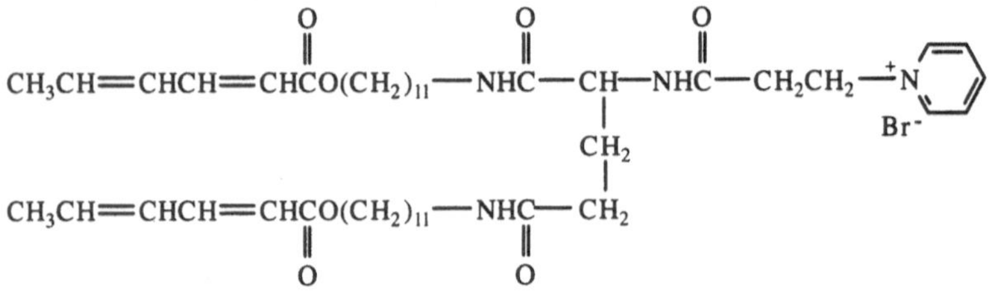

$$CH_3CH=CHCH=CHCO(CH_2)_{11}-NHC-CH-NHC-CH_2CH_2-\overset{+}{N}\text{(pyridinium)} \quad Br^-$$
$$CH_2$$
$$CH_3CH=CHCH=CHCO(CH_2)_{11}-NHC-CH_2$$

27

Chiral aldonamines

D-Glu

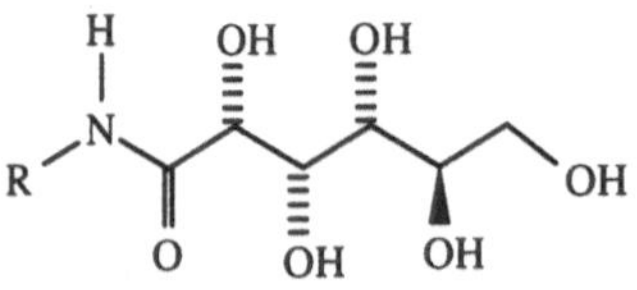

D-28

L-Glu

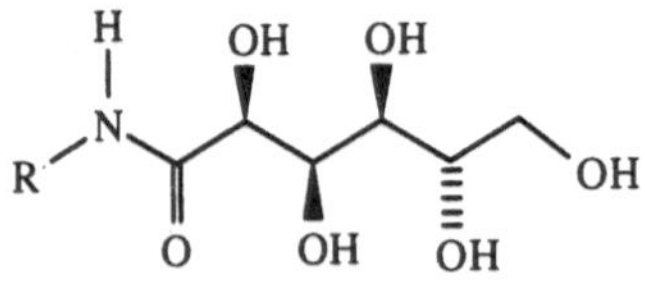

L-28

D-Man

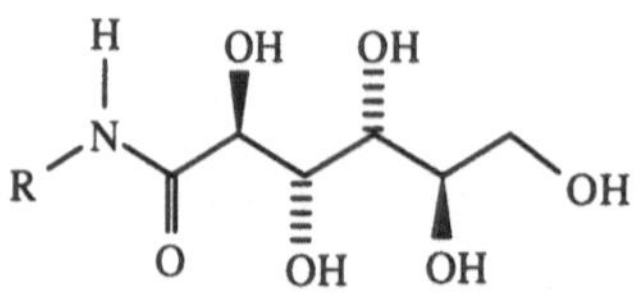

D-29

L-Man

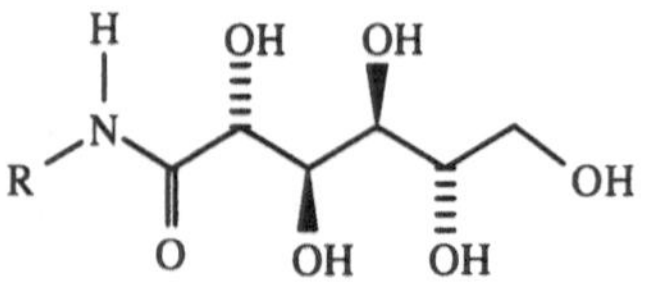

L-29

Table 1. Contd.

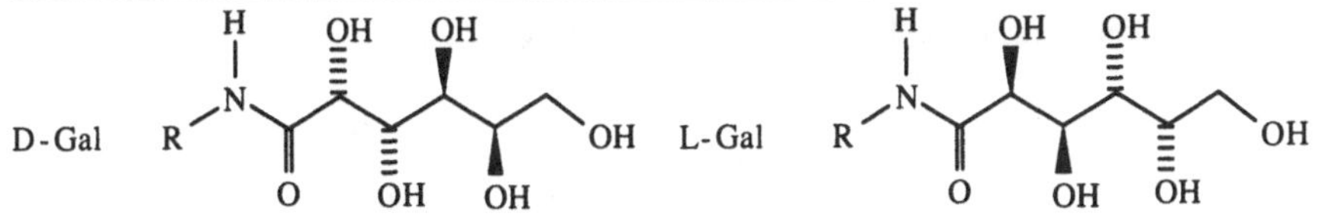

D-30   L-30

R = CH$_3$(CH$_2$)$_7$ = 8 = octyl-

R = CH$_3$(CH$_2$)$_{11}$ = 12 = dodecyl -

R = CH$_3$(CH$_2$)$_{17}$ = 18 = octadecyl-

N -Dodecyltartaric acid monoamides

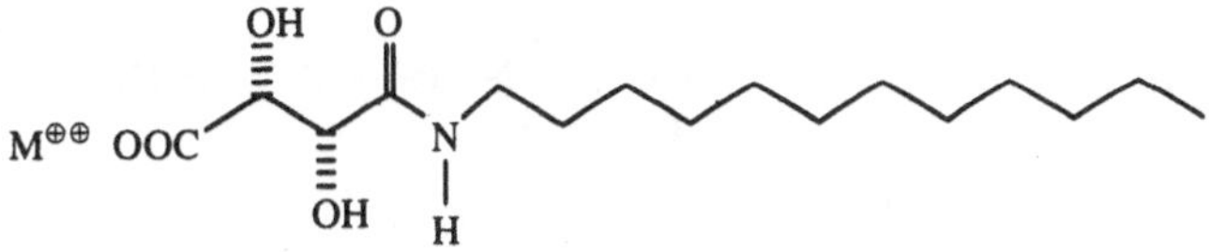

31a Na$^\oplus$, **31b** K$^\oplus$

A hyperbranched dendrimer

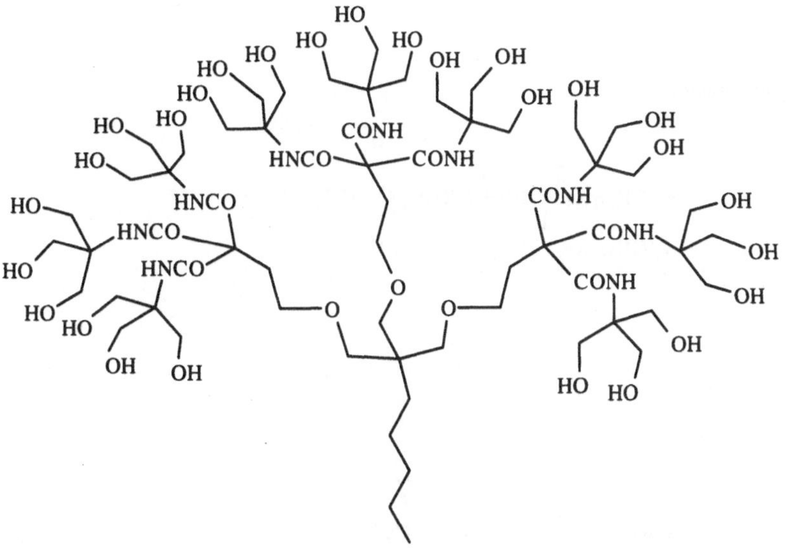

32

Table 1. Contd.

Alkylammonium surfactant with azobenzene functionaility

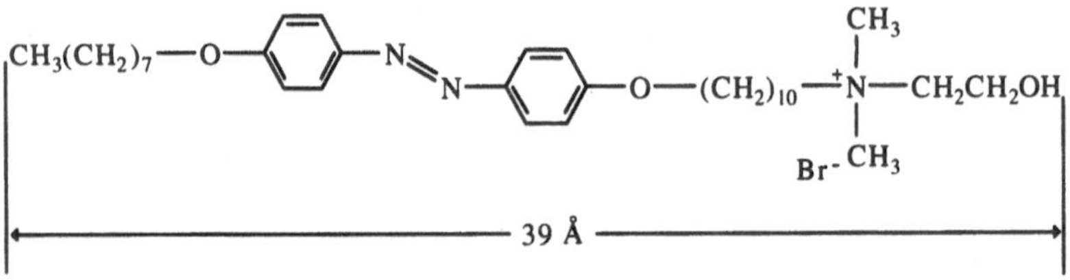

$$CH_3(CH_2)_7-O-\!\!\!\bigcirc\!\!\!-N\!\!=\!\!N-\!\!\!\bigcirc\!\!\!-O-(CH_2)_{10}-\overset{+}{\underset{CH_3}{\overset{CH_3}{N}}}-CH_2CH_2OH$$

$$Br^-$$

$$\longleftarrow\qquad 39\ \text{Å}\qquad\longrightarrow$$

33

Alkylammonium surfactant with glutamate functionality

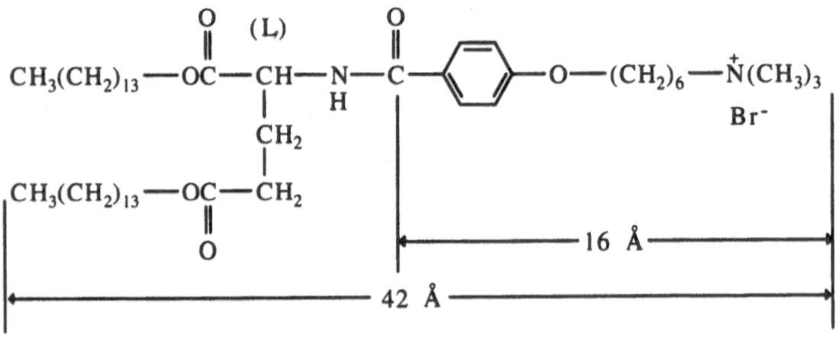

$$CH_3(CH_2)_{13}-O\overset{O}{\overset{\|}{C}}-\overset{(L)}{\underset{\underset{CH_2}{|}}{CH}}-\underset{H}{N}-\overset{O}{\overset{\|}{C}}-\!\!\!\bigcirc\!\!\!-O-(CH_2)_6-\overset{+}{N}(CH_3)_3$$

$$CH_3(CH_2)_{13}-O\overset{}{\underset{O}{\overset{|}{C}}}-CH_2\qquad\qquad Br^-$$

$$\longleftarrow 16\ \text{Å}\longrightarrow$$

$$\longleftarrow\qquad 42\ \text{Å}\qquad\longrightarrow$$

34

A dialkylammonium surfactant

$$CH_3(CH_2)_{15}O\overset{O}{\overset{\|}{C}}\overset{}{\underset{}{C}}HNH\overset{O}{\overset{\|}{C}}(CH_2)_{10}\overset{+}{\underset{CH_3}{\overset{CH_3}{N}}}-CH_3Br^-$$

$$CH_3(CH_2)_{15}O\overset{}{\underset{O}{\overset{|}{C}}}(CH_2)_2$$

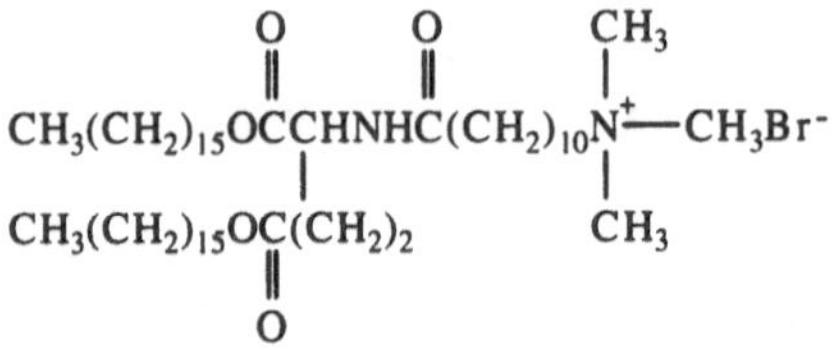

35

Bisacrylate monomer

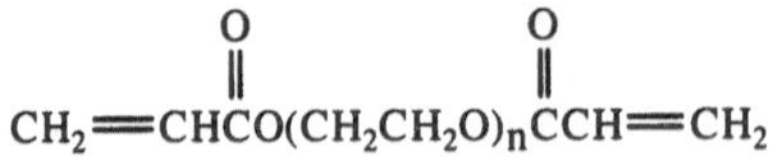

$$CH_2\!\!=\!\!CHC\overset{O}{\overset{\|}{}}O(CH_2CH_2O)_n\overset{O}{\overset{\|}{C}}CH\!\!=\!\!CH_2$$

36

Table 1. Contd.

$$CH_3 — (CH_2)_{11} — O \overset{\overset{O}{\|}}{C} \overset{\overset{H}{|}}{C} \overset{\overset{H}{|}}{N} \overset{\overset{O}{\|}}{C} — \langle benzene \rangle — O — (CH_2)_4 — N^+(CH_3)_3 \ Br^-$$

with branch:

$$\begin{array}{c} CH_2 \\ | \\ CH_3 — (CH_2)_{11} — O \overset{}{C} CH_2 \\ \overset{\|}{O} \end{array}$$

37

$$CH_3(CH_2)_{10}\overset{\overset{O}{\|}}{C}$$

carbazole ring system with N(CH_2)_4N^+(CH_3)_3   Br^-

$$CH_3(CH_2)_{10}\overset{}{C}$$
$$\overset{\|}{O}$$

38

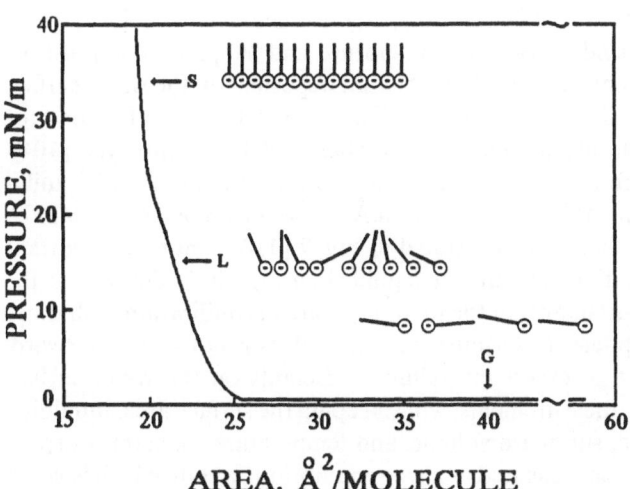

Fig. 6. Surface-pressure/surface-area isotherm of an AA monolayer floating on an aqueous 5.0×10^{-4} M Pb(NO$_3$)$_2$ solution at room temperature. Regions of gaseous (G), liquid (L), and solid (S) are indicated by *arrows*

2.1 Monolayers or Langmuir Films

Spreading an organic solution of a surfactant or phospholipid on an aqueous solution leads to monolayer formation at the water-air interface [68, 104–108]. Spreading is accomplished in a Langmuir trough equipped with a movable barrier which controls the surface pressure. Subsequent to evaporation of the organic solvent, the surfactant molecules lie on their backs on the water surface, relatively far from each other, in a two-dimensional gaseous state (Fig. 6). Careful examination reveals, however, the presence of a few aggregates in separate domains [109]. In the gaseous state, surfactant molecules occupy large areas (A) and have rather low surface pressures (π). Increasing the surface pressure by moving the barrier results in a two-dimensional gaseous-to-liquid phase transition which manifests itself in an increase in the gradient of the π vs A isotherm (Fig. 6). In the liquid phase, the surfactant molecules are beginning to "stand up" with their hydrophobic tails pointing away from the subphase. Other distinct phases have also been identified within the two-dimensional liquid phase of the monolayer; they are referred to as liquid expanded and intermediate phases and represent different degrees of surfactant compression in the monolayer. Further compression results in the transition of the surfactants to their two-dimensional solid state (also referred to as the liquid condensed, or LC, state). In the solid state, individual surfactant molecules are closely packed and are standing on their heads in an approximately upright confirmation. Consequently, changes in the surface pressure cause relatively minor changes in the headgroup area occupied by each molecule. Further increase in compression introduces mechanical instability and ultimately results in monolayer collapse. This event is accompanied by a decrease in π.

The pressure at which the monolayer collapses (the collapse pressure, π_c) and the corresponding area (the collapse area, A_c) are characteristic for a given surfactant and are dependent on the temperature and subphase composition. Knowledge of the concentration of the surfactant deposited on the water surface and the dimensions of the trough surface affords absolute values for the area occupied by each molecule in the different phases of the monolayer [104]. Langmuir isotherms are generally given as surface pressure (in $mN\,m^{-1}$, milli-Newton per meter) against molecular area (in $Å^2$/molecule) plots. A typical π-A isotherm for a thiol surfactant is illustrated in Fig. 7 [110]. Another important parameter is the spreading pressure at equilibrium (π_e). It is defined as the pressure at which the surfactants in the monolayer are in equilibrium with those in the bulk crystalline phase. Determination of π_e values is not straightforward; it requires the presence of excess crystalline surfactants on the water surface during surface-pressure measurements, while keeping the surface area, humidity, subphase concentration, subphase volume, and temperature constant. Further, π_e should, by definition, be measured at a true equilibrium, a state which is yet to be achieved.

Surface-pressure/surface-area isotherms provide valuable insight into the molecular packing of surfactants in monolayers. A steep slope in the π-A curve

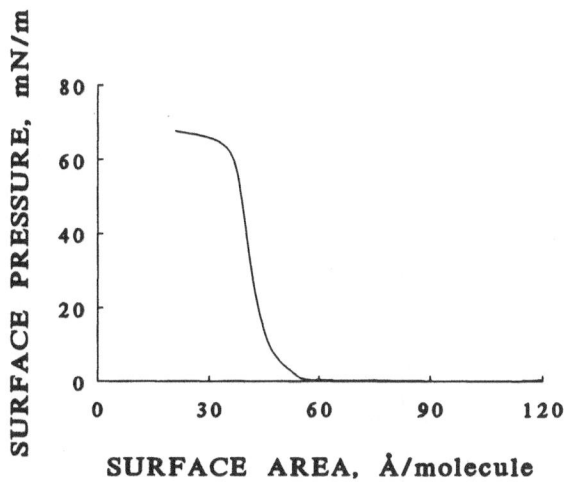

Fig. 7. Surface-pressure/surface-area isotherm of a thiol surfactant [110]

is indicative, for example, of strong chain-chain interactions and tight packing [104].

Precise determination of monolayer phase transition and π_e values allows, in fact, the thermodynamic treatment of these two-dimensional systems [105, 111–113]. Two examples illustrate this approach. In the first work, chiral discrimination between the packing of enantiomeric and racemic N-(α-methylbenzyl)stearamides into monolayers floating on aqueous acid subphases was observed [114]. Determination of π_e values at a range of temperatures allowed the evaluation of thermodynamic parameters for equilibrium spreading of the surfactants (Table 2). In all cases, the free energy of spreading (ΔG_s^0) was negative, while the enthalpies (ΔH_s^0) and entropies (ΔS_s^0) were found to be large and positive. This behavior indicates, of course, the spontaneity of spreading. Values of ΔG_s^0 for the racemates were always more negative than those for the enantiomers (Table 2), which was considered to signify a greater aggregation energy for the enantiomers than for the racemates [114, 115]. The examination of enantiomeric interactions in monolayers formed from chiral surfactants represented a significant contribution towards our understanding and utilization of molecular recognition [72, 115].

Investigation of thermodynamic behavior of monolayers formed from mixtures of bis[2-(n-hexadecanoyl)ethyl]methyl(4-vinylbenzyl) ammonium chloride (**1**) and dioctadecyl-dimethylammonium bromide (**5**) serves as the second example [116]. The determined surface-pressure/surface-area isotherms of monolayers prepared from (**1**) and (**5**) and from their mixtures are given in Fig. 8. The isotherms showed the expected behavior: following the liquid state there was a transition to the solid state in the region of 20–36 mN m^{-1}. This transition was characterized by a pressure, π_i, and by a corresponding area, A_i. All isotherms intersected in the vicinity of 30 mN m^{-1} and, at surface-pressure/surface-area

Table 2. Thermodynamic properties for spreading from crystals for the chiral stearamides on 10 and 6 N H_2SO_4 [a] [114]

	Subphase acidity								
	10 N H_2SO_4								6 N H_2SO_4	
	ΔG_s^0, kcal mol^{-1}		Q_s, kcal mol^{-1}		ΔH_s^0, kcal mol^{-1}		ΔS_s^0, cal deg^{-1} mol^{-1}		ΔG_s^0, kcal mol^{-1}	
temp, K	racemate	enantiomer	racemate	enantiomer	racemate	enantiomer	racemate	enantiomer	racemate	enantiomer
288	− 0.525	− 0.100	+ 8.6	+ 10.7	+ 8.1	+ 10.7				
298	− 0.776	− 0.461	+ 8.9	+ 10.7	+ 8.1	+ 10.3	28	35	− 0.515	− 0.066

[a] $\Delta G_s^0 = - \pi_e A_e$; $Q_s = TA_e d\pi_e/dT$; $\Delta H_s^0 = Q_s + \Delta G_s^0$; $\Delta S_s^0 = A_e d\pi_e/dT$

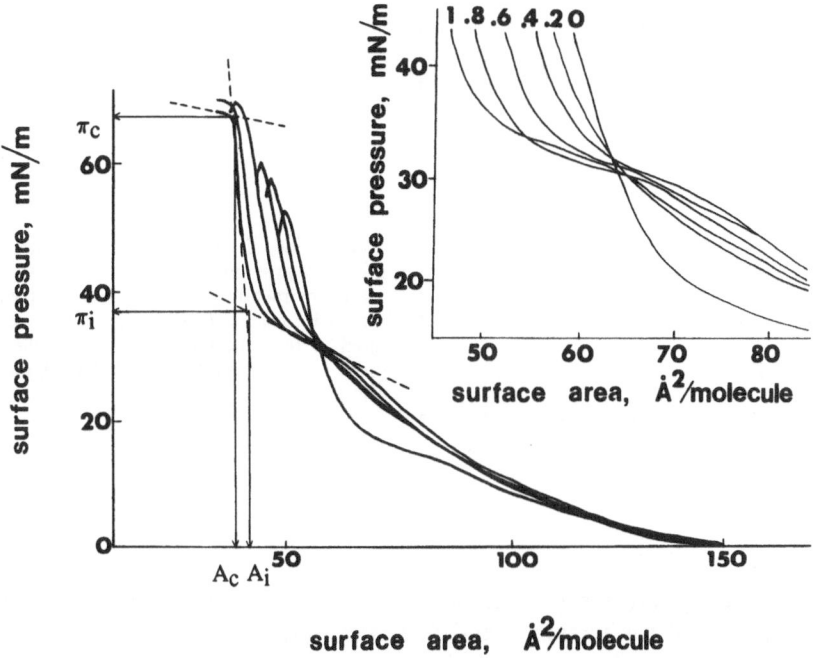

Fig. 8. Surface-area/surface-pressure isotherms for spreading **1**, **5**, and mixtures of **1** + **5** at 0.2, 0.4, 0.6, and 0.8 mole fractions of **1** on aqueous 5.0 mM NaCl. Areas (A_i) and pressures (π_i) associated with the transition to a compressed state were taken by projecting the intersection of straight lines drawn to the appropriate sections of the isotherm to the surface-area and surface-pressure axes. The collapse pressure (π_c) and collapse area (A_c) were taken by treating that transition, similarly. The *insert* shows an expansion of the isotherms between 20–40 mN/m. Temperature = 24.0 ± 0.5 °C [116]

isotherms above this pressure, monolayers arranged themselves according to their composition (see insert in Fig. 8). The equation of state of the solid monolayer was characterized by a relatively small change in the surface area with increasing surface pressure. This situation continued until the monolayers collapsed at a pressure, π_c, and at an area, A_c. The method of taking π_i and π_c, A_i and A_c values from surface-pressure/surface-area curves is shown in Fig. 8. The collapse pressure of the monolayer prepared from **5** (53 mN m^{-1}) was found to be smaller than that of the monolayer prepared from **1** (67 mN m^{-1}). Increasing the mole fraction of **1** (x_1) resulted in increased collapse pressures until a plateau value was reached, after which π_c decreased (Fig. 9). The obtained data could be treated in terms of a two-dimensional pressure (π_c) composition (x_1) phase diagram [104, 105, 117–120]. The collapse pressure of a monolayer containing two components (**1** and **5**) that are completely miscible is given by [117–120];

$$\chi_1^{Mf}_1 \exp\left(\frac{\pi_{c,m} - \pi_{c,1}}{\kappa T} A_1\right) + \chi_5^{Mf}_5 \exp\left(\frac{\pi_{c,m} - \pi_{c,5}}{\kappa T} A_5\right) = 1 \tag{1}$$

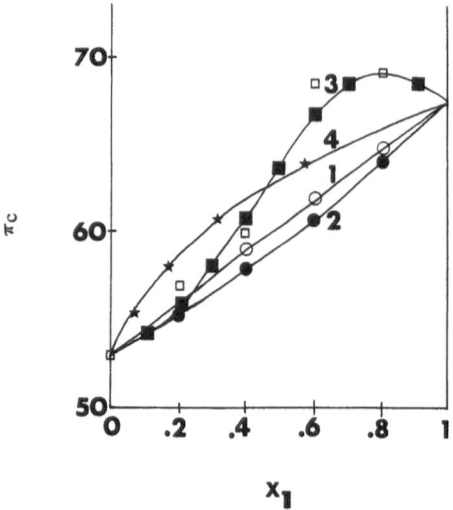

Fig. 9. Plot of the experimentally determined collapse pressures (π_c values) against x, the mol fraction of **1** in **1** + **5** (indicated by the *open squares*). *Lines 1, 2, 3, and 4* are fitted to theoretical values based on ideal (*lines 1 and 2*), modified ideal (*line 4*), and non-ideal (*line 3*) mixing [116]

where χ_1^M and χ_5^M are the mole fraction in the monolayer (indicated by the superscript M) of surfactants **1** and **5**, respectively; $\pi_{c,1}$, $\pi_{c,5}$, and $\pi_{c,m}$ are monolayer collapse pressures of pure **1**, pure **5**, and those of mixtures of **1** and **5** at a given composition (χ_1^M, χ_5^M, respectively); A_1 and A_5 are limiting surface areas at the collapse point; f_1 and f_5 are surface activity coefficients at the collapse point of components **1** and **5**, respectively; and κ and T are Boltzman's constant and the absolute temperature.

The activity coefficients are related to χ_1^M and χ_5^M by the interaction parameter, I

$$f_1 = \exp[I(\chi_5^M)^2] \tag{2}$$

$$f_5 = \exp[I(\chi_1^M)^2] \tag{3}$$

and hence to the interaction energy, $\Delta\varepsilon$:

$$\Delta\varepsilon = I\kappa T/6 \tag{4}$$

Substitution of Eq. (2) and Eq. (3) into Eq. (1) leads to

$$\chi_1^M \exp\left(\frac{\pi_{c,m} - \pi_{c,1}}{\kappa T}A_1\right)\exp[I(\chi_5^M)^2]$$

$$+ \chi_5^M \exp\left(\frac{\pi_{c,m} - \pi - c,5}{\kappa T}A_5\right)\exp[[I(\chi_1^M)^2] = 1 \tag{5}$$

In the absence of interactions, $f_1 = 1$, $f_5 = 1$, and

$$\pi_{c,m} = \chi_1^M \pi - c,1 + \chi_5^M \pi - c,5 \tag{6}$$

Changes in the collapse pressure of monolayers prepared from mixtures of **1** and **5** as a function of χ_1^M will, therefore, be ideal. In the presence of interactions, $\pi_{c,m}$ will deviate, of course, from ideal behavior. Expanding Eq. (1) (or Eq. 5) and neglecting the higher order terms led to

$$\pi_{c,m} = \frac{\chi_1^M A_1 \pi_{c,1} + \chi_5^M A_5 \pi_{c,5}}{\chi_1^M A_1 + \chi_5^M A_5} \tag{7}$$

which allows the approximation of $\pi_{c,m}$ values of different compositions of surfactants. Line 2 in Fig. 9 was calculated by substituting the appropriate parameters into Eq. (7). These results indicated a complete, but non-ideal, miscibility of **1** and **5** and the presence of a surface azeotrope.

Monolayers are best formed from water-insoluble molecules. This is expressed well by the title of Gaines's classic book: "Insoluble Monolayers at Liquid–Gas Interfaces" [104]. Carboxylic acids (**7–13** in Table 1, for example), sulfates, quaternary ammonium salts, alcohols, amides, and nitriles with carbon chains of 12 or longer meet this requirement well. Similarly, well-behaved monolayers have been formed from naturally occurring phospholipids (**14–17** in Table 1, for example), as well as from their synthetic analogs (**18, 19** in Table 1, for example). More recently, polymerizable surfactants (**1–4, 20, 21** in Table 1, for example) [55, 68, 72, 121], preformed polymers [68, 70, 72, 122–127], liquid crystalline polymers [128], buckyballs [129, 130], gramicidin [131], and even silica beads [132] have been demonstrated to undergo monolayer formation on aqueous solutions.

Composition of the subphase is of paramount importance for monolayer formation. Electrolytes, by virtue of their ability to salt-out organic molecules, often render surfactants insoluble and, thus, permit the formation of a well-behaved monolayer. A case in point is the previously cited example of optically active (N-(α-methylbenzyl)stearamides [114]. Stable monolayers could not be formed with these molecules on water. Stable monolayers readily formed, however, on strong aqueous acid solutions [114].

Scrupulous cleanliness is an absolute must in working with monolayers [104]: all materials used should be of the highest purity; solvents and solutions should be dust-free; glassware should be cleaned with a special cleaning solution and kept separate from other equipment; the film balance should be kept in a clean area on a vibration-free table; and surgical gloves should be worn at all times. These necessary precautions have been elaborated in frightening detail in a recent publication [114] and a review [107].

2.2 Langmuir–Blodgett Films

Monolayers and multilayers on solid substrates have come a long way since their initial preparation by Katherine Blodgett during the 1930s [133, 134]. Deservedly termed as "Langmuir–Blodgett (LB) films", they have flourished into a major discipline, advances in which are documented in review articles and books [55, 68, 104, 135–141].

Moving a clean plate (the substrate) through an aqueous solution-air interface results in the transfer of the monolayer onto the solid support during immersion (dipping) or withdrawal. Hydrophobic substrates preferentially attract the tails of surfactants and the monolayer is transferred during immersion. Conversely, polar substrates favor the surfactant headgroups and monolayer transfer preferentially occurs during the withdrawal process. Repeated withdrawal and dipping of a hydrophilic substrate through the monolayer leads to the buildup of a substrate-head-tail-tail-head-head Y-type multilayer LB film (Fig. 10). Alternatively, consecutive dippings (i.e. no alternating withdrawal through the monolayer) of a hydrophobic substrate results in a substrate-tail-head-tail-head, or X-type, multilayer deposition. Consecutive withdrawals (i.e. no alternative dipping through the monolayer) of a hydrophilic substrate produces a substrate-head-tail-head-tail Z-type multilayer LB film. Schematics of the different types of LB films are illustrated in Fig. 11.

The most common LB films are of the Y-type. Tail-to-tail hydrophobic and head-to-head electrostatic interactions stabilize these films. X- and Z-type LB films are more difficult to prepare and are less stable than Y-type LB films.

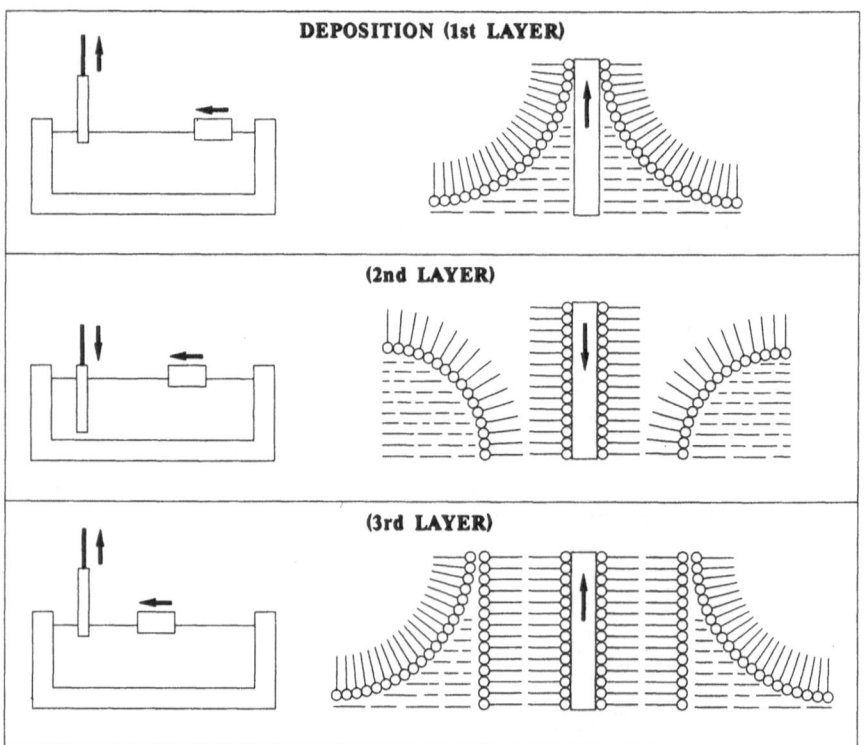

Fig. 10. Schematics of Y-type Langmuir-Blodgett (LB) formation on a hydrophilic substrate [164]

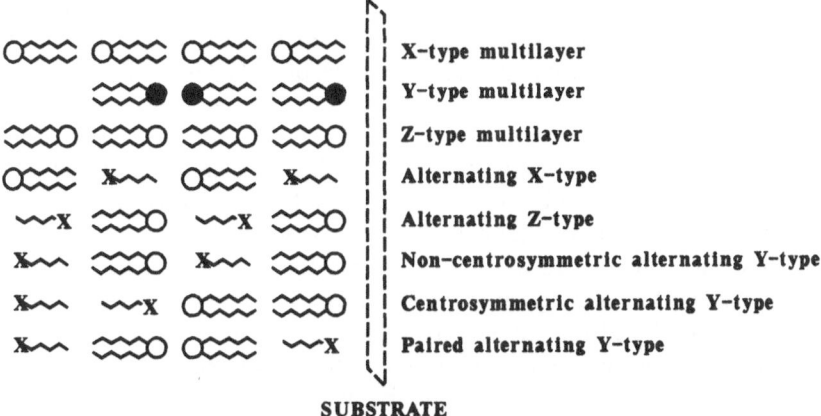

SUBSTRATE

Fig. 11. Schematics of different types of LB films [164]

Stable Z-type LB films have been prepared, however, by the alternative deposition of two different monolayers (A and B) in a head-to-head ABAB arrangement [142]. Two different monolayers can be layered a number of different ways (AABBAA, AABAAB, ABBABB, etc.) and there are a myriad of ways to arrange three or more different monolayers in LB films. These types of LB films have no plane of symmetry (i.e. they are non-centrosymmetric) and manifest non-linear optical behavior [108, 143]. Schematics of some of the different types of LB films are illustrated in Fig. 11.

The effectiveness of monolayer deposition onto a substrate is given by the transfer ratio $T_R = A_T/A_S$, where A_T is the decrease in surface area occupied by the monolayer at the aqueous solution-air interface resulting from transfer to the substrate at a constant surface pressure and A_S is the area of the substrate which becomes coated by the monolayer. Deposition is considered to be ideal when the determined T_R value is 1.00 ± 0.05 for both the downward and upward passage of the substrate through the monolayer [144]. Reaching an ideal transfer ratio is experimentally demanding. It requires, in the first place, a stable, well-compressed, and homogenous monolayer. Secondly, monolayer transfer should be performed at a constant, steady, and slow (on the order of millimeters per second) speed. Indeed, there is a critical velocity above which no monolayer transfer can be achieved. Satisfactory transfer ratios are most easily accomplished with Y-type deposition. Fluorescence microscopy [90, 93, 145], X-ray diffraction measurements [146–148], FTIR spectroscopy [149, 150], and X-ray photoelectron spectroscopy [151] have all been fruitfully used for the characterization of LB films. Unfortunately, in spite of extensive measurements of contact angles [152–155] and surface forces [156, 157], the mechanism of monolayer transfer to substrates is not completely understood [144].

Long-term mono- and interlayer stabilities, controllable morphologies, and defect-free structures are necessary requirements for the employment of LB films in device construction. Selective polymerization of monolayers provides a potentially viable approach for meeting these requirements [55, 57, 68, 121, 158–162]. Advantage has been taken, for example, of surfactants which contain styrene in their headgroups (1–3) or in their tails (4) to prepare polymerized monolayers and LB films [163, 164]. Ultraviolet irradiation of monolayers prepared from 1 and 2 resulted in decreased surface areas at constant surface pressures. Conversely, monolayers prepared from 4 showed increased surface areas at constant surface pressures during photopolymerization. It was considered that photopolymerization pulled the headgroups of 1 and 2 together thereby decreasing the average area occupied by each surfactant. Conversely, polymerization of 4 tied the surfactant tails together and, thus, increased the effective areas somewhat. These proposed interpretations of the data are illustrated in Fig. 12. Similar results were obtained in the polymerization of the corresponding LB films [164]. Although polymerized monolayers and LB films had greater stabilities, they contained more defects than their non-polymerized parents. Polymerization of monolayers prepared from mixtures of 1 and 4 is likely to remedy this situation.

LB films have also been prepared from simple (20, for example) [165] and functionalized (21, for example) [166] amphiphatic diacetylenes. Two different approaches were pursued. In the first approach, diacetylenes were polymerized as monolayers and subsequently transferred to substrates to generate LB films. In the second approach, LB films were formed from monomeric diacetylenes and were subsequently polymerized. Strong absorption of polydiacetylenes in

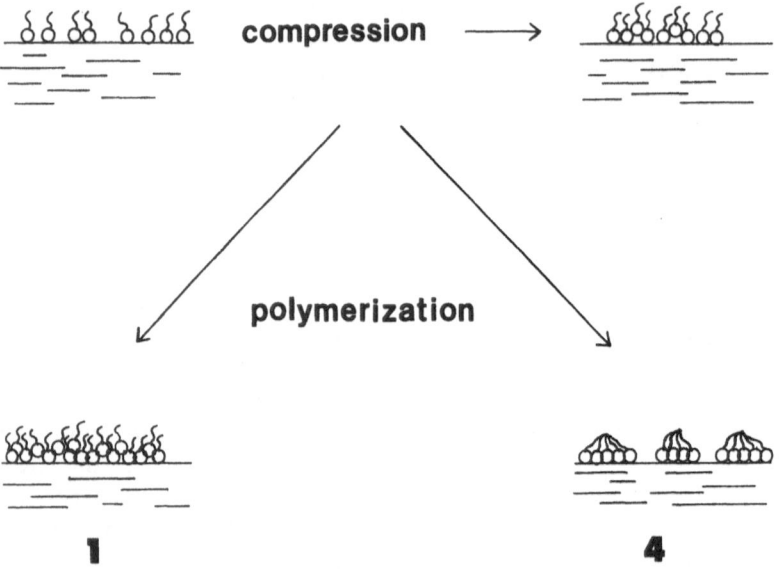

Fig. 12. An artist's conception of the compression and polymerization of monolayers prepared from 1 and 4 [163]

the visible spectrum afforded convenient monitoring of monolayer photopolymerization [167]. The degree of conjugation and conformation of the
polymer backbone determined the color (blue or red) of the polydiacetylenes
which were formed. Alternative double and triple bonds characterized the fully
conjugated polymer:

$$
R-C{\equiv}C-C{\equiv}C-R' \; + \; R-C{\equiv}C-C{\equiv}C-R' \;\rightarrow\;
\underset{R'}{\overset{R}{=}}C-C{\equiv}C-\underset{R'}{C}{=}\overset{R}{C}-C{\equiv}C-\underset{R'}{C}{=}
$$

(8)

Diacetylene monolayer photopolymerization was found to be topochemical; it
only occurred in the two-dimensional solid state of the surfactants. Polymerized
diacetylenes, both in monolayers and in LB films, were found to be rather rigid
and prone to cracking [160]. This undesirable property somewhat limits
the exploitation of polymerized diacetylene LB films for potential electronic
applications.

Large numbers of functionalized LB films have been prepared. Highly
ordered LB films have been formed by the inclusion of surface-active cobaltous
phthalocyanine [168]; amphiphilic TCNQ was assembled to function as conducting LB films [169]; liquid-crystalline LB films, potentially capable of
undergoing thermotropic or lyotropic phase transitions [170, 171], have also
been generated. Spacer groups introduced into polymeric surfactants (23) helped
to stabilize the LB films which they formed by decoupling the motion of pendant
polymers (see Fig. 13) [172].

In summary, LB films provide a route to precise two-dimensional molecular
architecture and, hence, to advanced nanostructured materials. However, LB
films do have disadvantages, such as the experimental difficulties associated with
the creation of defect-free, stable, and long-lasting structures in the dimensions
and scales required for device construction at an economically viable cost.

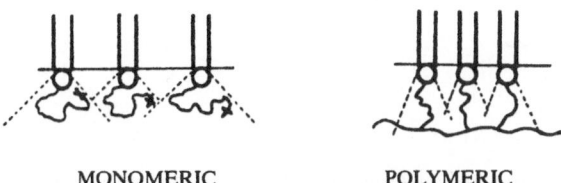

MONOMERIC   POLYMERIC

Fig. 13. Schematic representation of the influence of the hydrophilic spacer on the area occupied by
the lipids in the monolayer. High mobility of the spacer requires large areas in the fluid phase (*left*).
Reduced mobility of the spacer results in a decrease of the area occupied (*right*) [172]

2.3 Self-Assembled Monolayers and Multilayers

SA monolayers spontaneously form, by definition, upon the immersion of a substrate into an organic solution of a suitable surfactant. The method is attractive since it avoids the complex mechanical manipulation required for making LB films and it is amenable, at least in principle, to scale-up. Conditions for the reproducible formation of highly ordered, well-packed, and stable monolayers and multilayers have only been established during the last decade.

The rediscovery of wettability as an extremely sensitive and convenient surface characterization technique [47, 68, 101, 173–176] has considerably aided progress in this area. Qualitatively, it is immediately obvious that a hydrophilic surface (clean silicon, silicon dioxide, or gold, for example; i.e. the usual substrates for self assembly) is completely wettable by a polar solvent and has, therefore, a 0° solid-liquid contact angle. It is equally obvious that coating by a surfactant renders the surface to be more hydrophobic and, hence, decreases its wettability. This manifests in the incomplete spreading of the liquid droplet on the substrate and in an increase in the solid-liquid contact angle. Quantitatively, the contact angle, θ, is related to the solid-vapor (γ_S), the liquid-vapor (γ_{LV}), and the solid-liquid (γ_{SL}) interfacial tensions by Young's equation (Fig. 14):

$$\cos\theta = (\gamma_S - \gamma_{SL})/\gamma_{LV} . \tag{9}$$

This implies, of course, that wettability or contact-angle measurements only provide information on the differences and ratios of surface tensions (and, hence, on their relative surface-free energies) and not on their absolute values. Nevertheless, valuable information has been obtained from wettability measurements for a given SA system monitored as a function of changing one of its parameters (surfactant chain length or pH, for example) [177, 178]. Additional characterization by such techniques as FTIR [149, 150], X-ray photoelectron spectroscopy [151], scanning tunneling microscopy [179], X-ray specular reflectivity [180], and ellipsometry [181, 182] has also been fruitfully used to gain a better insight into the structures of SA systems.

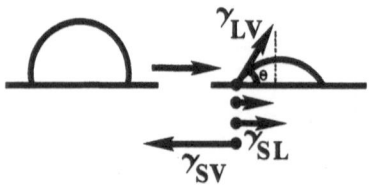

Fig. 14. Schematic illustration of a drop of liquid spreading in contact with a solid surface, showing the relations between the relevant parameters: the contact angle, θ; the solid/vapor interfacial free energy, γ_{SV}; the liquid/vapor interfacial free energy, γ_{LV}; and the solid/liquid interfacial free energy, γ_{SL}. Young's equation describes the relationship between these parameters for a stationary drop at thermodynamic equilibrium [175]

Adsorption of the surfactant headgroups to the substrate surface is the rate-determining step in the formation of SA monolayers. Van der Waals, electrostatic, and other weak interactions between the different regions of the surfactants are responsible for their reorganization into well-packed, ordered structures. Two different SA monolayer systems have been investigated to date. Self assembly, in the first system, involved the interaction of organosilanes with silicon-type substrates; their subsequent reaction resulted in the formation of polymer, network-like structures. The second system exploited chemisorption in order to locate organosulfur compounds onto gold and other metal surfaces. Preparation and properties of these two SA systems will be detailed in subsequent sections.

2.3.1 Self-Assembled Monolayers and Multilayers Derived from Organosilicon Derivatives

Formation of a true SA monolayer was first reported by Sagiv [183–185]. He immersed scrupulously clean glass, poly(vinyl alcohol), oxidized polyethylene, and evaporated aluminum substrates into millimolar solutions of n-octadecyl-trichlorosilane ($C_{18}H_{37}SiCl_3$, OTS) in an organic (8% $CHCl_3$ + 12% CCl_4 + 80% n-hexadecane, v/v/v) solvent. Exposure to OTS was quick, with the substrates being immersed and withdrawn at an approximate speed of 2.7 mm/s. The mode of self assembly was discussed in terms of chemisorption and hydrolysis of the Si–Cl bonds at the substrate surface and of the subsequent formation of a network of Si–O–Si bonds (Fig. 15). The availability of three reactive hydroxyl groups was considered to be sufficient for the formation of at least one covalent bond per surfactant molecule with the substrate [183].

The mechanism of self assembly was explored by the exposure of substrates to organic solutions which contained two different surfactants [183]. Mixtures of long-chain carboxylic acids + long-chain cyanine dyes, OTS + long-chain cyanine dyes, and long-chain carboxylic acids + OTS were used as pairs of surfactants. The cyanine dyes used had high extinction coefficients and distinctive fluorescence, and their transition moments aligned along the longitudinal axis of the chromophores. These properties permitted convenient spectroscopic measurements in addition to the determination of the orientation of the cyanine dye surfactants with respect to the substrate by linear dichroic measurements [137].

SA OTS was established to be a true monolayer (rather than a bi- or multilayer) by conductivity and Förster-type energy transfer measurements [183]. Energy transfer experiments were carried out on a composite system which consisted of a mixed OTS and donor cyanine dye (D) monolayer on a glass slide, on top of which a mixed cadmium arachidate acceptor cyanine dye (A) monolayer was deposited (by the LB technique) in such a manner that it covered only half of the glass slide. The other half was covered by a pure

Chemisorption of n-Octadecyl-trichlorosilane (OTS) on glass

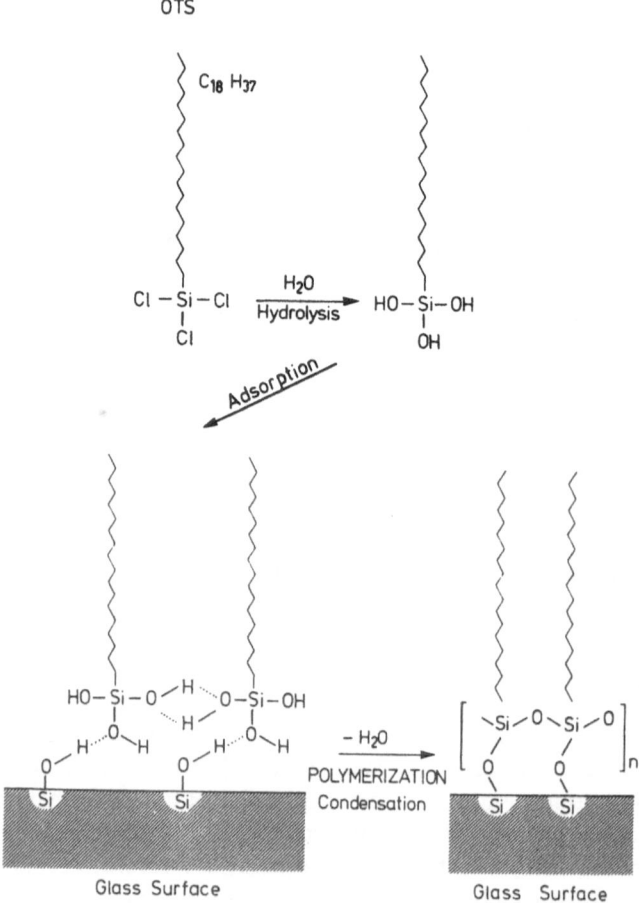

Fig. 15. Chemisorption of *n*-octadecyltrichlorosilane (OTS) on glass [183]

cadmium arachidate monolayer. A third layer of cadmium arachidate mono-layer (also deposited by the LB technique) covered the entire glass slide for protection. A similar system was prepared entirely by the LB technique in which the OTS was replaced by cadmium arachidate. Florescence energy transfers from the donor to the acceptor were found to be identical for both systems (Fig. 16), indicating identical monolayer thicknesses [183].

Preferential adsorption of a surfactant from a mixture depends on the structures of the amphiphiles and the substrate. Self assembly by physisorption is reversible, while that by chemisorption is irreversible. Thus, surfactants physisorbed in monolayers can be replaced by surfactants which are able to chemisorb. Such behavior was demonstrated by allowing a donor cyanine surfactant, D (capable of physisorption), and OTS (known to chemisorb) to

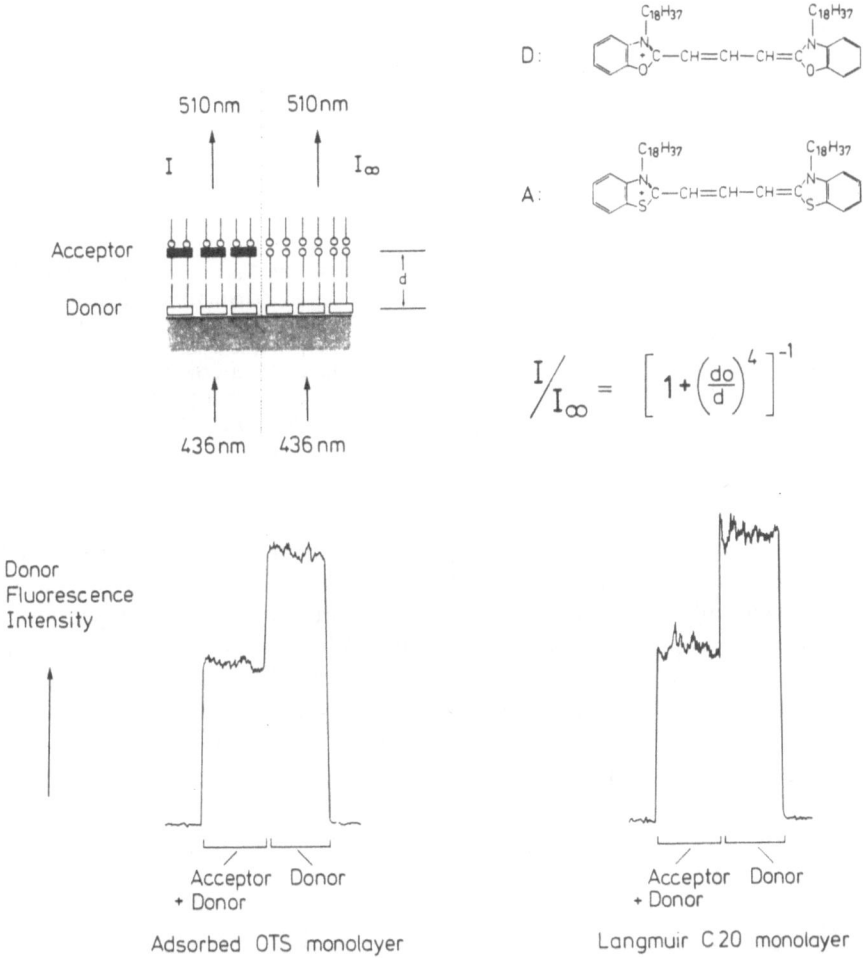

Fig. 16. Energy transfer experiment used for estimating the thickness of an octadecyltrichlorosilane (OTS) film adsorbed on glass. The layer containing the donor dye (D) is either a mixed adsorbed film of OTS + D or a transferred Langmuir monolayer of C_{20} + D. The acceptor layer (A) is a transferred Langmuir monolayer of C_{20} + A. A narrow excitation light beam scans the system from left to right and the corresponding donor fluorescence is recorded. The fluorescence intensity is quenched in the sample section containing the acceptor layer (*left part*) to about half of its value in the section where only the donor is present (*right part*). Approximately the same quenching effect is observed with either Langmuir or adsorbed donor layers [183]

compete for selected substrate. The ratio of OTS to D (R_M) present in the SA monolayer was found to increase with increasing time of substrate immersion into an organic solution containing fixed concentrations of D and OTS (Fig. 17) [183]. This behavior indicated kinetically controlled physisorption of the cyanine-dye surfactant and thermodynamically controlled chemisorption of OTS.

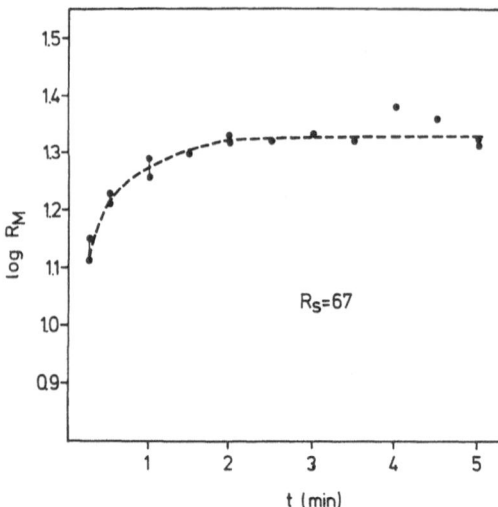

Fig. 17. Varying composition of mixed octadecyltrichlorosilane (OTS) + dye (D) monolayers adsorbed on glass as a function of adsorption time: R_M and R_S are the molar ratios OTS/D in the monolayer and in solution, respectively [183]

Removal of reversibly adsorbed (physisorbed) cyanine-dye molecules, D, led to skeletonized, SA silane monolayers having pinholes in the shape of D (Fig. 18) [185]. Such pinholes can, in turn, serve as templates for similarly shaped guest molecules. This approach opens the door for molecular recognition and device construction based on molecular recognition [186–190]. The idea relies upon the construction of a well-packed, SA monolayer from a mixture of OTS molecules and a solvent removable guest species. Removal of the guest molecules leaves pinholes with precise dimensions which only accept molecules with

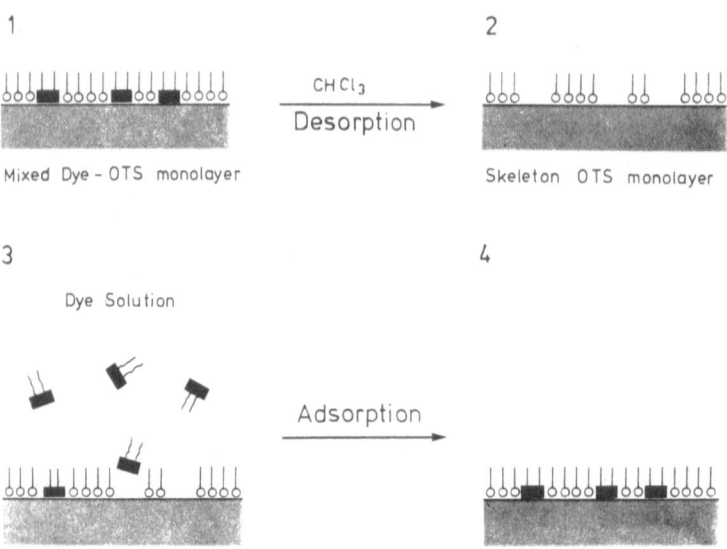

Fig. 18. Sequence of adsorption-desorption steps, assuming skeleton octadecyltrichlorosilane (OTS) monolayers to be perfectly stable in $CHCl_3$ and other solvents of low polarity [185]

the right configuration and reject others. Setting up the substrate as a transduction device (electrochemical or spectroscopic, for example) permits recognition and quantification of molecules which fit into the vacancies in the SA
monolayer.

Solvent molecules were found to be suitable template-forming guests in SA
monolayers [189, 191]. SA monolayers, constructed from a mixture of OTS and
hexadecane, provided sites for long-chain molecules subsequent to solvent
removal [189]. Template-forming reactions were also carried out by using
chlorodimethyloctadecylsilane with toluene and hexadecane as solvents [191].
Subsequent washing with chloroform removed the organic solvents and created
vacancies for octadecanoic acid and octadecanamine (see Fig. 19 for the
scheme).

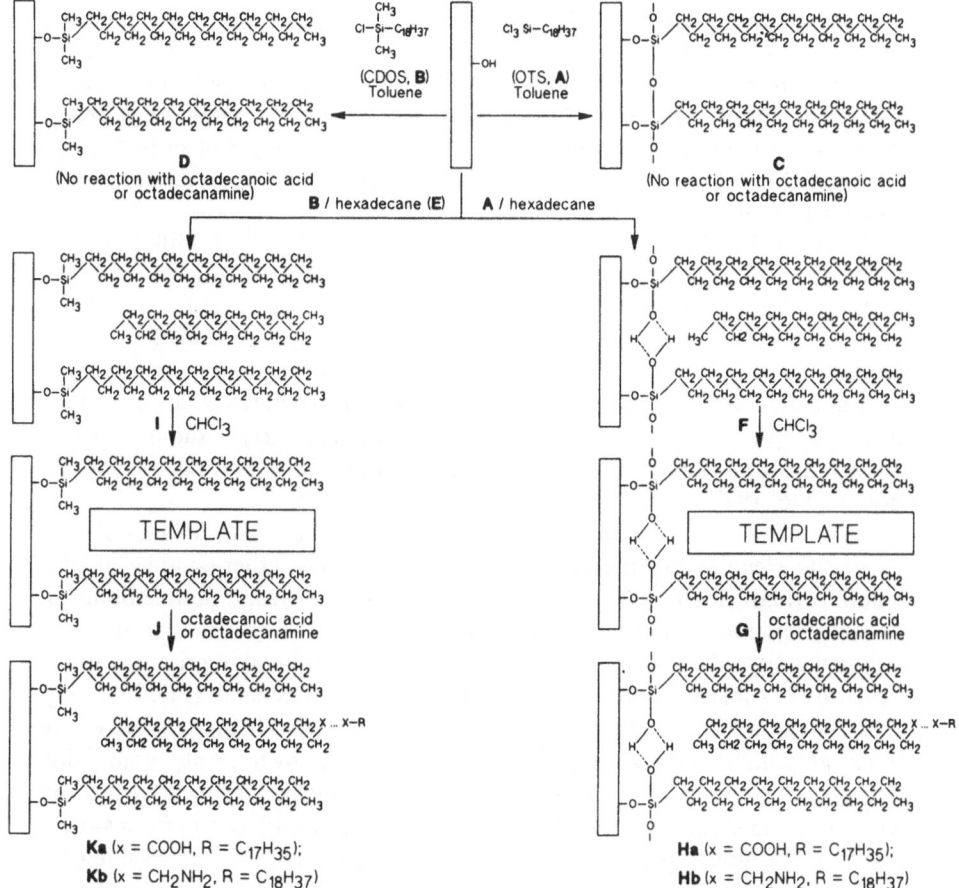

Fig. 19. Silanization of silicon substrates with octadecyltrichlorosilane (OTS) and chlorodimethyloctadecylsilane (CDOS) and reactivity of the silanized surfaces with octadecanoic acid and
octadecanamine (it is noted that the exact details of the interaction between OTS and CDOS and the
silicon surface are not defined) [191]

Linear-dichroic spectra of SA monolayers prepared from mixtures of OTS and a cyanine-dye surfactant established the absence of dimerization and the orientation of the chromophore parallel to the substrate [183]. In contrast, the same cyanine dye underwent sandwich-type dimer formation in LB films and had its chromophore oriented perpendicular to the water surface [192]. These results highlight an important difference between LB and SA monolayers. Parameters which determine monolayer formation on an aqueous subphase are also responsible for the orientation and organization of the surfactants therein. Furthermore, the configuration of the surfactants is retained regardless of the structure of the substrate to which the floating monolayer was subsequently transferred to by the LB technique. Conversely, in SA monolayers, surfactant organization is primarily dependent upon the nature of the substrate [183].

The first report on the construction of a SA multilayer appeared in 1983 [193]. A terminally bifunctional surfactant, 15-hexadecenyltrichlorosilane, was the initial building block for the formation of a SA monolayer. The trichloro-silane functionality reacted, as did OTS, with both of the hydroxyl groups on the substrate surface and with adsorbed water to form a network of Si–O–Si bonds in the SA monolayer. Conversion of the terminal double bonds to hydroxyl group functionality allowed the chemisorption of a new layer of surfactants to produce a SA bilayer (Fig. 20) [193]. The process could be repeated to form subsequent multilayers.

A similar strategy has been employed in construction of SA multilayers from methyl 23-(trichlorosilyl)tricosanoate, $H_3CO_2C-(CH_2)_{22}SiCl_3$ [194]. Chemi-sorption of this surfactant onto substrates resulted in a well-behaved SA monolayer whose exposed ester moieties could be reduced to alcohol groups which, in turn, could serve as the reactive sites for chemisorption of a subsequent layer of surfactant. Repetition of this process led to structures which contained up to 25 equally thick layers in a SA multilayer (Fig. 21); although surface characterizations indicated an increasing disorder in the surface hydroxyl groups [176, 194].

Desired chemical functionality may also be introduced into SA monolayers via in situ cleavage of unsaturated surfactants. Crown-ether-complexed pot-assium permanganate was shown to break the carbon-carbon double bonds in the substrate-immobilized, unsaturated surfactants and to convert them to carboxyl groups. Slow lateral propagation of this bond breaking resulted in the formation of shorter monolayers with exposed terminal moieties (Fig. 22) [195].

SA monolayers and multilayers derived from organosilicon derivatives are remarkably stable. They are highly resistant to washing by hot water containing 1% detergent, to prolonged extraction by organic solvents, or to treatment by acids [176, 183–185, 195–199]. Thermal stability was found to vary from system to system. SA monolayers and multilayers formed from OTS remained stable up to 110 °C, while those constructed from cadmium arachidate underwent irre-versible rearrangement [200]. The stability of OTS is derived from the presence of a polysiloxane network which aids in the immobilization of the surfactant headgroups. Water chemisorbed at the substrate surface has an important role

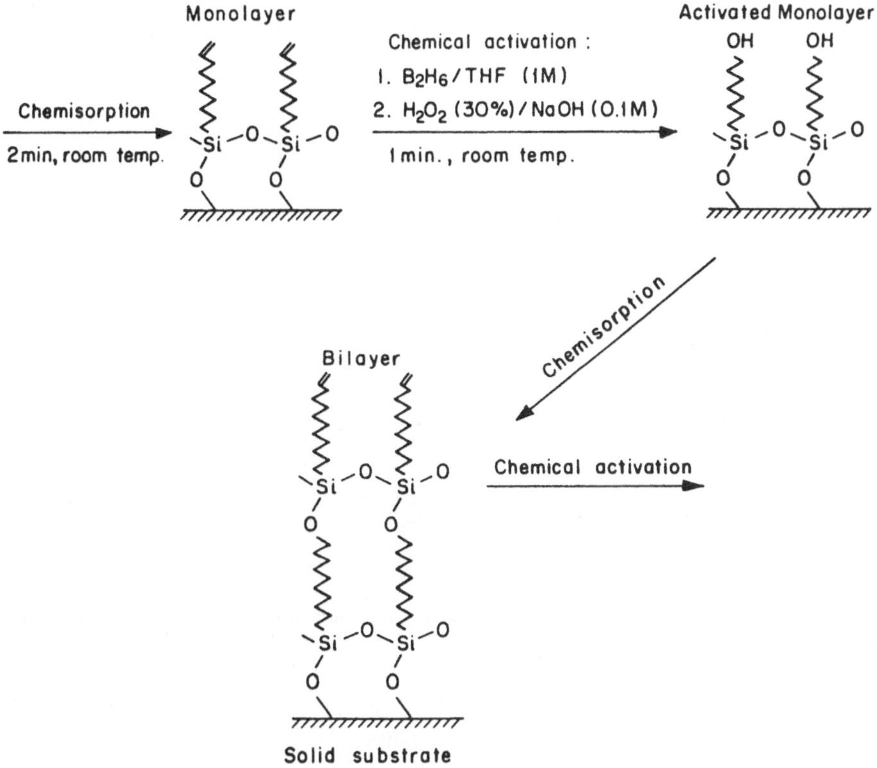

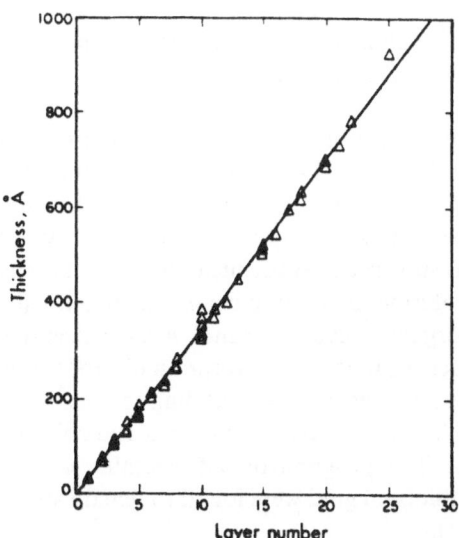

Fig. 20. Multilayer formation by adsorption, using bifunctional silane surfactants as monolayer building units [193]

Fig. 21. Film thickness, determined by ellipsometry, vs layer number, measured on eight different multilayer samples [194]

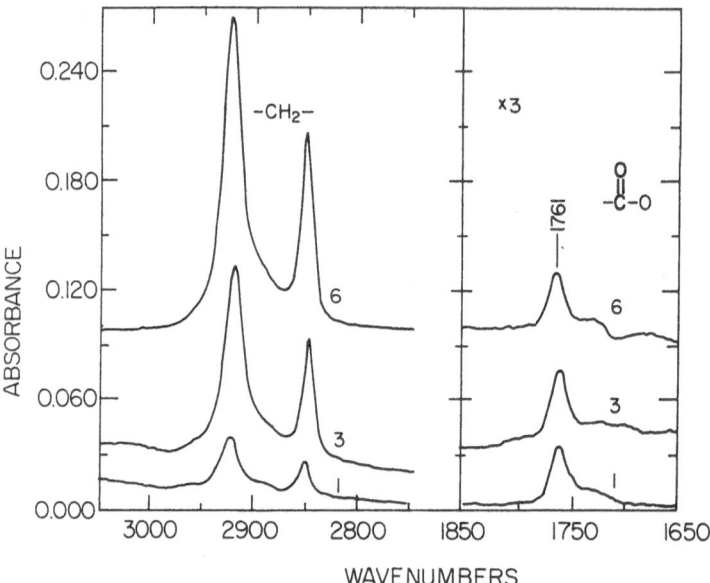

Fig. 22. FTIR-ATR spectra on silicon taken after deposition of one, three, and six monolayers [195]

in OTS stabilization. Ellipsometry and FTIR-ATR spectroscopy revealed the formation of closely packed SA OTS monolayers only on hydrated silicon oxide substrates; dry substrates could not be fully coated by OTS [201]. These results substantiated the involvement of trace amounts of water, both in the covalent bond formation between the monolayer and the substrate and in the subsequent polymerization (Fig. 23).

2.3.2 Self-Assembled Monolayers and Multilayers Derived from Organosulfur Compounds Chemisorbed on Gold and Other Metals

The initial report on the spontaneous assembly of oriented monolayers of organic disulfides on gold surfaces [202] prompted vigorous activity in this area [203–246]. Preparation of SA monolayers on gold is deceptively simple. Thin films of gold (about 2000 Å, > 99.999% purity), vapor deposited onto clean (> 99.999% purity) silicon, are used as substrates. Subsequent immersion of the substrate into a 10^{-3} to 10^{-1} M solution of a long-chain alkanethiol (or dialkyl disulfide) in a highly purified organic solvent (ethanol, hexane, acetone, and methylene chloride, for example) results in the chemisorption of surfactants onto the gold surface. Selected immersion times varied, according to the surfactant used, from minutes to days. Finally, the residual surfactants and solvents are removed by rinsing with alcohol. The progress of self assembly can be monitored by ellipsometry. SA alkanethiols on gold were found to remain stable indefinitely at room temperature [212].

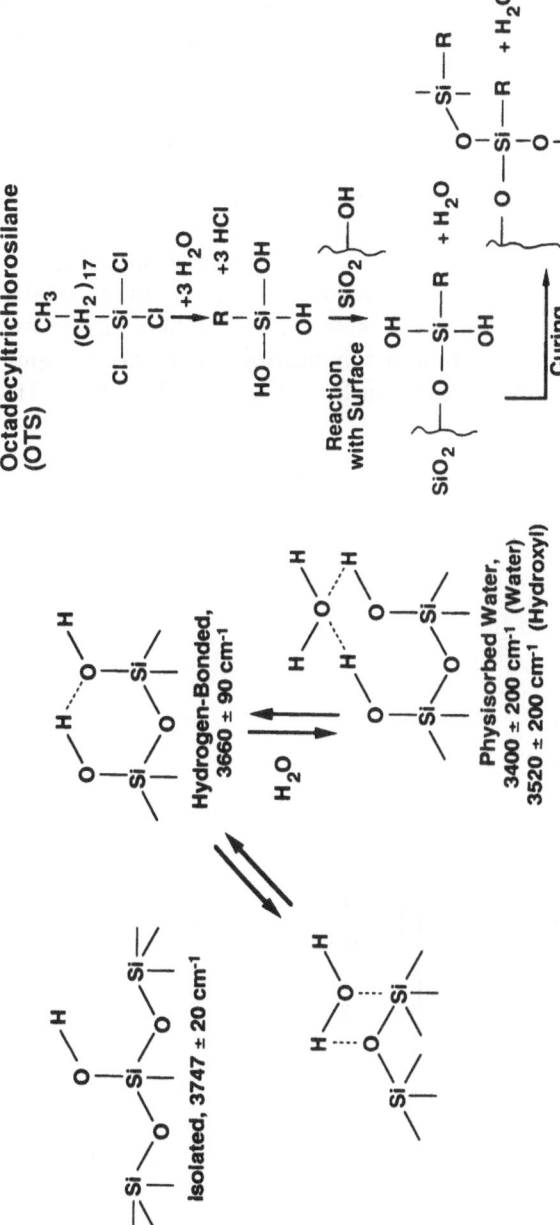

Fig. 23a. Hydration and dehydration reactions of the silica surface and the IR absorption frequencies of the surface species. Chemisorption of water produces surface silanols, which serve as adsorption sites for water. **b** Reaction of octadecyltrichlorosilane (OTS). Hydrolysis of the chloride group by trace amounts of water in solution to silanol is followed by condensation with surface silanols, resulting in covalent bond formation between the monolayer and the substrate. OTS molecules can also cross-link to form polymeric species during film curing [201]

Self assembly is governed by the strong interaction between the thiol (or disulfide) headgroups of the surfactant and the gold surface, as well as by the weak forces between the surfactant tails. Once again, useful insight has been gained into the assembly process by the examination of mixed SA monolayer formation in solutions where two surfactants were allowed to compete for the gold surface [224–227]. A case in point is the formation of SA monolayers from mixtures of different chain length α,ω-hydroxyalkanethiols – $SH(CH_2)_{11}OH$ and $SH(CH_2)_{19}OH$. Identical wettability was established for SA monolayers prepared separately from either the short- or from the long-chain surfactants, since both contained ordered arrays of hydroxyl groups at their surfaces. Monolayers self assembled from mixtures of these two surfactants had, however, different composition-dependent wettabilities, indicating the absence of significant surfactant segregation on the gold surface (Fig. 24) [226, 227]. This

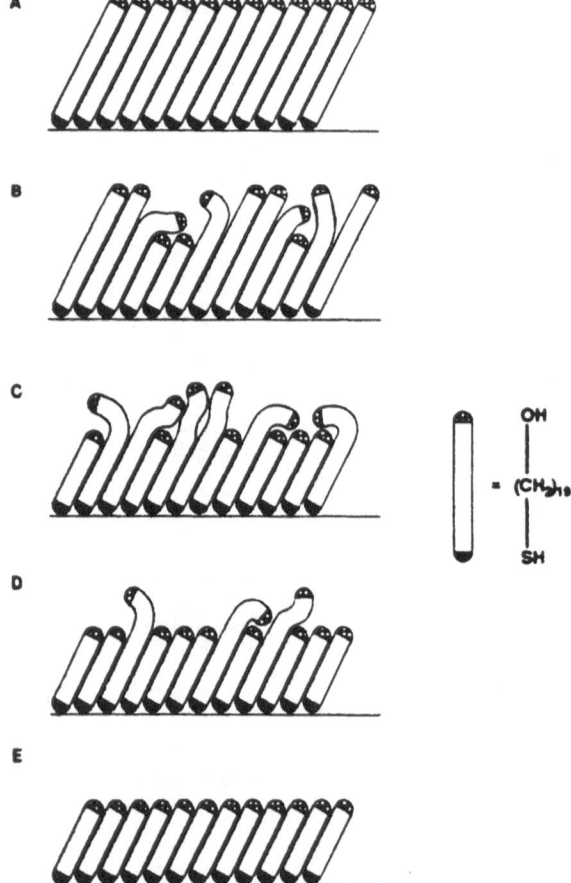

Fig. 24a–e. Schematic illustration of monolayers formed by adsorption of mixtures of $HS(CH_2)_{11}OH$ and $HS(CH_2)_{19}OH$ onto gold: **a** pure monolayer of $HS(CH_2)_{19}OH$; **b–d** monolayers containing different mixtures of the two components; **e** pure monolayer of $HS(CH_2)_{11}OH$ [175]

behavior is another manifestation of the differences between SA monolayers and those floating on aqueous solutions. Domain formation, an event prevalent in aqueous monolayers and LB films, is a rarity in SA systems.

Self assembly of monolayers on gold electrodes provided a viable approach to sensor construction (see below), as well as permitted electrochemical characterization of the mechanism involved in chemisorption [233, 236–239]. Electrochemical investigation of the spontaneous adsorption of *n*-alkanethiols onto a gold electrode and the subsequent desorption of the monolayer were rationalized in terms of oxidation and reduction of the sulfur atom [233]. More intimate details concerning the chemical fate of all species involved in the chemisorption process have not yet been elucidated.

Structures of SA alkanethiol monolayers generated on gold substrates are better understood than the chemistries involved in their formation. Electron diffraction and FTIR measurements, together with computer simulations, have provided a picture of a SA monolayer in which the axes of the alkyl chains of the surfactants are tilted approximately 30° with respect to the surface normal of the substrate and the sulfur atoms reside in the three-fold hollows of the gold (111) surface (Fig. 25) [68, 207–210, 230, 240]. The use of scanning tunneling micro-

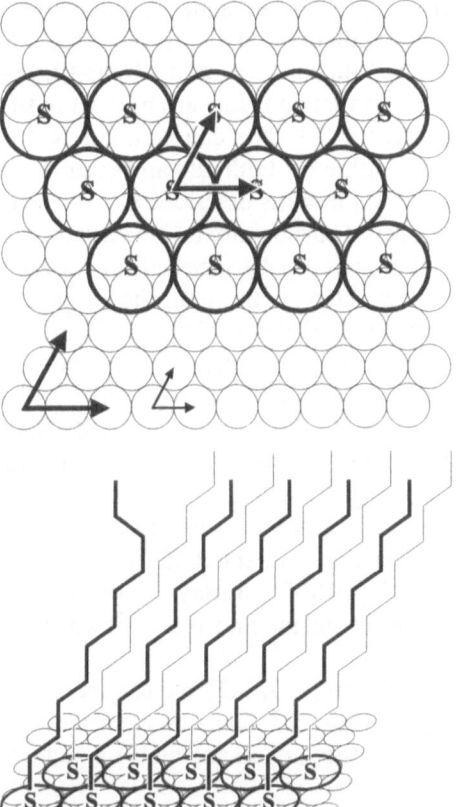

Fig. 25a, b. Schematic illustration of an alkyl thiolate monolayer on Au(111): **a** a top view, shows the inference from electron diffraction experiments: the adsorbed thiolates are epitaxially located on the gold surface. The sulfur atoms are located in the threefold hollow sites. The *circles* surrounding the sulfur atoms are used to suggest the area parallel to the gold surface occupied by the alkyl chain; **b** a side view, shows the orientation of the alkyl chains inferred from polarized infrared external reflectance spectroscopy (PIERS). These alkyl chains exist largely in an ordered, *trans*, zigzag conformation canted at an angle of approximately 30° from the normal to the surface. The presence of gauche conformations in the alkyl chains cannot be ruled out [175]

scopy has confirmed this picture. Atomically resolved images of ethanethiol and
n-octadecanethiol on the (111) surface of epitaxially grown gold revealed a
hexagonally packed array of adsorbates with respective nearest-neighbor and
next-nearest neighbor spacings of 0.50 ± 0.02 nm and 0.87 ± 0.04 nm
[179, 234].

Scanning tunneling microscopic techniques have also been used for etching
SA octadecanethiol monolayers on atomically flat gold (111) substrates on mica
[219]. The observed image of the substrate in a 3.5 nm × 3.5 nm area clearly
showed the hexagonal close packing of Au (111) with respective nearest-neighbor and next-nearest-neighbor spacing of 0.29 ± 0.02 nm and 0.49 ± 0.02 nm
(Fig. 26). The STM image of an octadecanethiol monolayer on this gold
substrate prior to etching is shown in Fig. 27a [219]. The effects of 10 s, 10 min,
and 35 min of etching (at a tip bias of 10 mV and tip current of 10 nA) are
illustrated in Figs. 27b, 27c, and 27d, respectively. The surfactant can be seen to
have been scattered off and the underlying gold surface has been disrupted. The
potential of this approach for nanofabrication is evident [219].

Structures of SA alkanethiol monolayers have also been examined on silver
and copper substrates [218, 241]. Orientation of the alkyl chains of the surfactants on gold, silver, and copper substrates have been assessed by reflection,
infrared, and X-ray photoelectron spectroscopies and by numerical simulation
[218]. The coordinate system used to define chain orientation is given in Fig. 28.
Here, the tilt angle, α, is defined as the cant of the long axis of the trans-zig-zag
polymethylene chain relative to the surface normal direction, z. The twist of the
plane of the chain around its axis is given by β. The conservation of the
metal-S-C geometry was adopted as a convention for both odd and even chain
lengths. As drawn, the direction of the C–CH$_3$ bond is described in Fig. 28 as
a positive tilt for a chain with an odd number of carbon atoms (even number of
CH$_2$ groups). The convention requires that α and β be specified for an even
chain in which the initial C–CH$_3$ bond is tilted in the opposite (negative)
direction. The proposed structures for the alkyl chains in SA monolayers on
silver and copper substrates were all-trans zig-zag, canted (in the opposite
directions alternating between $+\alpha$ and $-\alpha$ for odd and even numbers of
methylene groups) by about 12° from the normal to the surface, and had about
a 45° twist of the plane containing the carbon backbone (Fig. 29) [218]. The
proposed structures for SA alkanethiols on gold substrates were also trans-
zig-zag extended, had a cant angle of about 27° from the normal to the surface,
and had a twist of the plane of the carbon backbone of about 53° (Fig. 29). On
gold substrates, sulfur atoms, in an $(\sqrt{3} \times \sqrt{3})$R30° overlayer with a fixed
Au–S–CH$_2$ angle, were considered to force the chains into the same absolute
orientation, regardless of the chain length.

Formation of SA monolayers of phospholipids [217, 242] and surface-active
viologens [244, 245] on gold substrates, those of long-chain zirconium-
phosphonates on silicon substrates [243], and those of alkanethiols on gallium
arsenide substrates [246] have also been reported.

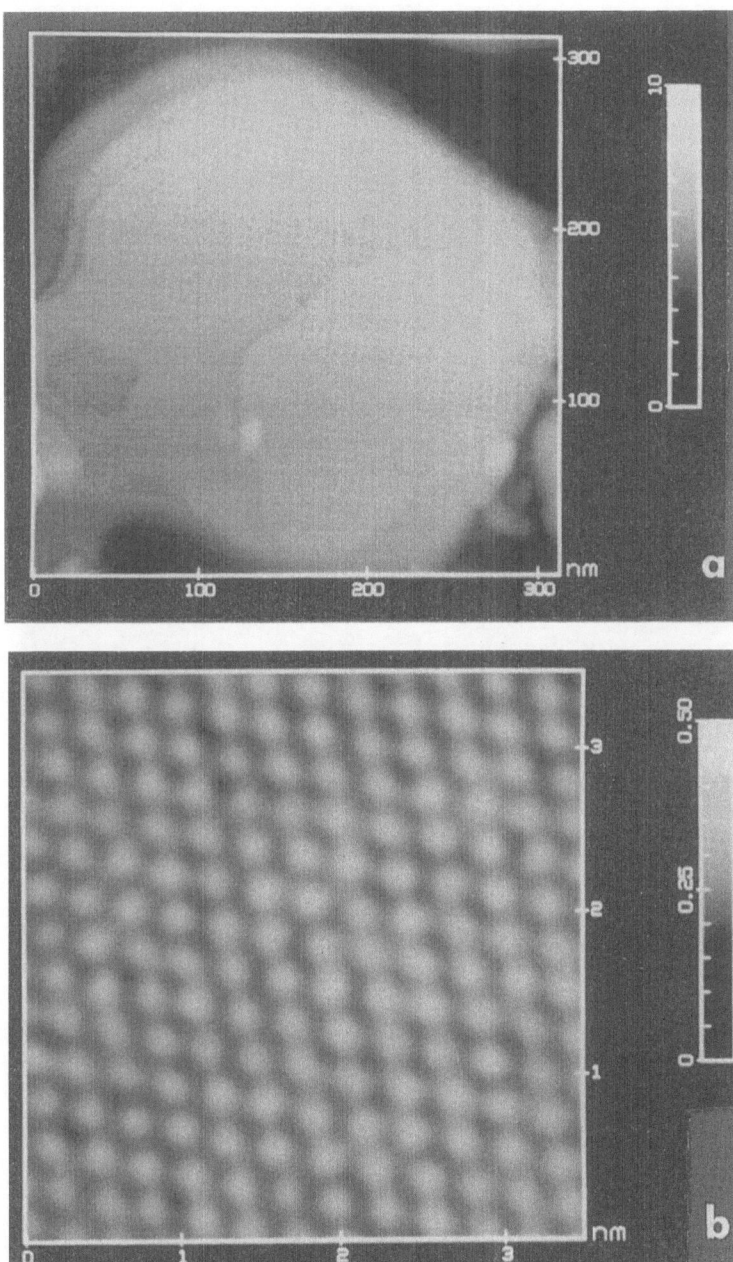

Fig. 26a. Scanning tunneling microscopic image of a 310 nm × 310 nm area of bare Au film epitaxially grown on a mica substrate (Au; mica) measured in air: constant current mode; bias (V_b) of + 50 mV, tunneling current (I_t) of 10 nA. **b** STM image of a 3.5 nm × 3.5 nm area of a bare Au/mica film: constant current mode; $V_b = + 4.9$ mV; $I_t = 3.0$ nA [219]

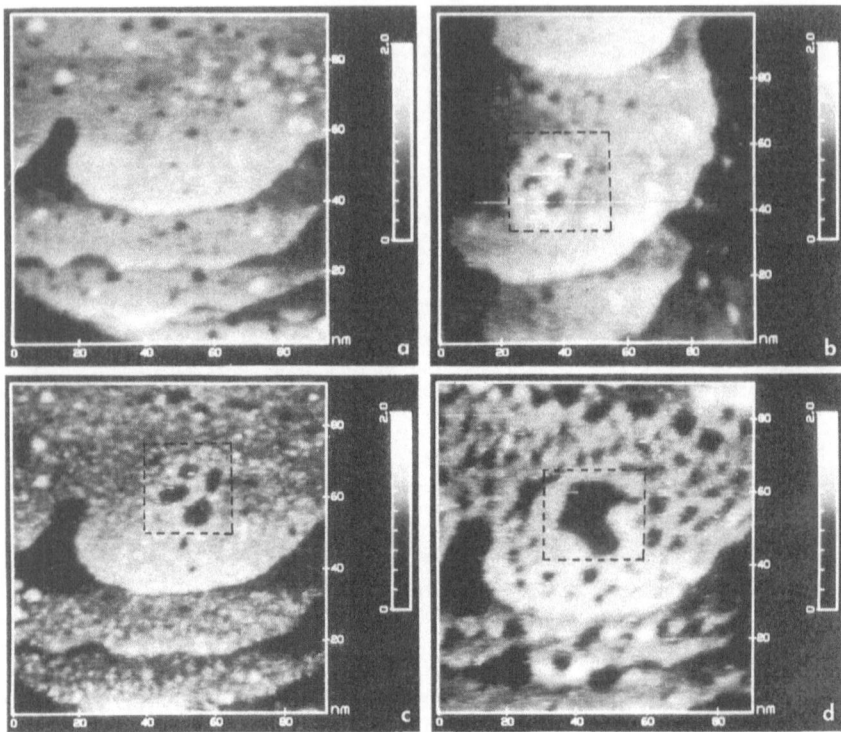

Fig. 27a–d. STM image of a 100 nm × 100 nm: **a** *n*-octadecanethiol film coated on Au/mica; **b** same after 10 s of etching mode ($V_b = 10$ mV, $I_t = 10$ nA) over a 10 nm × 10 nm area (within area shown by *broken line*); **c** same area after 10 min of continuous normal scanning; **d** after 35 min normal scanning. All images were taken in constant current mode, $V_b = +1$ V, and $I_t = 1$ nA [219]

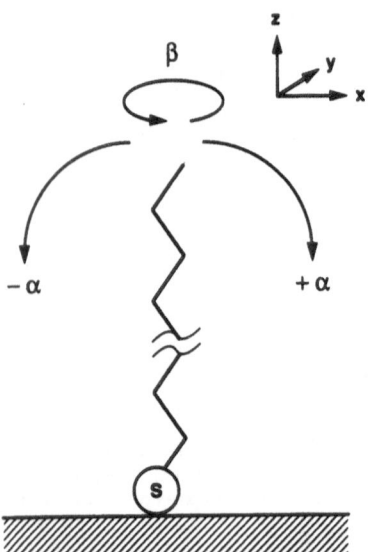

Fig. 28. Schematic diagram of an all-*trans* chain in an *n*-alkanethiols monolayer on a surface. The co-ordinate system used to define the orientations of the chain, the cant angle α, and the chain twist β are shown along with their relationship to the surface coordinates [218]

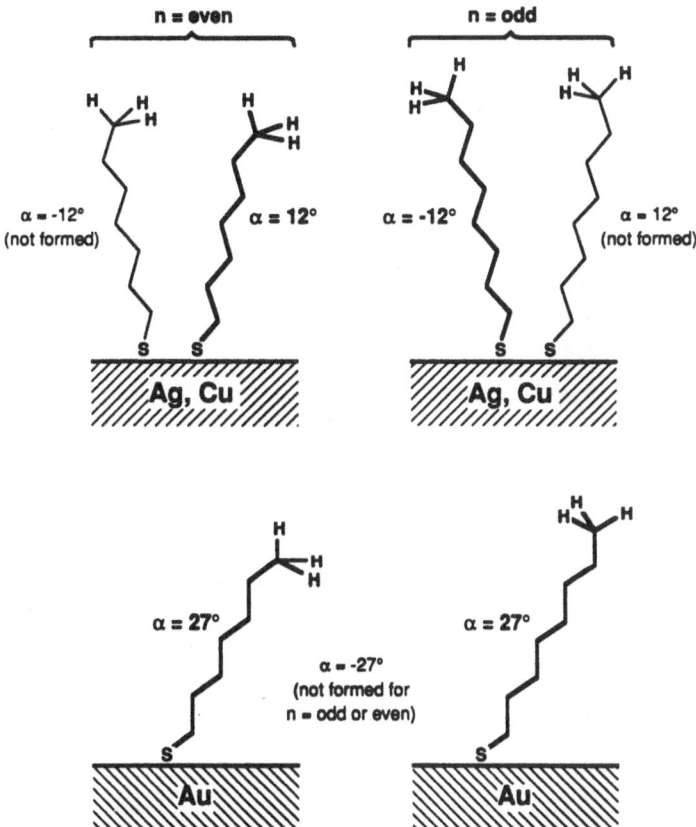

Fig. 29. Illustration of the canted structures formed upon adsorption of *n*-alkanethiols, $CH_3(CH_2)_nSH$, on copper, silver, and gold. The structures that form on copper and silver do not exhibit any change in the intensities of methyl modes with incremental increase in the number of methylene groups. This observation is consistent only with the formation of a structure in which the cant angle, α, is positive for an even number of methylene groups and negative for an odd number of methylene groups; the geometry of the metal-thiolate interaction on these metals is not fixed as *n* is varied (*upper panel*). The structures formed on gold exhibit an odd-even modulation in the intensities of the methyl modes with increasing chain length. This modulation reflects the formation of a structure in which variations in the chain length of the adsorbate do not perturb the geometry of the gold-thiolate interaction (*lower panel*) [218]

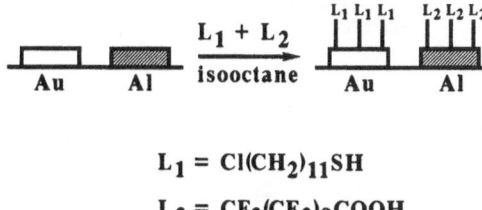

$$L_1 = Cl(CH_2)_{11}SH$$

$$L_2 = CF_3(CF_2)_8COOH$$

Fig. 30. Schematic illustration of the simultaneous formation of two independent (orthogonal) self-assembled (SA) monolayers on gold and alumina surfaces by exposure to a solution containing both a thiol and a carboxylic acid [175]

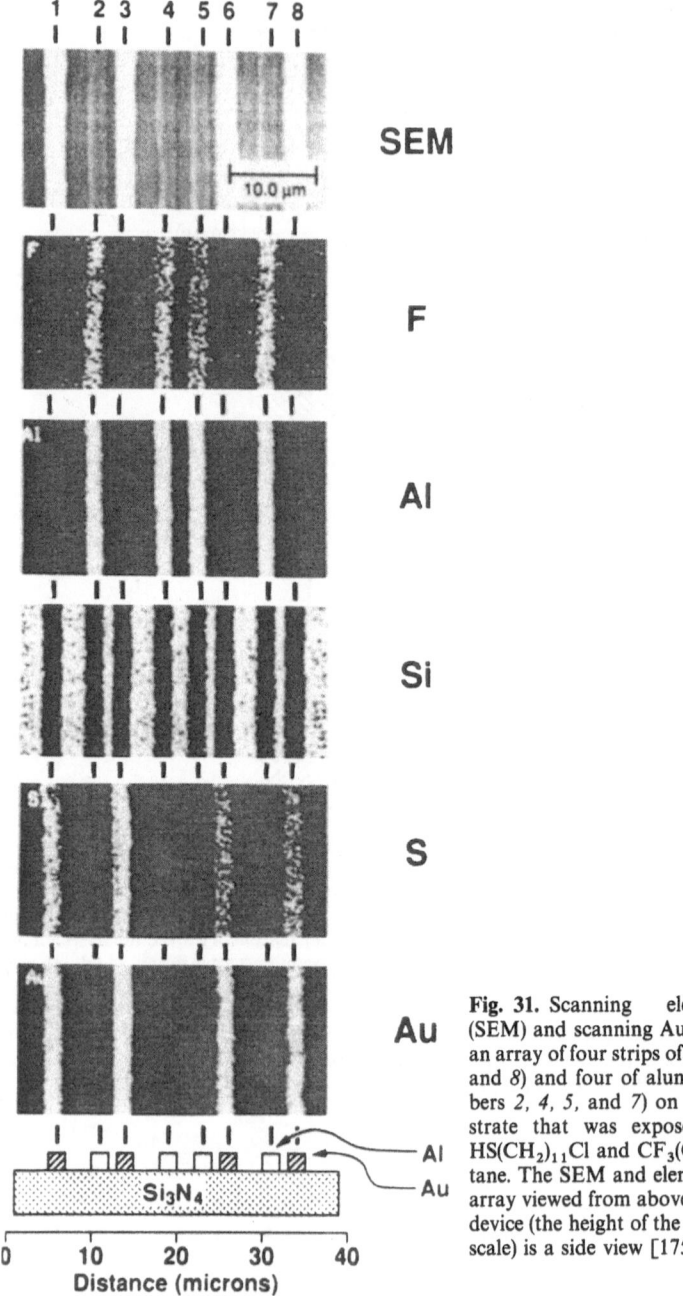

Fig. 31. Scanning electron micrograph (SEM) and scanning Auger element maps for an array of four strips of gold (numbers *1, 3, 6,* and *8*) and four of aluminum/alumina (numbers *2, 4, 5,* and *7*) on a silicon nitride substrate that was exposed to a mixture of $HS(CH_2)_{11}Cl$ and $CF_3(CF_2)_8CO_2H$ in isooctane. The SEM and element maps are for the array viewed from above; the schematic of the device (the height of the strips is not drawn to scale) is a side view [175]

All of the SA systems discussed so far permitted a two-dimensional and organizational control perpendicular, but not parallel, to the substrate. Micronscale organizational control in the plane of the substrates has been recently described [247]. Advantage was taken to the selective adsorption of an alkanethiol (L_1) on gold and of a fluorine-labeled alkanecarboxylic acid (L_2) on aluminum. Exposure of a substrate patterned with gold and aluminum to a solution which contained a mixture of L_1 and L_2 resulted in the expected selected adsorption, as confirmed by scanning electron microscopy and scanning Auger spectroscopy (Figs. 30, 31) [247]. This work represents a major milestone in the chemical approach to nanofabrication. Related is the very recently reported procedure of using a pen, filled with hexadecanethiol, to write 1–100 μm features on a gold substrate [248]. Similarly, mixtures of hydrophobic (methyl-terminated) and hydrophilic (hydroxyl-, maltose-, and hexa(ethylene glycol)-terminated) alkanethiols were reported to self assemble onto gold and, thus, to provide good matrices there for protein recognition [249].

This section can be ended on a highly optimistic note. The relative ease of formation and the high degree of molecular level organization, coupled with physical and chemical stability, render SA monolayers and multilayers to be eminently suitable matrices for advanced materials construction.

2.4 Aqueous Micelles, Reversed Micelles, and Microemulsions

The aggregates termed "aqueous micelles" are generated from the spontaneous and dynamic association of ca. 50 to 100 surfactant molecules above a characteristic surfactant concentration, labeled the critical micelle concentration (CMC) (Fig. 32) [55–58, 101, 250]. Hydrodynamic diameters of aqueous micelles are of the order of 20–50 Å. Formation of aqueous micelles is a cooperative process. The opposing forces of repulsion between the polar headgroups and attraction between the hydrophobic chains of the surfactants are responsible for micellization. In general, the longer the hydrophobic tails are, the lower the CMC is. Aqueous micelles are the most disordered aggregates among the organized

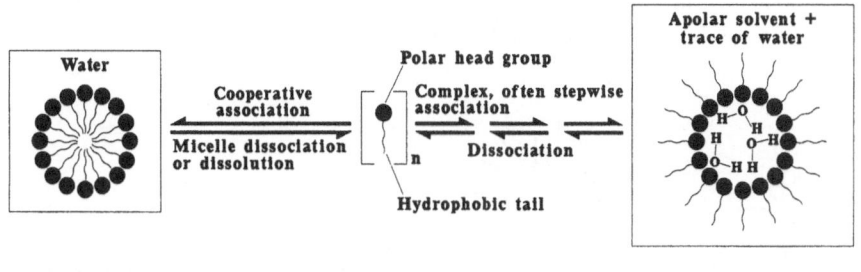

Fig. 32. Surfactant aggregates in both water and apolar solvents

surfactant assemblies characterized. They rapidly break up and reform by two known processes. The first one, occurring in the microsecond time scale, releases a single surfactant from the micellar ensemble and reincorporates it. The second process, taking place in the millisecond time frame, corresponds to the dissolution and subsequent reformation of the micelles. The dynamic nature of aqueous micelles has precluded, to date, their use as compartments for advanced materials.

Increasing the concentration of surfactants in water to a level above the CMC leads to the formation of rod-like micelles and, subsequently, liquid crystals [251]. Both liquid crystals and liquid-crystalline polymers [252] have been used as media for small particle generation [253, 254] and have also acted as piezoelectric devices [255]. Of particular interest are metallomesogens, the metal complexes of organic ligands which exhibit liquid crystalline behavior [255].

Surfactants having an appropriate hydrophobic/hydrophilic balance (sodium bis(-2-ethylhexyl)sufosuccinate, or AOT, for example) undergo concentration-dependent self association in apolar solvents to form reversed or inverted micelles (Fig. 33) [256–262]. Reversed micelles are capable of solubilizing a large number of water molecules (AOT reversed micelles in hexane are able to take up 60 water molecules per surfactant molecule, for example). Reversed-micelle-entrapped water pools are unique; they differ significantly from bulk water. At relatively small water-to-surfactant ratios (w = 8–10, where w = $[H_2O]/[Surfactant]$), all of the water molecules are strongly bound to the surfactant headgroups. Substrate solubilization in the restricted water pools of reversed micelles results in altered dissociation constants [256, 257, 263–265], reactivities [256, 258, 266], and reaction products [267].

Proteins and enzymes have been successfully entrapped in surfactant-solubilized water pools in organic solvents [268–278]. Furthermore, many reversed-micelle-entrapped enzymes retained their activity and could be used for peptide synthesis [273, 274]. That the water pools corresponding to very small w-values exhibited freezing points below − 50°C enabled both the enzyme structures and the rates of enzyme-catalyzed reactions to be investigated at low temperatures. These studies much aided the development of cryoenzymology [279, 180].

Increasing the concentration of surfactant-entrapped water results in the formation of larger aggregates which eventually become water-in-oil (w/o)

reversed
micelle

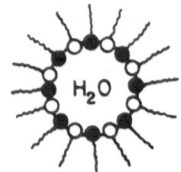

w/o microemulsion

Fig. 33. Oversimplified representations of a reversed micelle and a water-in-oil (w/o) microemulsion formed from surfactants

microemulsions [281]. Addition of gelatin or suitable polymers to AOT w/o microemulsions produced organogels in the water pools [282–287]. Alternatively, reversed micelles prepared from lecithin gelled, even in the absence of an added polymer [283]. Organogel reversed micelles are highly viscous and contain large, flexible cylindrical aggregates [284–288]. Evidence has also been presented for the formation of bilayer aggregates in organic solvents, called reversed vesicles [289, 290]. Structures formed from surfactants in organic solvents are, unlike aqueous micelles, amenable to hosting incipient advanced materials. Indeed, catalytic, magnetic, and semiconducting particles have been in situ generated in these systems (see below).

Interestingly, self replication has been demonstrated in reversed micelles [291, 292]. Reaction between cetylbromide and trimethylamine in an organic solvent gave hexadecyltrimethyl-amine which formed, in the presence of small amounts of water, reversed micelles. This system was considered to self replicate since an endogenous growth of the micelles corresponded to the in situ appearance of surfactants [291].

2.5 Surfactant Vesicles, Polymerized Vesicles, and Polymeric Vesicles

2.5.1 Surfactant Vesicles

Closed bilayer aggregates, formed from phospholipids (liposomes) or from surfactants (vesicles), represent one of the most sophisticated models of the biological membrane [55–58, 69, 72, 293]. Swelling of thin lipid (or surfactant) films in water results in the formation of onion-like, 1000- to 8000-Å-diameter multilamellar vesicles (MLVs). Sonication of MLVs above the temperature at which they are transformed from a gel into a liquid (phase-transition temperature) leads to the formation of fairly uniform, small (300- to 600-Å-diameter) unilamellar vesicles (SUVs; Fig. 34). Surfactant vesicles can be considered to be spherical bags with diameters of a few hundred Å and thickness of about 50 Å. Typically, each vesicle contains 80,000–100,000 surfactant molecules.

SUVs can also be prepared by injecting an alcohol solution of the surfactant through a small bore syringe into water, by detergent dialysis, by ultracentrifugation, and by gel, membrane, or ultrafiltration [55]. Spontaneous formation of SUVs by swelling slightly charged phospholipids on a small surface [294], by swelling mixtures of single-tail cationic and anionic surfactants [295], or by subjecting MLVs to pH-jump [296] have also been reported. SUV formation has been rationalized by assuming disk-like transition structures and delicately controlled equilibria between the forces responsible for maintaining the various micellar and bilayer geometries in the solution [297–299].

Publication of reports on the formation of well-characterized SUVs from long-chain dialkylammonium salts [300, 301] has prompted the syntheses of

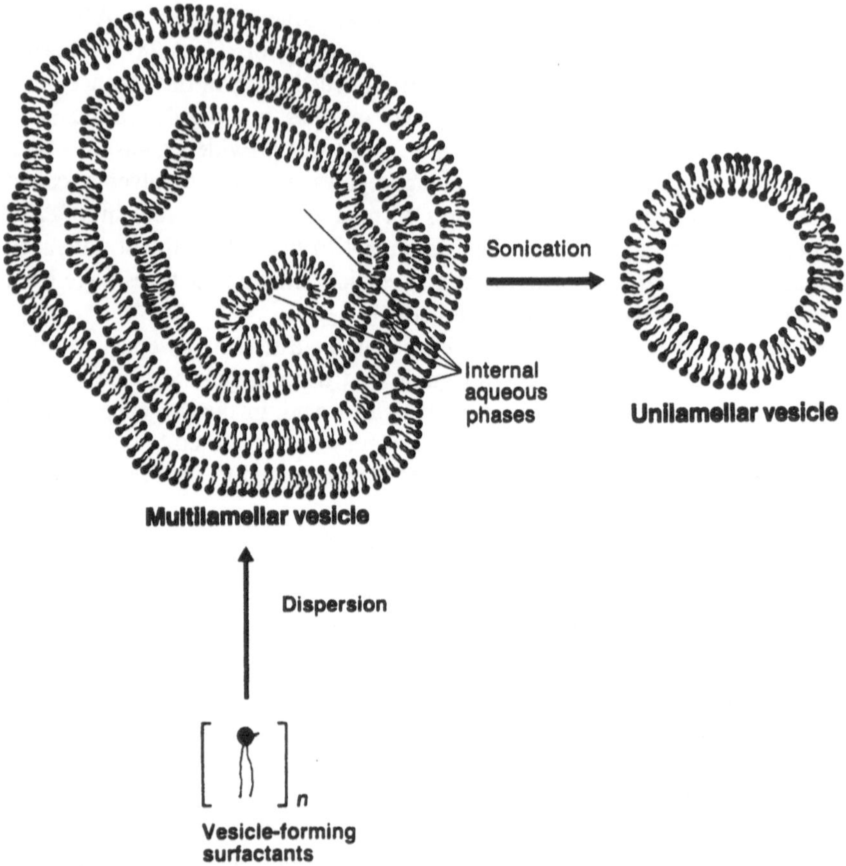

Sonication

Internal
aqueous
phases

Unilamellar vesicle

Multilamellar vesicle

Dispersion

$$\left[\; \right]_n$$

**Vesicle-forming
surfactants**

Fig. 34. Formation of surfactant vesicles

numerous single- and multiple-chained vesicle-forming surfactants [55–58, 69, 72, 293]. A great deal of versatility has been accomplished by the introduction of rigid segments, spacers, and connectors into surfactants (Fig. 35) [69]. Advantage has been taken of X-ray structural analysis and molecular modeling in

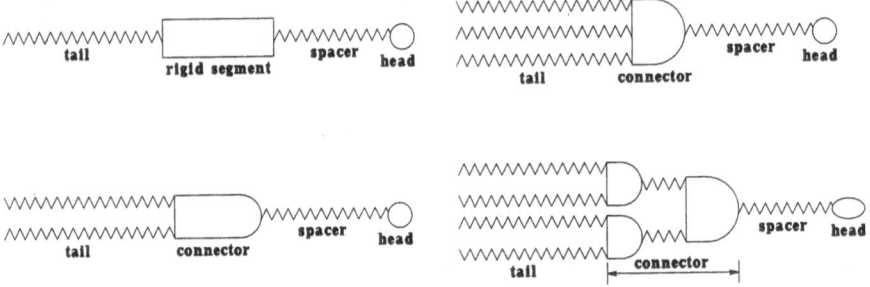

Fig. 35. Structural elements (modules) of bilayer-forming amphiphiles [69]

designing vesicle-forming surfactants. For example, the crystal structure of dioctadecyldimethylammonium bromide (**5**) indicated a 45° tilting of the molecules relative to the bilayer surface, the trans zig-zag conformation of one of the octadecyl chains, the folded conformation near the polar group, the trans zig-zag conformation towards the alkyl group of the other octadecyl chain, and the presence of the charged N^+ and Br^- in the hydrophilic plane parallel to the bilayer surface (Fig. 36) [302, 303]. This structural arrangement is responsible, at least in part, for restricting the conformational mobility of the surfactants in the bilayers and, hence, for maintaining the stability of SUVs prepared from di-alkylammonium surfactants. Introduction of rigid segments into single-chain surfactants has restricted alkyl chain mobilities and, thus, has facilitated SUV formation [30]. Appropriate molecular architecture has allowed the formation of SUVs from triple-chain [305], quadruple-chain [306], and fluorocarbon [307] surfactants. Vesicle formation from amphiphiles containing polar groups at both of their ends (bolaform surfactants) required, however, the introduction of curvature-creating moieties or co-surfactants [308].

Once formed, SUVs, unlike aqueous micelles, do not break down upon dilution; there is no equivalent of CMC for SUVs. Additionally, depending on their chemical composition, vesicles remain stable for days to weeks. SUVs, like membranes, are osmotically active; addition of electrolytes shrinks the vesicles, while placing them in solutions more dilute than their internal electrolytic concentrations causes swelling. SUVs, like membranes, are destroyed (lysed) by the addition of detergents or alcohols.

Below their phase transition temperatures, surfactants in SUVs are arranged in tilted, one-dimensional lattices; they are in highly ordered "solid" states. Above the phase transition temperature, the surfactants are somewhat separated

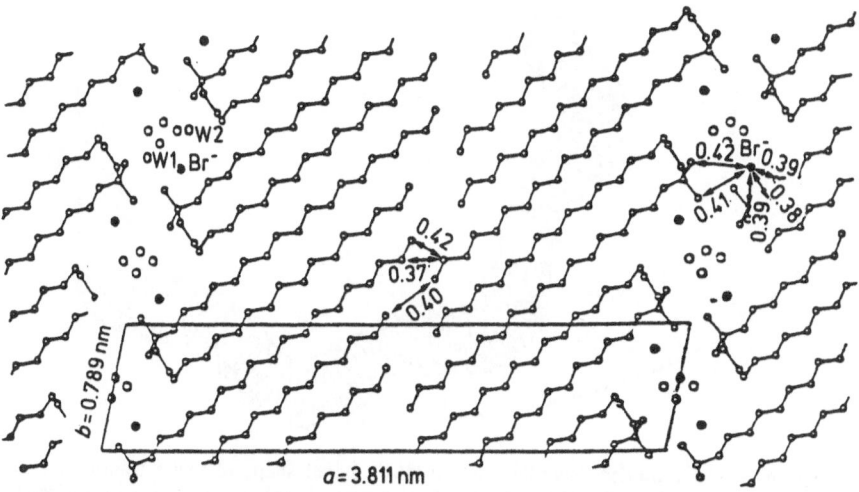

Fig. 36. The packing structure of a single crystal of the monohydrate of **5** [69]

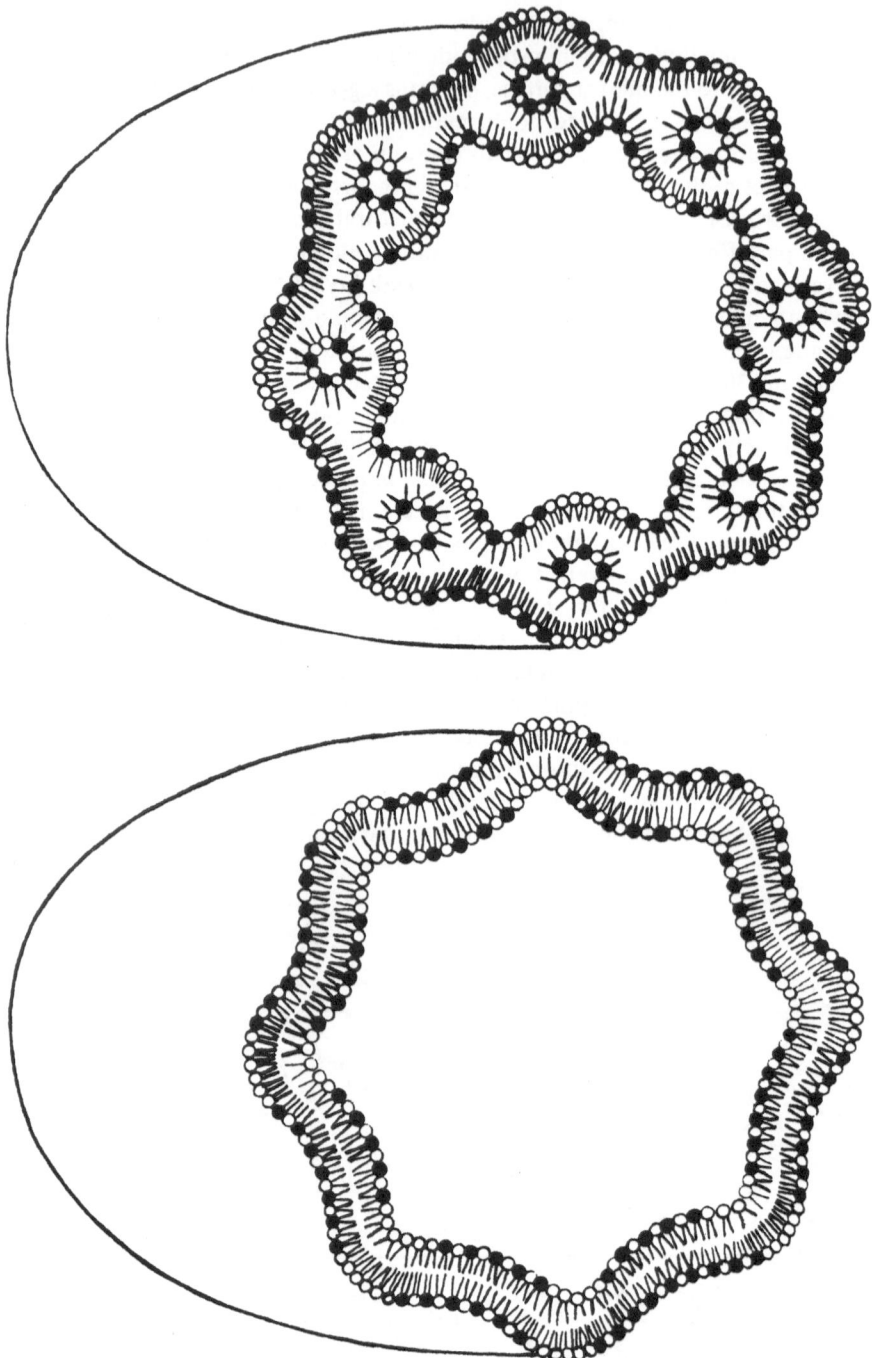

Fig. 37a, b. Model proposed according to ^{31}P-NMR signal shape of phase transition range. Membrane structure of mixed vesicles prepared from oppositely charged vesicles: **a** inverse micelle model; **b** bulge model of clusters with different spontaneous membrane bending [310]

from each other and assume liquid-like configurations; they are in their "fluid" states. Temperature-induced phase transition and phase separation are important properties of vesicles, offering routes to permeability control and molecular recognition. Indeed, the phase behavior of over 9500 surfactants and lipids have been investigated and the obtained data has been compiled [309]. Phase separation and domain formation can be realized by constructing SUVs from surfactant mixtures. Cosonication of positively charged 5 and negatively charged 4 led, for example, to the formation of a single-population of long-lasting, closed-bilayer vesicles [310]. Observed ^{31}P NMR and absorption spectroscopy, as well as phase transition behavior have been accommodated in terms of two models (Fig. 37). The first model involved the formation of reversed micelles which contained domains of 4 + 5 ion pairs. The second, and more likely, model has been discussed in terms of hemispheric bulges on the vesicles which contained the 4 + 5 ion-pairs [310]. These phase-separated domains are potentially exploitable as sites for the in situ growth of metallic, semiconducting, and magnetic particles.

Vesicles can organize many guest species in their compartments. Hydrophobic molecules can be distributed among the hydrocarbon bilayers and polar molecules may move relatively freely in vesicle-entrapped water pools, particularly if they are electrostatically repelled from the inner surface. The binding of small charged ions to the oppositely charged outer and/or inner vesicle surfaces is facile. Species with charges identical to those on the vesicles can be anchored onto the surfaces by long hydrocarbon tails.

Cell-sized "giant" liposomes [311–313] have been prepared and their mechanical properties have been extensively investigated [312–315]. Utilization of these systems as containers for advanced materials awaits the interested researcher.

2.5.2 Polymerized Vesicles

The need for increased stabilities and for controllable permeabilities and morphologies led to the development of polymerized surfactant vesicles [55, 158–161]. Vesicle-forming surfactants have been functionalized by vinyl, methacrylate, diacetylene, isocyano, and styrene groups in their hydrocarbon chains or headgroups. Accordingly, SUVs could be polymerized in their bilayers or across their headgroups. In the latter case, either the outer or both the outer and inner surfaces could be polymerized separately (Fig. 38). Photopolymerization links both surfaces; selective polymerization of the external SUV surface is accomplished by the addition of a water-soluble initiator (potassium persulfate, for example) to the vesicle solution.

Dissymmetrical SUVs can be formed by limiting reactions to the outer surfaces of polymerized surfactant vesicles (Fig. 39). Chemical dissymmetry has been created, for example, in polymerized vesicles prepared from surfactants containing ester-linked viologen moieties in their headgroups. Cleavage of the

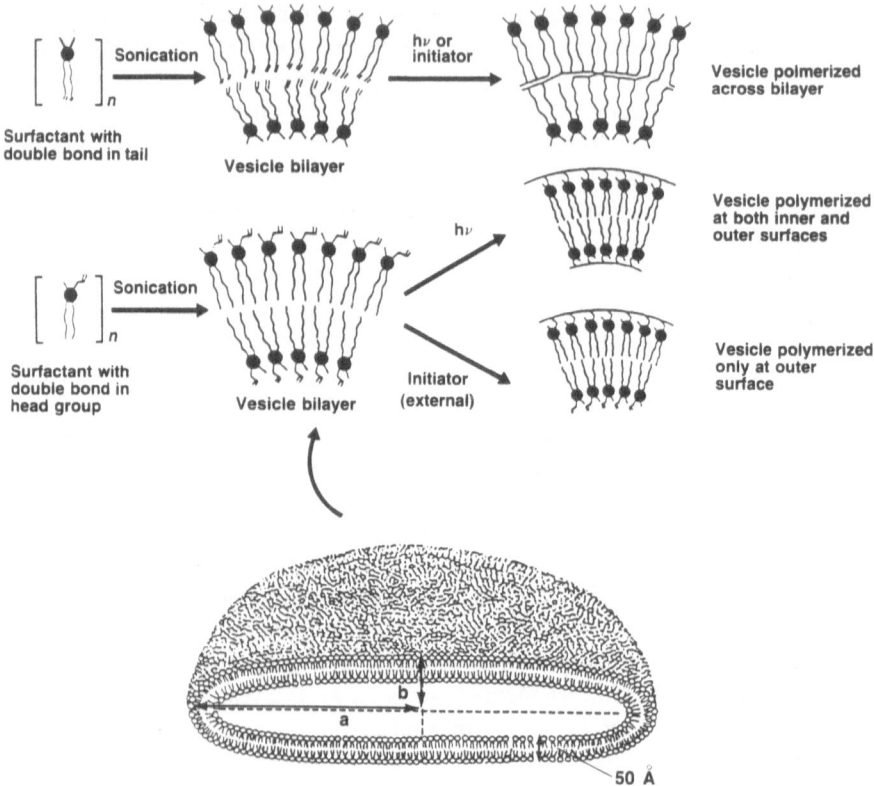

Fig. 38. Vesicle formation and polymerization

labile ester groups on the outer surface led to viologen functionalities in the inner, and carboxylate groups in the outer, surfaces of SUVs [316].

SUVs prepared from surfactants containing thiol and dithiol moieties have been shown to undergo reversible polymerization/depolymerization via S–S linkages [317–319]. Polypeptide vesicles have also been prepared from long-chain α-amino acid esters [72, 160]. The inherent high density of well-oriented reacting groups ensured polypeptide formation in SUVs (Fig. 40) [317]. In contrast, bulk condensation of α-amino acid esters leads to 2,5-diketopiperazins. Biodegradability renders polypeptide vesicles to be suitable drug carriers.

The correct alignment of surfactants in some, but not all, SUVs is an essential requirement for polymerization. Polymerization of diacetylenes is topochemically controlled and only occurs below the phase transition temperature of the surfactant. In contrast, SUVs prepared from styrene-containing surfactants could be polymerized in their fluid states [55]. The degree of polymerization varied from very low (10–20 for SUVs prepared from styrene containing surfactants) to rather high (several hundred for SUVs prepared from diacetylene-containing surfactants).

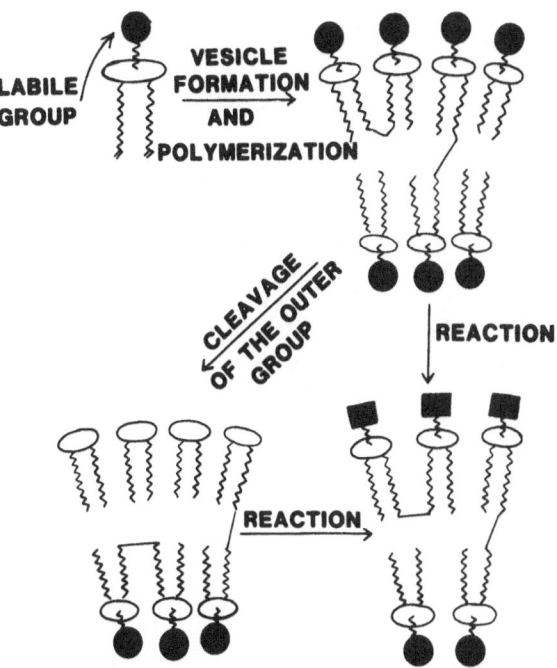

Fig. 39. Schematics of dissymmetrical polymerized vesicle formation [158]

Polymerized vesicles, unlike their non-polymerized counterparts, remain stable for months, are not destroyed by the addition of a few percent of alcohol or detergents, and, in the absence of additives, do not undergo size changes. This latter property can be advantageously utilized for stabilizing very small SUVs. While 300-Å dipalmitoylphosphatidylcholine (18) SUVs remained stable above their phase transition temperature, below it their mean diameter grew spontaneously over a period of several days to 700 Å [320, 321]. Similar spontaneous growth was observed in SUVs prepared from mixtures of 18 and a polymerizable cationic surfactant having a styrene moiety in one of its alkyl chains (2). Polymerization completely stabilized these SUVs; their hydrodynamic diameters remained at 300 Å for weeks (Fig. 41) [322]. Polymerization of SUVs prepared from surfactants which contained styrene in their headgroups (1 and 2, for example) resulted in the pulling together of approximately 20 monomers and thereby creating 15-Å-diameter surface clefts [323, 324].

Substrate entrapment, retainment, and ion permeabilities are important properties of polymerized SUVs. Indeed, substrates entrapped in polymerized SUVs display much lower leakage rates than those encapsulated in non-polymerized SUVs [158–160, 325, 326]. Selective polymerization allows a fine control of acid and base transport from the bulk solution to the vesicle interiors, or vice versa. These species permeate unpolymerized SUVs almost instantaneously. In partially polymerized SUVs, acid and base transfer occurs on the

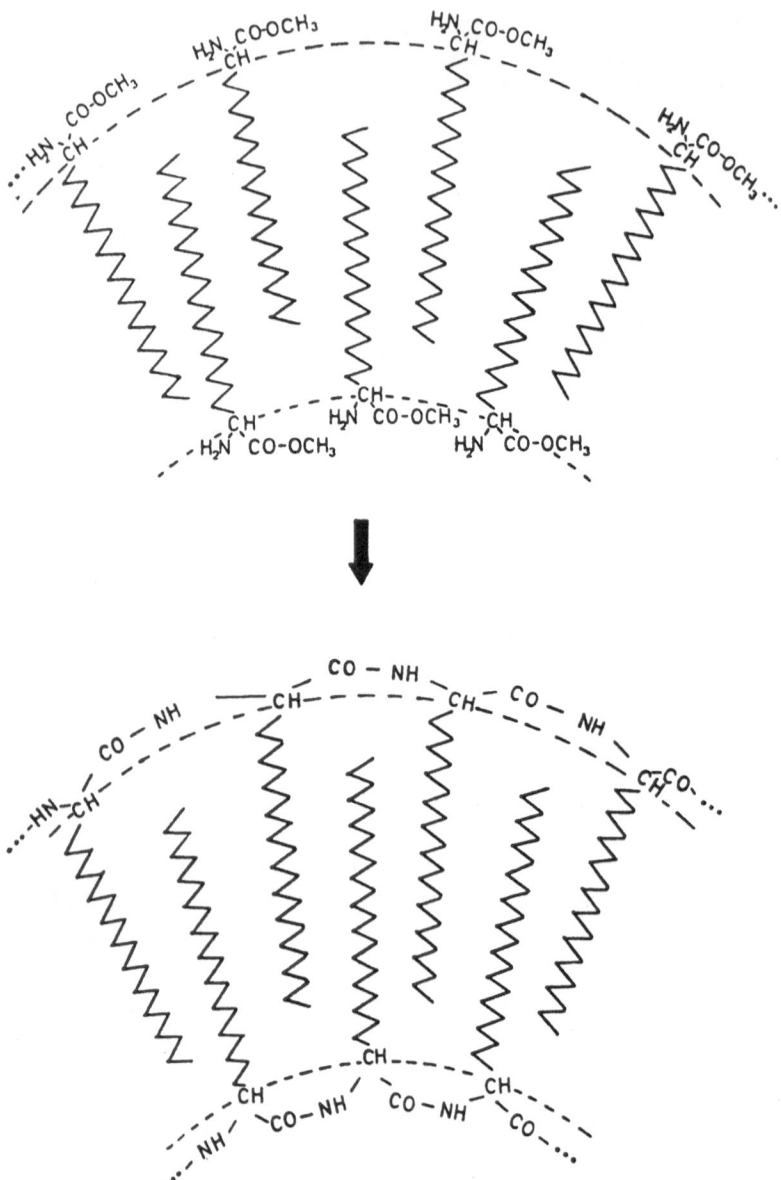

Fig. 40. Formation of polycondensed vesicles [160]

minute-to-hour time scale; a pH gradient of several units can be maintained across completely polymerized vesicles [158].

Polymerized SUVs prepared from surfactants containing styrene or vinyl moieties retained their phase transition behavior. Conversely, polymerized SUVs prepared from diacetylenic surfactants lost their phase transitions and those of methacrylamide surfactants had lower phase transition temperatures

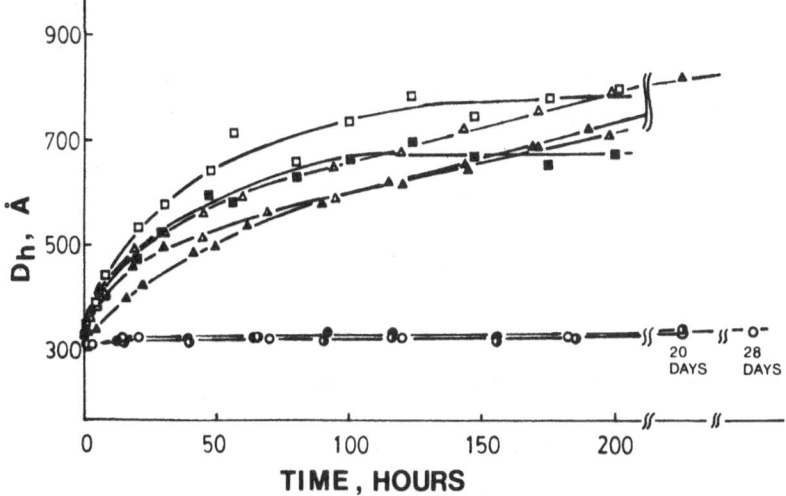

Fig. 41. Spontaneous growth of **18** (10.2×10^{-4} mol dm^{-3} (*open square*), 5.0×10^{-4} mol dm^{-3} (*solid square*)) and non-polymerized **18** = 7.4×10^{-4} mol dm^{-3} + **2** = 4.4×10^{-4} mol dm^{-3} (*open triangle*), **18** = 4.7×10^{-4} mol dm^{-3} + **[2]** = 2.8×10^{-4} mol dm^{-3} (*solid triangle*), **[18]** = 3.4×10^{-4} mol dm^{-3} + **[2]** = 2.3×10^{-4} mol dm^{-3} (*half-filled triangle*) vesicles as a function of incubation time. Polymerized **18** + **2** vesicles (**[18]** = 5.3×10^{-4} mol dm^{-3} + **[2]** = 3.5×10^{-4} mol dm^{-3} (*half-filled circle*), **[18]** = 3.6×10^{-4} mol dm^{-3} + **[2]** = 2.5×10^{-4} mol dm^{-3} (*solid circle*), **[18]** = 2.6×10^{-4} mol dm^{-3} + **[2]** = 1.6×10^{-4} mol dm^{-3} (*open circle*)) are seen to retain their sizes for extended periods. Plotted are the hydrodynamic diameters (D_h) of the vesicles, determined by dynamic light scattering, against incubation time at 23 °C [322]

than their non-polymerized couterparts [160]. Partially polymerized phospholipid vesicles were shown to undergo reversible structural changes upon cooling [327]. These results indicate the danger of making generalizations concerning structurally different polymerized SUVs. Both the packing of the surfactants in the vesicles and the functional-group chemistry involved in polymerization govern the behavior of polymerized SUVs.

SUVs have also been stabilized by "coating" their outer surfaces with chitin [328], polylysine [329–332], polyelectrolytes [333], and polysaccharides [334]. Advantage has also been taken of electrostatic interactions to attract oppositely charged polyelectrolytes to outer SUV surfaces and subsequently polymerize them in situ [335–339]. These systems have been referred to as "liposomes in a net" [72]. A particularly telling example is the attachment of a two-dimensional polymeric network either to the inner or to the outer surfaces of SUVs by ion exchanging the vesicle counterions with oppositely charged polymerizable short-chain counterions and their subsequent polymerization (Fig. 42) [340–342].

Surfactant vesicles have, in fact, been used as templates for the synthesis of ultrathin, spherical polymer membranes [337]. The approach involved the formation of SUVs from mixtures of diallylammonium dihexadecylphosphate and sodium dihexadecylphosphate, polymerization of the counterions coating

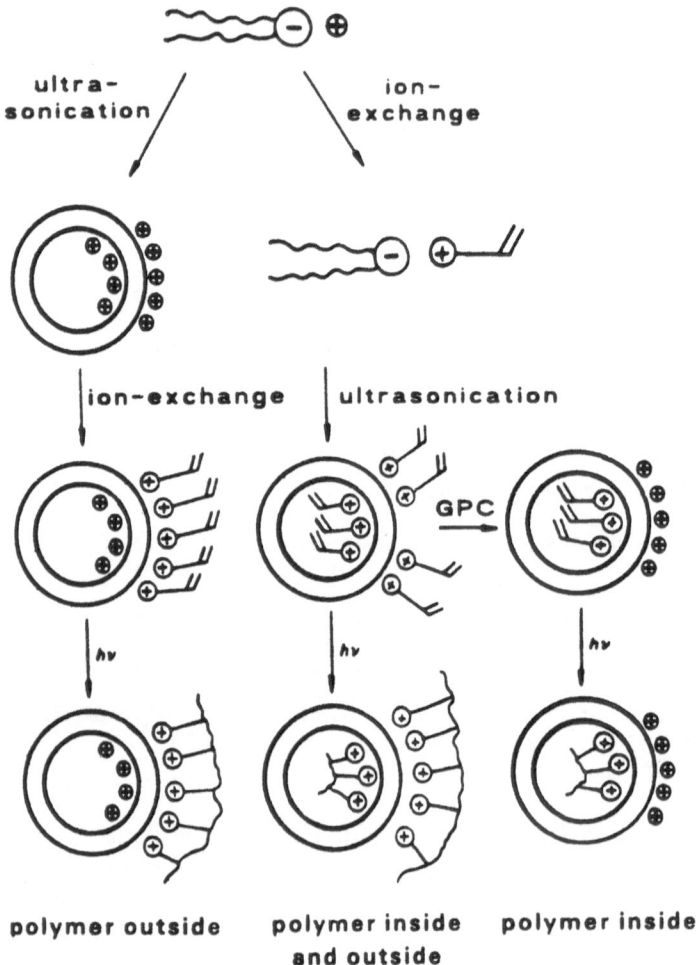

Fig. 42. Preparation of vesicle membranes with asymmetrically or symmetrically bound poly-electrolytes from lipids with polymerizable counterions [340]. GPC = gel permeation chromatography [72]

the vesicles by ultraviolet radiation, removal of the surfactants (by treatment with HCl and dialysis), and redispersal of the remaining polymeric bags (termed "ghost vesicles"). Electron microscopy and (^{14}C) sucrose entrapments indicated the presence of 200- to 2000-Å-diameter, ghost vesicles which were closed spherical and relatively porous in character [337].

A related system is that of the lipid-bilayer corked capsule membranes which are formed from ultrathin (about 1 µm thick), spongy, 2.0- to 2.5-mm-diameter, more-or-less spherical nylon bags in which multiple bilayers are immobilized (Fig. 43) [343–345]. They were considered to combine the advantages of mechanical and chemical stabilities of polymeric membranes with the controllable permeabilities of surfactant vesicles. Polymerization of the bilayers, in situ,

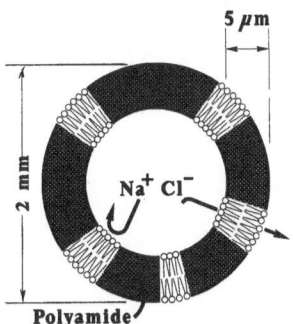

Fig. 43. A capsule nylon membrane [448]

alleviated the undesirable and premature release of surfactants from the matrices of the capsule membranes. The surfactants used contained polymerizable groups either in their tails, in their headgroups, or in their counterions. While all three types of polymerized bilayers led to stable lipid-bilayer-corked capsule membranes, the beneficial properties of vesicles (temperature-dependent phase transitions and controllable ion permeabilities) were retained only in those which had been polymerized at their counterions [345]. Although lipid-bilayer-corked capsule membranes have been primarily used in biotechnology related separations and molecular recognition processes [343], they can clearly play an important role in the construction of advanced materials.

2.5.3 Polymeric Vesicles

SUVs can, in principle, be prepared from pre-polymerized surfactants. In practice, a three-dimensional polymer has to be fitted, two-dimensionally, around the vesicle surface. This can be accomplished by using relatively small oligomers or through the introduction of hydrophilic spacers into the polymerizable surfactants (Fig. 44) [346, 347]. The function of the spacers is to obviate the interference of the polymer main chain with the alignment of the pendant surfactant groups in the vesicles (i.e. to decouple the motion of the polymer chain and thereby permit self assembly). Thus, SUVs were obtained from copolymers, prepared from hydrophilic (acrylamide or N-acetamidoacrylamide, for example) and hydrophobic (dialkyl acrylamide, for example) monomers [348]. Each vesicle was estimated to contain between 100 and 200 polymeric molecules. In an alternative approach, SUVs were prepared from unpolymerized and polymerized **23–26** [349]. Interestingly, the phase transition temperatures of SUVs prepared from prepolymerized **23–26** were higher than those formed from their non-polymerized counterparts. The situation for polymerized SUVs is quite different; their transitions are, generally, more broad and shifted to lower temperatures. Apparently, polymeric surfactants having appropriate spacers show greater cooperativity than their non-polymeric counterparts. As expected, substrates entrapped in polymeric SUVs leaked at slower rates than those encapsulated in the corresponding non-polymeric vesicles.

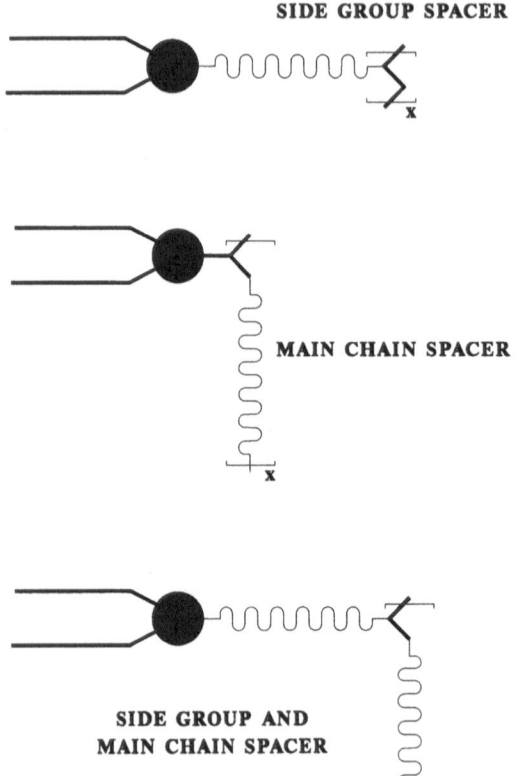

SIDE GROUP SPACER

MAIN CHAIN SPACER

**SIDE GROUP AND
MAIN CHAIN SPACER**

Fig. 44a. Side group spacer (hydrophilic segments). **b** Main chain spacer (hydrophilic comonomers). **c** Main chain spacer and side group spacer (spacer combination) [346]

Polymers and polymeric vesicles have been used for destroying SUVs [172, 319, 350, 351].

2.6 Tubules, Rods, Fibers, and Related Self-Assembled Structures

The appearance of tubular myelin-like structures in swollen lecithin was observed by light microscopy well before the systematic investigation of liposomes [351–352]. Similarly, it was also demonstrated some time ago that the addition of calcium ions converted phospholipid liposomes to cochleate cylinders [353]. Subsequent studies have, however, revealed that the system is extremely complex. For example, examination of the phase-transition behavior of synthetic sodium di-n-dodecyl phosphate [$(C_{12}H_{25}O)_2PO_2^-Na^+$ or NaDDP] and calcium di-n-dodecyl phosphate [$Ca(DDP)_2$] showed the presence of many diverse structures [354]. In particular, hydrated NaDDP crystals were shown to form lyotropic liquid-crystalline phases which transformed, upon heating to 50 °C, to myelin-like tubes. Structures of the tubes formed were found to depend upon the ionic strength and the shear rate. MLVs and/or SUVs could only be formed from NaDDP tubes or from NaDDP lyotropic liquid crystals upon the input of mechanical energy (sonication, for example). A variety of different fragmented structures (windmills, bananas) were observed in the intermediate steps. NaDDP vesicles and the tubular structures were found to transform to lamellar Ca(DDP)$_2$ crystals by the addition of calcium ions. Finally, Ca(DDP)$_2$ could be reconverted to NaDDP lyotropic liquid crystals by ion exchange with Na$_2$EDTA (Fig. 45) [354]. The important lesson from this work is that judicious variation of the experimental conditions (type of surfactant, type of counterion, presence of electrolytes, concentrations of the components, and temperature) can bring about the stabilization of a desired structure.

The realization that tubules may be formed on temperature reduction of polymerized SUVs, prepared from polymerizable diacetylenic phosphatidylcholines (21; where n = 7–16 and m = 5–11), represented a major breakthrough in obtaining the desired supramolecular structure [355–360]. In the initial experiments, 0.4- to 1.0-μm-diameter and 10- to 1000-μm-long tubules were prepared by the gradual lowering of the temperature (to about 38 °C) of 21 (m = 8, n = 9) SUVs [358]. The walls of the tubules had thickness of 10–40 nm and were coated by spiral ripples and helical bilayer strips. Many tubules contained trapped SUVs. Polymerization of the acetylenic moieties greatly enhanced the mechanical and thermal stabilities of the tubules [355–360].

More recently, tubules have been produced by precipitation from solution of 21 in an organic solvent at temperatures 10–30 °C *below* the phase transition of 21 (T$_m$ of 21 (m = 8, n = 9) = 43 °C); that is, via a route which *did not* involve

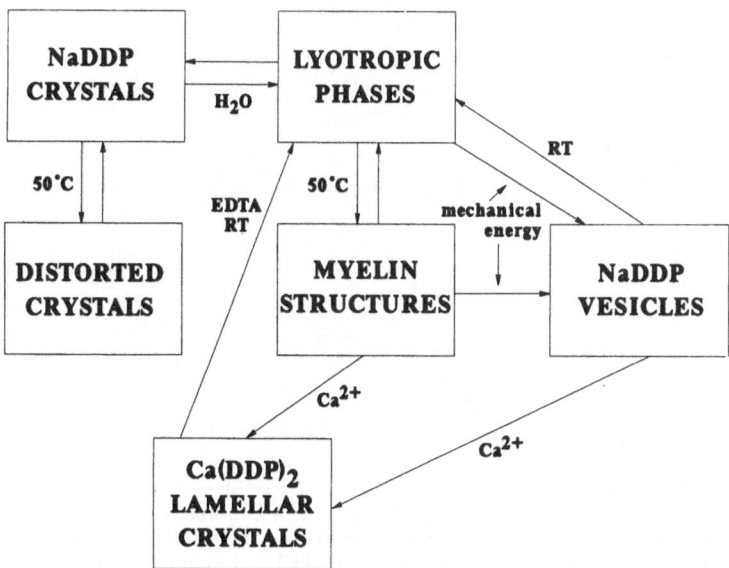

Fig. 45. Sodium di-*n*-dodecyl phosphate (NaDDP) phase transitions [354]

SUVs [361]. The length of the tubules was found to depend on the amount of water added to the alcoholic solution of **21** and on the incubation time (Fig. 46). In general, the more water that was added, the faster precipitation occurred and the smaller the tubules became. Those prepared from **21** (m = 8, n = 9) had diameters and lengths in the ranges of 0.3–3.0 µm and 5.0–1000 µm, respectively. Precipitated tubules, unlike those formed from SUVs, were open hollow cylinders (i.e. they contained trapped SUVs) and contained a large proportion of helical structures (Fig. 47). The helices were right-handed if prepared from chiral **21**. That the tubules had highly ordered structures was further demonstrated by their successful polymerization; topochemical alignments of the molecules is a requirement for diacetylene polymerization. Polymerization, by light or γ-radiation, stabilized the tubules without a loss of their helicity [361]. Tubule formation upon precipitation has been rationalized in terms of oriented helical growth of the initially precipitated, quasi-two-dimensional (i.e. no thicker than two bilayers) crystallites. The growth was considered to produce a helix (with a pitch of about 1 µm and a diameter of about 0.4 µm) which widened into a spiral ribbon, extending in width to the point of fusing to a seamless tubule and growing at both ends to a finite length [361].

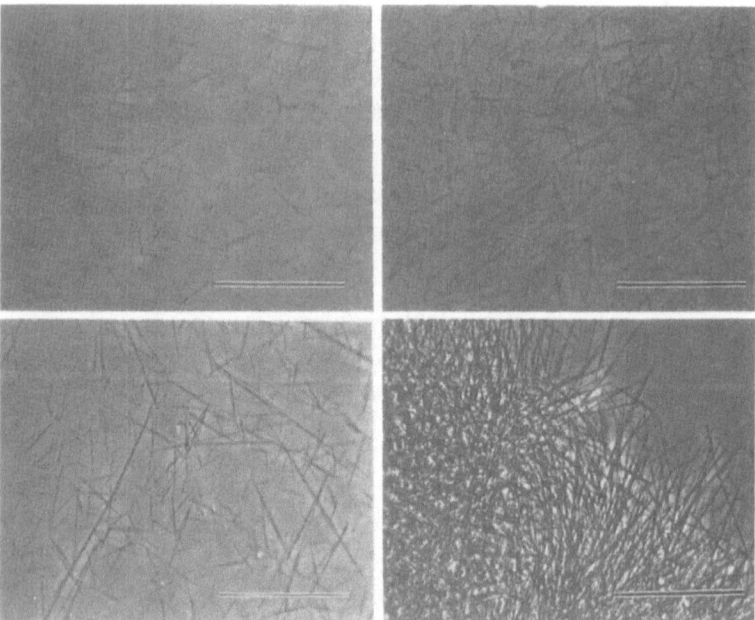

Fig. 46a–d. Darkfield optical micrographs of tubules formed by addition of water to ethanol solutions of **21** (m = 8, n = 9). The micrographs were taken at room temperature. With little water the tubules were uniformly long – some in (d) are longer than 300 µm. The final concentration of the lipid was 0.5 mg/ml in all cases. The final concentrations (v/v) of ethanol were: **a** 50%; **b** 55%; **c, d** 70%. The incubation times in the ethanol-water mixtures were 10 h (a, b), 144 h (c), and 6 months (d). *Scale bars* = 100 µm [361]

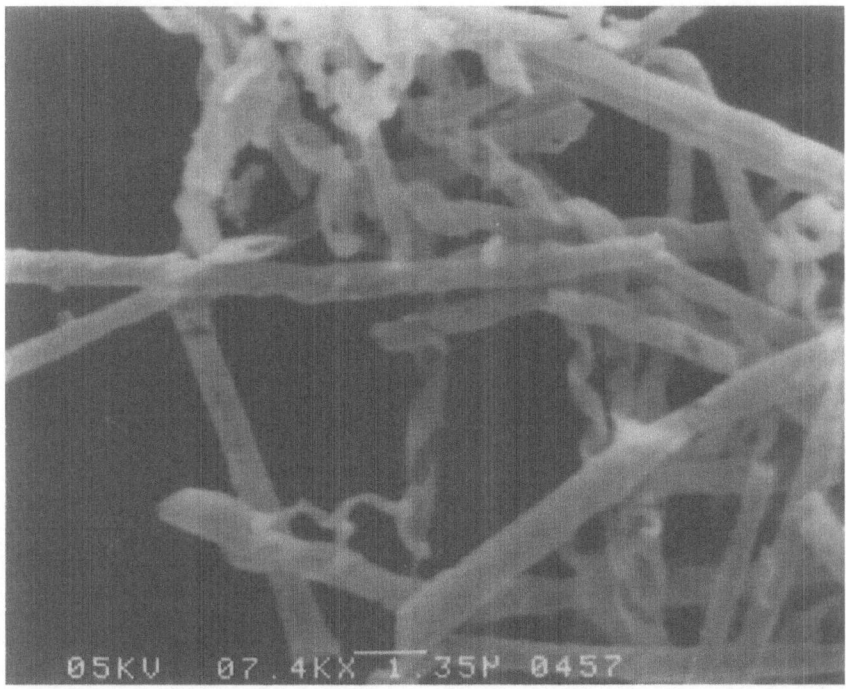

Fig. 47. SEM of tubules and helices formed from **21** (m = 8, n = 9) at 50% 2-propanol in water that were subsequently coated with copper metal as described in the text. Note that all helical structures are right-handed and that the pitch of the helices is somewhat variable. *Scale bar* = 2.48 μm [361]

Tubules have also been prepared by swelling thin films of polymerizable diacetylenic phosphatidylhydroxyethanol (choline functionally in **21** is replaced by hydroxyethanol) in aqueous metal ion solutions *above* the phase transition temperature of the lipid. Various cylindrical structures were observed upon swelling the lipid in the presence of mono- and divalent cations. In contrast, no definable microstructures were noted in the absence of cations [362].

Videomicroscopy provided a means for the in situ observation of the different structures which were dynamically formed in dispersions of **21** (n = 9, m = 10) in ethanol-water (70%–30% v/v) mixtures [363]. Thin filaments were observed upon cooling the solution from 55–60 °C to room temperature (Fig. 48). Some of these filaments had bulb-like termini from which they appeared to grow. The filaments also underwent thermal fluctuations which appeared to be both temperature and morphology dependent; the thinner the filaments were, the faster they undulated. At constant temperatures, the morphological changes of **21** were quite different. The filaments in close proximity to each other underwent fusion, becoming thicker, and ultimately coiling up to generate toroidal structures and rod-like vesicles (Fig. 49). Although such structures have been theoretically predicted [297–299, 364, 365], the observation in Fig. 49 represents the first experimental evidence of their existence.

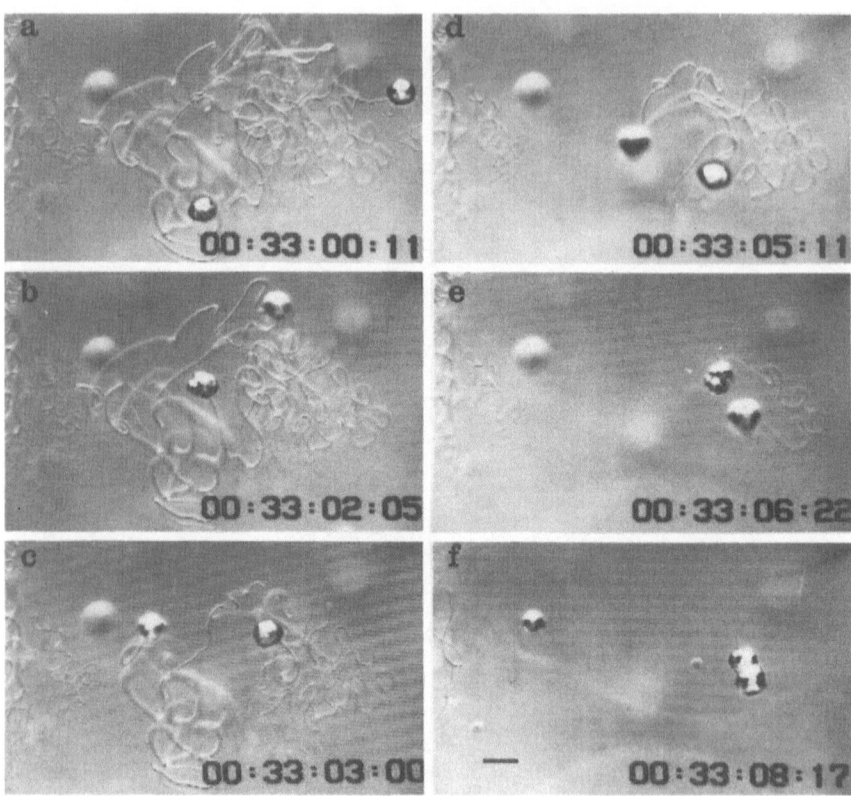

Fig. 48a–f. Representative video images of a single long filament during retraction, a typical single filament formed on cooling (cooling rate is 1 °C per min) from the isotropic phase. Temperature is 40.4 °C. The two ends of the filaments are attached to balls of lipid. As the temperature is lowered (a–f), the two ends move along the length of the filament, engulfing them in the process. The phenomenon occurs very rapidly, the whole sequence takes ~ 8 s (time-code is shown at the bottom of each photograph). Scale bar, 20 μm [363]

Evidence has been presented for the aggregation of tubules, prepared from **21** (m = 8, n = 9) in ethanol-water mixtures (70%–30% v/v) [366]. Cooling the lipid solution from 60 °C to room temperature resulted in the appearance of needle-like structures with lengths of the order of several tens of micrometers. Ethanol was removed from the solution by repeated centrifugation in water and the tubules were allowed to equilibrate in $H_2O + D_2O$ mixtures (at a solvent density of 1.093) at 27 °C. Within the first couple of hours, local turbidities ("dust-ball" formation) were observed, indicating the aggregation of tubules to "log jams". No such aggregation was noted when the pH of the solution was kept in the range of 2.6 < pH < 3.0, indicating the need for delicately balanced dispersive and Columbic repulsive forces for maintaining well-separated tubules [366].

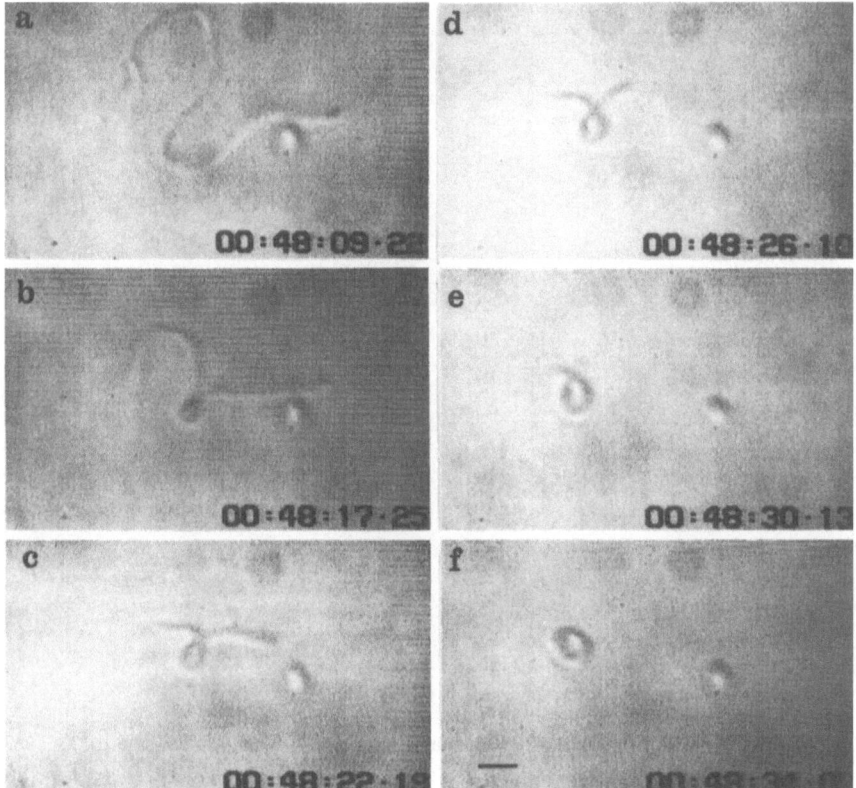

Fig. 49a–f. Formation of the torus from a filament. Temperature is 46 °C: **a** thicker filament formed after the fusion of thinner filaments; **b** if the temperature is still held constant, an initial loop forms; **c–f** as time elapses, the rest of the filament is drawn into the loop to form a torus (f). *Scale bar*, 10 µm [363]

Helical tubular structures have been observed in dispersions of surface-active L-glutamates [367–369]. Once again, polymerization has been found to stabilize the supramolecular structures formed, Single-bilayer helical aggregates were observed to form spontaneously upon the hydration of N',N''-bis [11-(sorboyloxy)undecyl] (pyridinium-N-ylpropionyl)-L-glutamide bromide (**27**), both below and above the phase transition ($T_c = 51$ °C) temperature (Fig. 50) [367]. The helical rods, seen in Fig. 50, had 40- to 60-Å-thick strands and were 250–300 Å in diameter across the helices, regardless of temperature. The helical structures of **27** showed a high degree of exciton coupling at 15 °C, having a 200-fold greater θ value than that for the monomeric **27** (Fig. 51). The high degree of helicity was rationalized in terms of head-to-head stacking of the sorboyl groups in **27** in an S-chiral arrangement (c in Fig. 52) [367].

Photoirradiation of supramolecular **27** decreased the absorption due to the sorboyl groups, indicating that polymerization had occurred [367]. Polymerization at 70 °C was accompanied by morphological transitions to well-

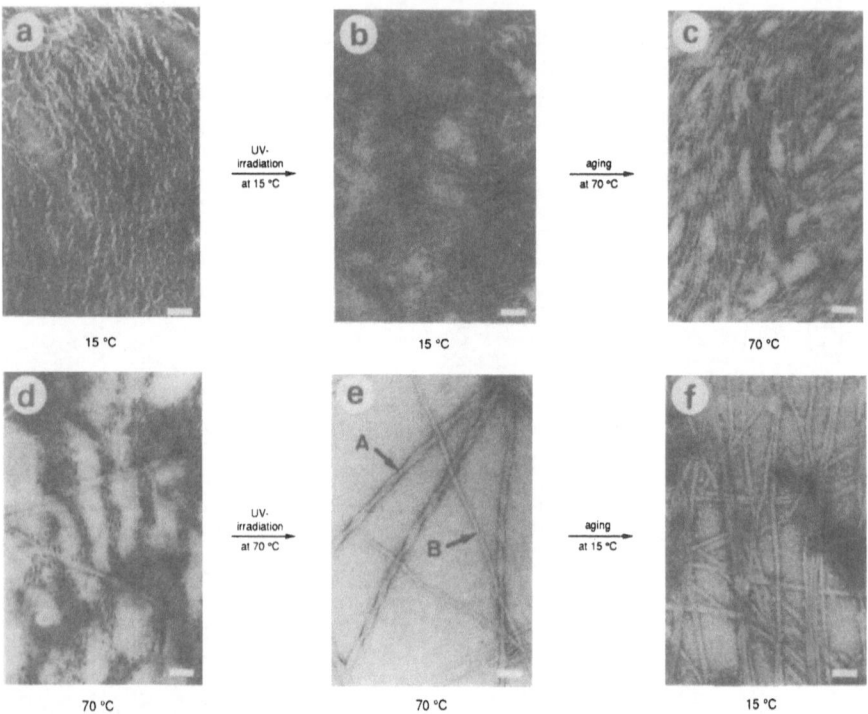

Fig. 50. Typical electron micrographs of aqueous **27** aggregates. *Scale bars* indicate 1000 Å [367]

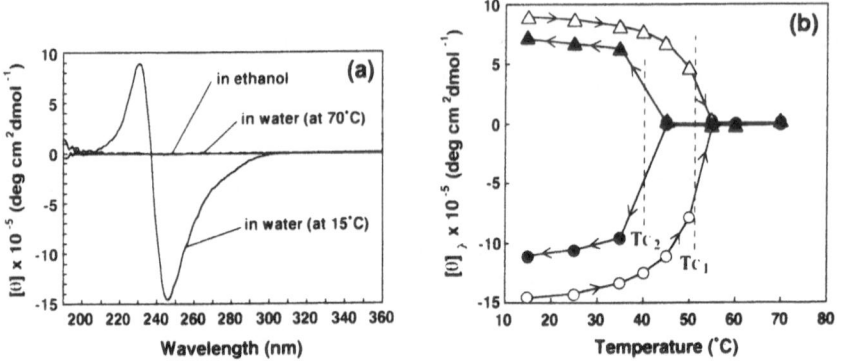

Fig. 51a. Circular dichroism spectra for hydrated lipid **27** (0.24 mM). **b** Temperature dependencies of $[\theta]_\lambda$ values for hydrated lipid **27** (0.24 mM). The *circles* and *triangles* show the molecular ellipticity at 246 and 231 nm, respectively [367]

developed, twisted (A of Fig. 50e) and untwisted (B of Fig. 50e) fibrous aggregates. In contrast, irradiation, below phase transition temperatures produced tubular aggregates. Incubation of the fibrous aggregates at 15 °C (Fig. 50f) or the tubular aggregates at 70 °C (Fig. 50c) did not induce morphological changes.

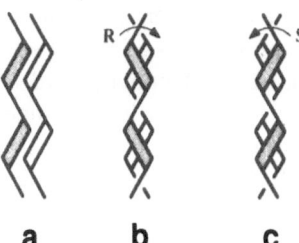

a   b   c

Fig. 52. Schematic illustrations of three possibilities of head-to-head stacking of sorboyl groups in aggregates of lipid **27** [367]

Morphologically diverse supramolecular assemblies have been formed from long-chain chiral aldonamides [370–378]. Typical compounds include **28–30**. Formation of 200- and 300-Å-diameter rods and helical rope-like structures in aqueous *n*-octyl- and *n*-dodecyl-ᴅ-gluconamide (D-**28**-8 and D-**28**-12) gels has been recognized for some time [379]. In a subsequent work, helical double-

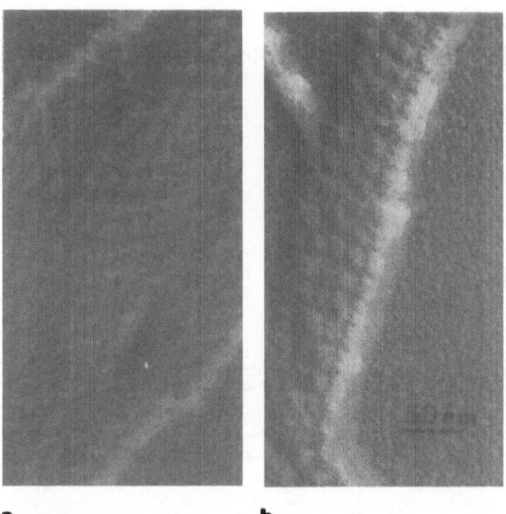

a   b

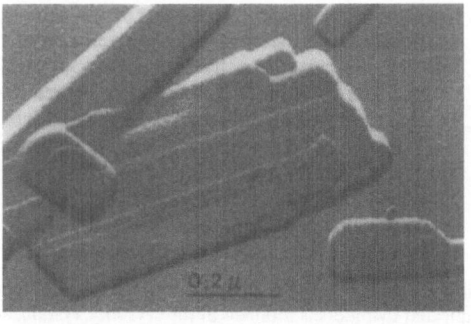

Fig. 53a–c. Electron micrographs of the platinum shaded bulgy helices of: **a** D-**28**-8; **b** L-**28**-8; **c** the racemic D-**28**-8 + L-**28**-8 platelets. Magnification 56 000 [373]

c

strained fibers of biomolecular thicknesses were formed after heating aqueous dispersions of D-**28**-8 or D-**28**-12 or L-**28**-8 to 100 °C and then cooling to room temperature [373] These helices contained wide bulges and knots placed at regular intervals along their lengths. The constituent molecules in the single strands were arranged tail-to-tail to each other. In contrast, anhydrous crystals of *N*-alkyl-D-gluconamides were formed by a head-to-tail molecular packing, generating sheets [380]. Electron microscopy revealed the presence of right-handed helices in aqueous samples prepared from *N*-octyl-D-gluconamide (D-**28**-8) and left-handed helices in those formed from *N*-octyl-L-gluconamide (L-**28**-8). More significantly, mixing of the enantiomeric solutions in equal proportions produced only non-fibrous, non-twisted platelets (Fig. 53) [373]. Apparently, chirality is an important requirement for helical fiber formation from surfactants.

The chiral-bilayer-effect hypothesis has been evoked for the rationalization of the helical fibers formed from enantiomeric or diastereomeric surfactants (Fig. 54) [373]. Different packing of the chiral surfactants in the crystals (head-to-tail) and in bilayer or micellar aggregates (tail-to-tail) is the basis for this postulate. Crystallization from aggregates requires an energetically costly, 180°

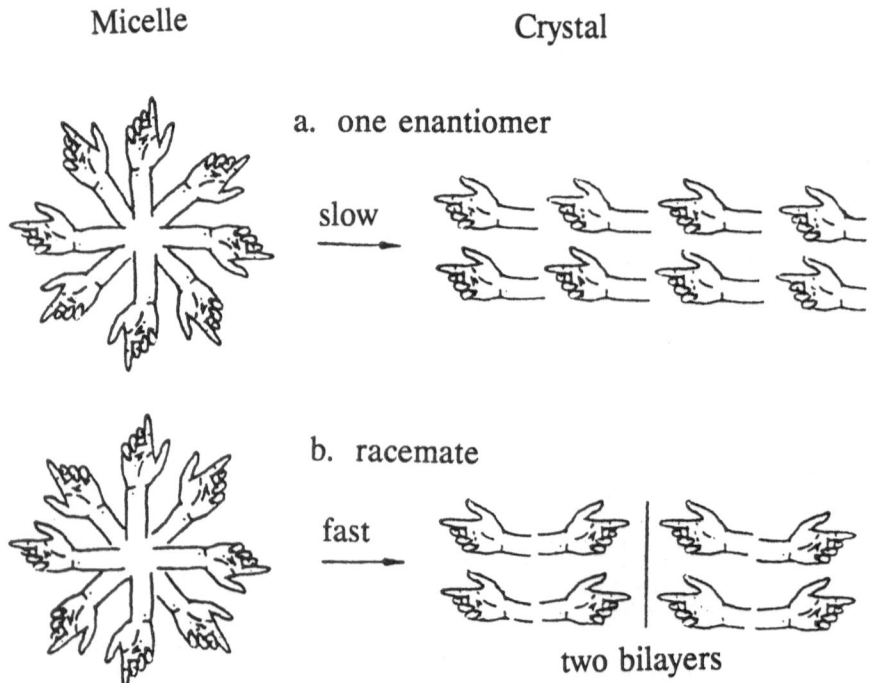

Fig. 54a–b. The chiral bilayer effect: **a** chiral micellar cylinders rearrange slowly to enantiopolar crystals; **b** the hydrophobic bilayer of achiral micellar cylinders is retained in the crystal. Crystallization is fast [373]

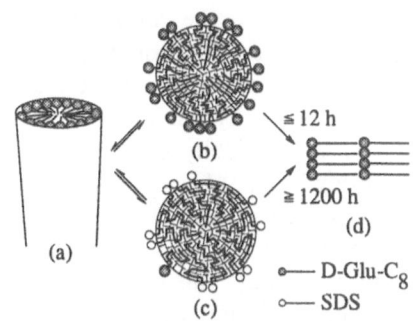

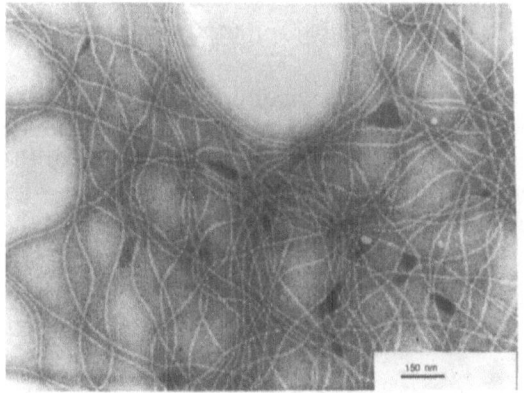

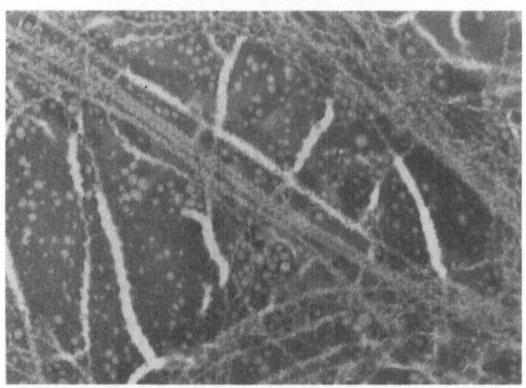

Fig. 55a. In the presence of detergents, e.g. SDS, micellar fibers do *not* rearrange to crystals, because crystallization nuclei with head-to-tail sheets cannot be formed. **b** Electron micrograph of a 2-month-old gluconamide D-**28**-8 gel, which was kept at 60 °C in the presence of SDS (molar ratio 10:1). Micelles and double helices occur (PTA 2% post-strained, bar = 100 nm). **c** Electron micrograph of a gel, which was kept at 20 °C and contained more SDS (molar ratio 2.5:1). Vesicles and multiple helices are apparent (PTA 2% poststained, *bar* = 100 nm) [377]

molecular realignment and resisting it, according to the chiral-bilayer-effect
postulate, is the predominating factor in the preferential formation of helical
fibers. Once formed, the helical fibers are stabilized, in a relatively shallow
potential-energy well, by amide hydrogen bonds and hydrophobic interactions.
In the absence of additives, the initially clear or slightly opaque gels which
contained the fibers decomposed within a day [373, 381].

Racemates and achiral bilayers do not have to reorganize for crystallization.
Platelet formation was observed upon cooling 1:1 mixtures of aqueous solutions of D-**28**-8 + L-**28**-8 and D-**29**-8 + L-**29**-8 [382]. This behavior appeared
to be general with the exception of recemic octylgalactamide, i.e. a 1:1 mixture
of D-**30**-8 + L-**30**-8, which gave tubules instead of platelets upon cooling. The
low solubility of galactamides in cold water and the presence of initially formed
racemic whiskers was suggested to be responsible for the tubule formation
[382].

Addition of micelle-forming surfactants and sodium dodecyl sulfate (SDS),
in particular, dramatically increased the stabilities of N-alkalygluconamide
aggregates (Fig. 55) [377]. It was considered that the rate of dissolution of the
incipient head-to-tail sheets of crystallization nuclei by the micellar aggregates
exceeded the growth rate of the crystals. This rationalization is supported by the

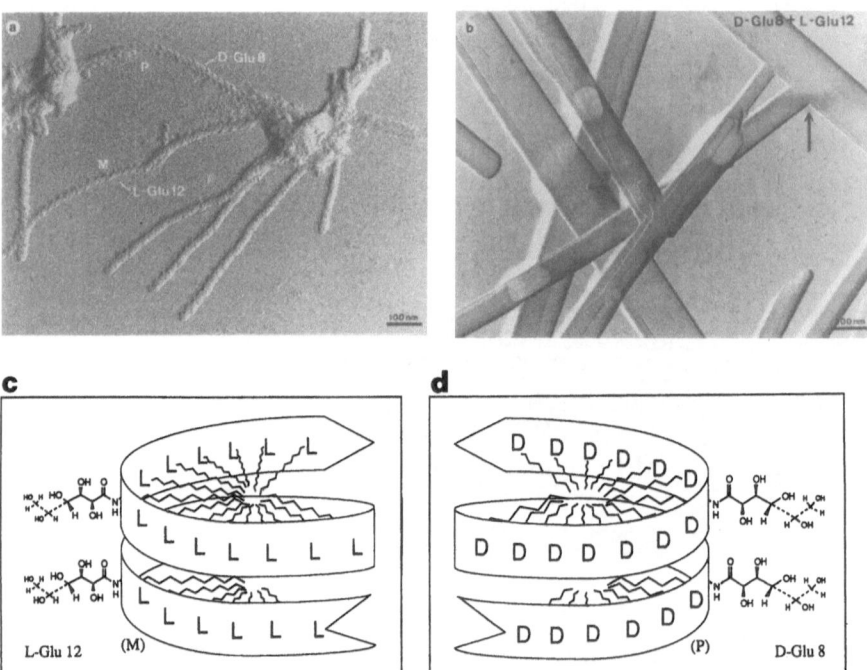

Fig. 56a–d. P-Helices (D-**28**-8) and M-helices (L-**28**-12) first separate (a, c) (bar = 100 nm) and then
unite (b, d) (bar = 300 nm) to form elongated "racemic" platelets. Notice smooth branchings (*arrow*)
[374]

outcome of the reverse experiment in which precipitation occurred upon the addition of a few gluconamide crystals to a 10%, SDS-containing, one-month-old D-**28**-8 gel.

Morphologies of the structures formed upon cooling mixtures of two N-alkyl-aldomates have also been examined [374]. Surfactant mixtures which differed only in their chain lengths (D-**28**-8 + D-**28**-12, for example) gave large spherical aggregates of ill-defined structures. Platelets were formed from racemic mixtures (see above) and mixed gels of different enantiomers and chain lengths initially formed short lived separate fibers, which subsequently converted to platelets (D-**28**-8 + L-**28**-12; Fig. 56). The behavior of the different mixtures examined is summarized in Table 3 [374].

Tubules have been formed from diacetylenic gluconamides, although UV irradiation had no apparent effect on the appearance of the aggregates [370, 371]. Chiral fibers have also been formed from protoporphyrin IX glucosamides [378]. Lowering their pH to 4.1–4.9 and cooling hot aqueous solutions of N-dodecyltartaric acid monoamides (**31**) led to the formation of cloth like-aggregates (Fig. 57) [376].

Table 3. List of examined mixed-gel systems and their aggregation behavior[a] [374]

Gel system	Aggregation behavior[b]	Molar ratio	Dry weight [mg]	Vol [ml]
1. D-**28**-8 + L-**28**-8	P	1:1	25 + 25	5.0
2. D-**28**-12 + L-**28**-12	P	1:1	29 + 29	5.8
3. D-**28**-8 + D-**28**-12	H	1:1	25 + 29	5.4
4. D-**28**-8 + L-**28**-12	S, later P	1:1	25 + 29	5.4
5. D-**29**-8 + D-**29**-12	H	10:1	25 + 2.9	5.6
6. D-**29**-8 + L-**29**-12	P	10:1	25 + 2.9	11.2[c]
7. D-**28**-8 + D-**29**-8	P	1:1	25 + 25	5.0
8. D-**28**-8 + L-**29**-8	S	1:1	25 + 25	5.0
9. D-**28**-8 + D-**29**-12	S	1:1	25 + 29	5.4[d]
10. D-**28**-8 + L-**29**-12	S	1:1	25 + 29	5.4
11. D-**28**-12 + D-**29**-8	B	1:1	29 + 25	5.4
12. D-**28**-12 + L-**29**-8	B	1:1	29 + 25	5.4
13. D-**28**-12 + D-**29**-12	H	1:1	29 + 29	5.8[e]
14. D-**28**-12 + L-**29**-12	P	1:1	29 + 29	5.8
15. D-**28**-8 + D-**30**-8	S	1:1	25 + 25	10.0
16. D-**28**-8 + L-**30**-8	S	1:1	25 + 25	10.0
17. D-**28**-12 + D-**30**-8	S	1:1	29 + 25	10.8
18. D-**28**-12 + L-**30**-8	S	1:1	29 + 25	10.8
19. D-**29**-8 + D-**30**-8	P	1:1	25 + 25	10.0
20. D-**29**-8 + D-**30**-8	S	1:1	25 + 25	10.0

[a] Solvent is double-distilled water.
[b] S = separation, function of individual aggregates for each component; B = heterogeneous mixed structures for "bistructures" (e.g. platelets with fringes); H = homogeneous mixed structures, except for platelets; P = mixed bilayer platelets.
[c] At lower volumes, the material remains insoluble.
[d] Clear solution after refluxing.
[e] Partly undissolved material

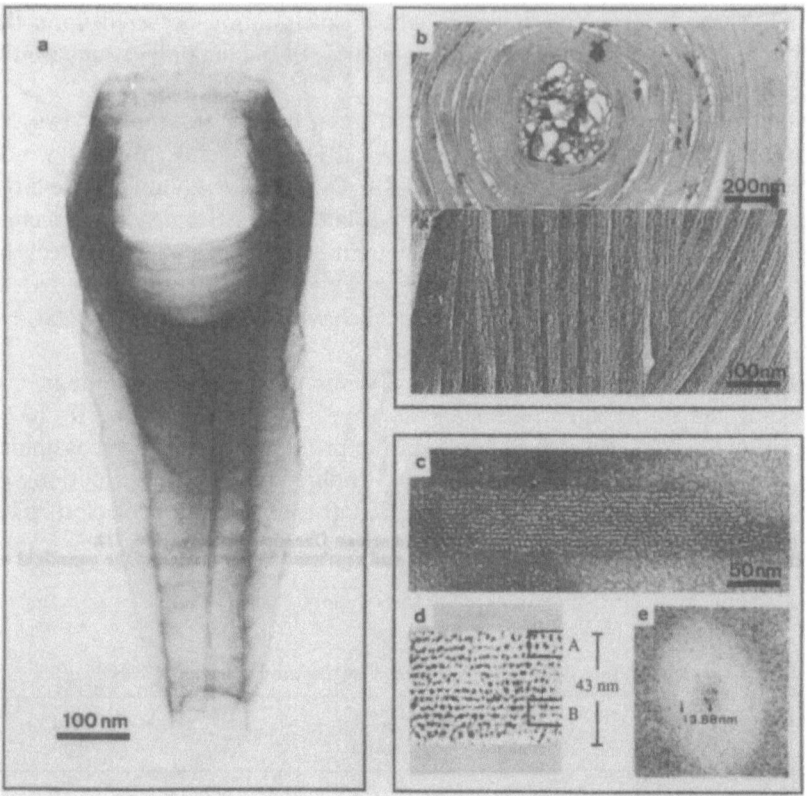

Fig. 57a. A typical multilayered cloth made of potassium tartaric amide **31b** at pH 5. Its observed physical shape is fortuitous. (Negative stain, uranyl acetate 1%). **b** Freeze-etching of a similar multilayer made of the sodium salt **31a**. At higher magnification (*below*), the bilayer profiles become visible (Pt/C shadowed). **c** Fiber pattern of **31a**. (Negative stain, uranyl acetate 1%). **d** Digitized area of the fiber bundle taken from (c). **e** Fourier transform of the input image from (d), as obtained by calculating the reciprocal space frequencies (x-y exchanged). Two intense spots yield a periodical pattern of 38.78 Å [376]

Polymers have provided an alternative approach to self assembly. Particularly exciting are the recently developed hyperbranched dendrimers [383] and arborols [384] **(32)**, both of which can be synthesized in a relatively simple step-wise manner. Pairs of arborols, connected with proper-length spacers, have been found to self assemble to form gels and long fibrous rods similar to those formed from chiral aldonamides [384].

Tubules, rods, and fibers assembled either from surfactants or from polymers are, in many ways, analogous to numerous biological structures formed from proteins and polynucleotides. Morphological versatility, together with the relative ease of large-scale production, render SA structures to be highly promising supramolecular organizers.

2.7 Bilayer Lipid Membranes

BLMs are formed at small teflon orifices (diameters of 2.00 mm or smaller) separating two compartments containing aqueous solutions, typically by painting the surfactant (or lipid), dissolved in a hydrocarbon solvent, across the pinhole (Fig. 58) [58, 385–389]. The initially formed film is rather thick and reflects the light with a gray color. Within a few minutes, the film thins and the reflected light exhibits interference colors that ultimately turn black (Fig. 59). At this point, the film is considered to be bimolecular and to be 5.0 ± 1.0 nm thick. The BLM owes its existence to the presence of a reservoir. The Plateau–Gibbs (or PG) border at its perimeter. Surfactant and solvent molecules are in dynamic equilibria between the PG border and the plane of one of the monolayers constituting the BLM and, consequently, some hydrocarbon-solvent molecules remain in the matrix of the BLM. The types of solvent and surfactants (or lipids) used and the size and geometry of the pinhole, as well as the method of

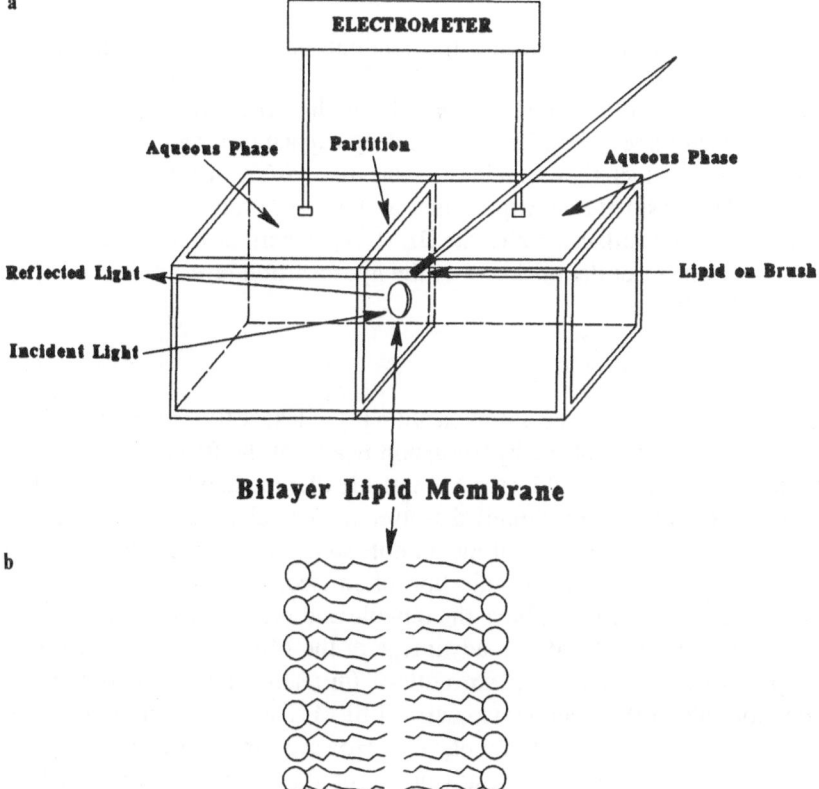

Fig. 58a. Formation of BLMs at a teflon pinhole which separates two compartments. **b** A schematic diagram of the molecular organization of the surfactant molecules in a bilayer lipid membrane (BLM)

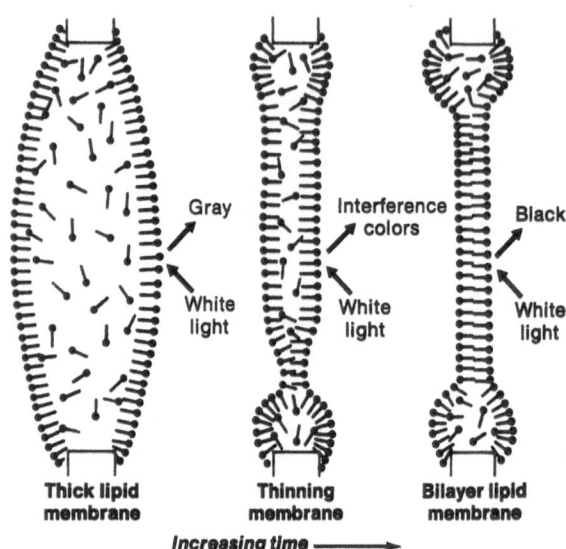

Fig. 59. Time-dependent thinning of a BLM in aqueous media indicating the patterns of reflected light

preparation (see below) determine the amount of solvent molecules which remain in the BLM.

Thinning of the BLM can also be monitored by capacitance measurements across the membrane [390]. Time-dependent capacitance values increase to a plateau beyond which they remain the same (to $\pm 0.2\%$) within the lifetime of the BLM. The measured total capacitance at its plateau value, C_t, together with a knowledge of the surface area of the BLM, A_m (determined microscopically), allow an assessment of the thickness of the hydrocarbon layer in the BLM, δ_h, from the relationship:

$$C_t = (\varepsilon_0 \varepsilon / \delta_h) A_m \tag{10}$$

where ε_0 is the dielectric constant of vacuum ($\varepsilon_0 = 8.85 \times 10^{-12}$ Fm^{-1}) and ε is the relative permittivity of the hydrocarbon region of the BLM ($\varepsilon = 2.1$).

BLMs can also be formed by the Montal-Mueller method [391, 392]. In this procedure, the surfactant (or liquid), dissolved in an apolar solvent, is spread on the water surface on both sides of the pinhole so as to form monolayers below the level of the pinhole. Careful injection of an electrolyte solution below the surface raises the water level above the pinhole and brings the monolayers into apposition to form the BLM. An advantage of the Montal-Mueller method is that it permits the formation of "solventless" (in reality, containing only a few solvent molecules) [387] and dissymmetrical BLMs (i.e. those containing different surfactants in the apposed monolayers). However, the necessity of a rather small pinhole (> 0.5 mm) is a disadvantage of the Montal-Mueller method. BLMs have also been prepared from surfactant vesicles (SUVs) via the Montal-Mueller method [391–399]. SUVs injected into an aqueous solution formed monolayers which, in turn, could be converted into BLMs (Fig. 60).

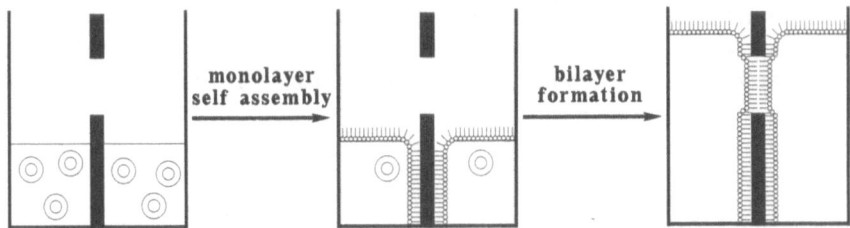

vesicle solutions

Fig. 60. Transformation of vesicles to planar bilayers via spontaneous monolayer formation [394]

Separation of two aqueous solutions by the BLM allows electrical measurements by using macroscopic electrodes. Precise capacitance, conductance, and impedance measurements, both in the absence and in the presence of ionophores, have contributed to our understanding of impulse and ion-transfer mechanisms [400, 401]. Particularly significant has been the development of voltage clamping (i.e. holding the BLM at a predetermined potential and measuring the current flow) and single-channel recording [401, 402].

Early investigations of BLMs suffered from two major drawbacks. In the first place, BLMs were notoriously unstable. Very rarely did they survive longer than a couple of hours. Secondly, voltage clamping provided information only on the transition from an open state to a closed state in ion channels and not on events in the closed state. These difficulties were overcome, respectively, by stabilizing the BLMs by polymerization or polymer coating [403–408] and via the development of simultaneous electrical and spectroscopic characterization of BLMs [409–414]. Polymerized BLMs, prepared from dialkylammonium surfactants, had day-long lifetimes [407]. Reflectivity [403], laser-intracavity absorption [409], steady-state and time-resolved fluorescence [412], holography [411], and interferometry [413, 414] have been utilized for the simultaneous characterization of BLMs with electrical measurements.

Two layers of closely packed surfactants are apposed tail-to-tail in the BLM. Their polar headgroups are in contact with, and are hydrated by, the aqueous solution bathing the two sides of the membrane. A delicate balance between the opposing forces is responsible for maintenance of the BLM. The repulsive force between headgroups is counteracted, in part, by the interfacial pressure of the aqueous solution on the surfactants. Since this attractive force (by unit length) is acting in the plane of the monolayer (i.e. one-half layer of the BLM), it is called interfacial surface tension and is designated by γ_m for monolayers and by γ for BLMs ($\gamma = 2\gamma_m$). Thermodynamically, γ is defined as the force (by unit length) required to increase the volume of the spherical cup (formed by applying a small hydrostatic pressure to one side of the BLM) by an infinitesimally small value, dV_c, upon stretching it by an infinitesimally small area, dA_c, at a hydrostatic pressure P_{hp}:

$$P_{hp}dV_c = \gamma dA_c . \qquad (11)$$

BLM surface tensions have been determined by the application of an ultrasmall hydrostatic pressure to one side of the membrane and measuring the resultant curvature changes by optical interferometry (Fig. 61). Typical values of 0.2–0.3 mNm^{-1} were obtained for γ for glyceryl monooleate and phosphatidylserine BLMs [413].

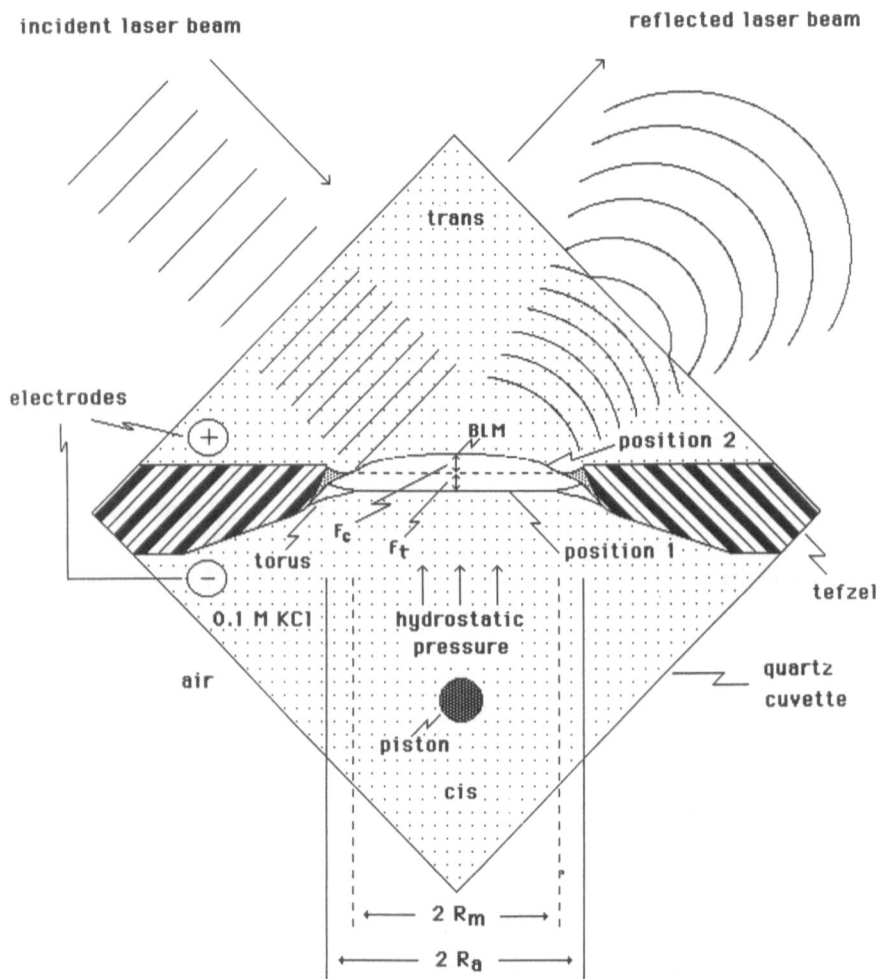

Fig. 61. Schematics of pressure-induced and applied-potential-induced BLM deformations. Application of hydrostatic pressure P_{hp} (by lowering a piston into the aqueous solution bathing the cis side of the BLM) displaces the BLM from *position 1* to *position 2*. The displacement involves both translational (lateral) motion (F_t) and curvature increase (F_c). As indicated, deformation of the BLM is accompanied by a change in its torus (Plateau-Gibbs border). $2R_a$ and $2R_m$ represent the diameters of the aperture of the pinhole in the Tefzel film and that of the membrane (excluding the torus). The object laser beam, incident upon the *trans* side of the BLM and reflected by it at 45° at a shortened wavelength produces concentric optical interference fringes with the reference laser beam. Ag/AgCl electrodes, placed in the cis and trans sides of the BLM, allow for continuous electrical measurements [413]

Consequences of BLM bending at the molecular level, investigated by measuring fluorescence lifetimes of diphenylhexatriene (DPH) incorporated into BLMs, have been discussed in terms of two effects [412]. First, bending is likely to be accompanied by a dynamic exchange of surfactant and solvent molecules between the Plateau–Gibbs border and the bilayer of the BLM. Minimizing both the Plateau–Gibbs border area and the presence of solvent molecules in the BLM tends to diminish this exchange and, hence, increases the stability and reproducibility of individual BLM preparations. The second effect of bending BLMs manifests itself in the reorientation of surfactant molecules. The headgroups are somewhat compressed at the side of the applied pressure and somewhat expanded at the opposite side. The net result of this is increased penetration of the aqueous solution between the hydrophobic parts of the BLM and a concomitant greater mobility of DPH, which is recognized by increased fluorescence lifetimes (Fig. 62) [412].

Generation of microvolt-range transmembrane potential differences through the periodic bending of BLMs has been termed "curvature-induced electricity" or "flexo-electricity" [415–424]. Flexoelectricity has been treated theoretically by considering the electrical multipole moments of the surfactant molecules [418, 420, 421], the effect of charge exchange across the BLM on the curvature elastic modulus [423], and electrostatics [424]. Experimentally, the magnitude of flexoelectricity has been shown to depend on the structure and charge of the

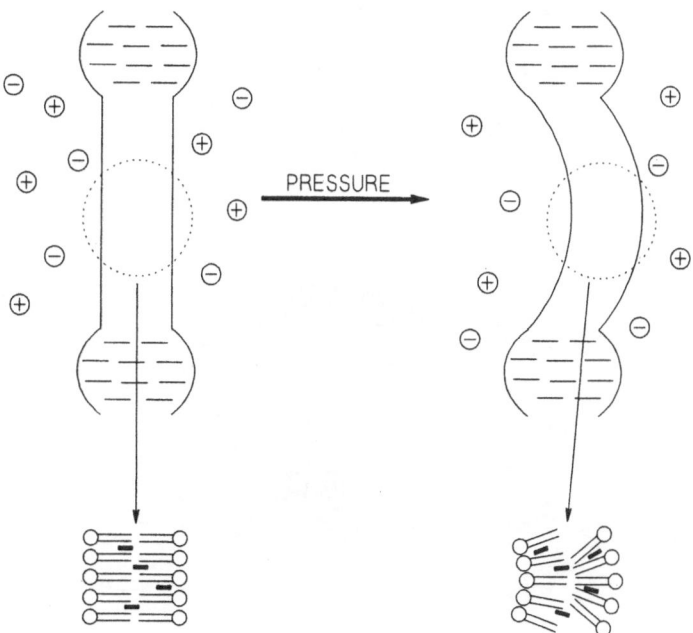

Fig. 62. A schematic representation of the hydrostatic-pressure-induced bending of a BLM. Distribution of DPH (indicated by ■) in the BLMs should only be considered to be schematic. In reality, the ratio of surfactants: DPH was 1 : 300 [412]

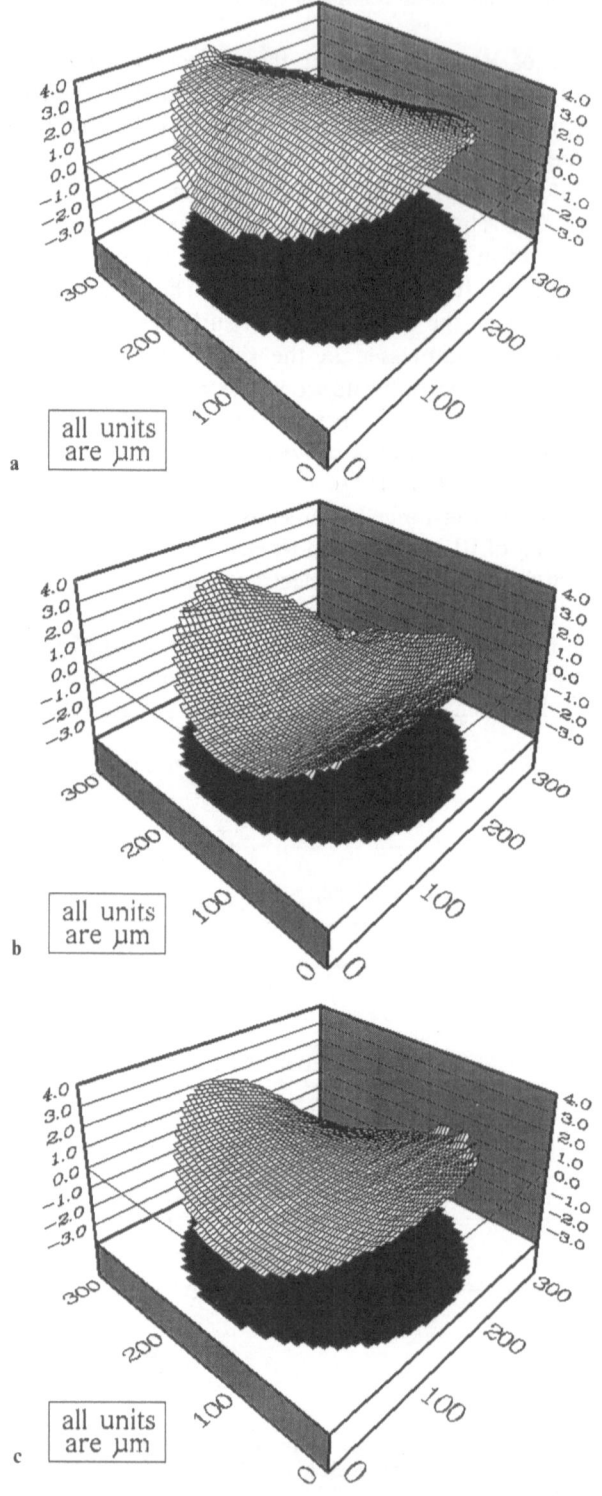

BLM, the type and concentration(s) of the electrolytes bathing the BLM, the ions adsorbed on the BLM surface, and the extent to and frequency with which the BLM is bent [419]. These experimental observations have led to a phenomenological definition of the flexoelectric coefficient, f, as the ratio between the bending-induced transmembrane potential, U_f, and the change of curvature, c, that accompanies the bending of the membrane:

$$f = U_f \varepsilon \varepsilon_0 / 2c . \tag{12}$$

An evaluation of f requires electrical measurements of U_f, together with a knowledge of BLM deformation during its subjection to oscillating hydrostatic pressure.

Precise movements of oscillating BLMs have been determined directly by real-time stroboscopic interferometry using laser pulses synchronized to the pressure of oscillation [422]. Stroboscopy slowed down the "apparent" rate of BLM oscillation so that optical interferograms could be continuously video-recorded at the normal speed of a commercial VHS recorder (a BLM interferogram was recorded every 33 ms). A time sequence of images was read from the videotape into a computer by use of a framegrabber. From the computer-analyzed frames, a time series of surface plots was generated and stored as images for rapid recall. By consecutive display of these images on screen, a movie was produced which clearly showed the three-dimensional movement of the BLM. Typical three-dimensional images taken from a hydrostatic-pressure-induced oscillating BLM are shown in Fig. 63 [422].

Flexoelectricity involves two degrees of freedom of the BLM: electrical and mechanical. The system is amenable to simultaneous electrical and optical investigation of mechanical-to-electrical-energy conversion mediated by a bi-molecularly thin membrane. BLMs themselves are unlikely to be used as device components. They offer, however, an eminently suitable means for conducting the fundamental studies which are necessary for the full potential of the membrane-mimetic approach to advanced materials to be realized.

2.8 Cast Multibilayers

Controlled evaporation of SUVs and MLVs on substrates has been shown to result in the formation of ultrathin films which retained the regular bilayer structure of vesicles [69, 425–427]. These immobilized bilayers, termed as "cast multibilayers", "cast multibilayers", or "ordered cast (ultrathin) films", have provided an alternative to LB films [425–446]. Alkylammonium surfactants with azobenzene (33) and glutamate (34) functionalities have been used, for example, in the preparation of cast-film-forming SUVs. X-ray diffraction

---

Fig. 63a–c. Three-dimensional presentations of the oscillating glyceryl monooleate (GMO) (22) BLM at: **a** $t = 0$ ms; **b** $t = 0.91$ ms; **c** $t = 1.67$ ms. Notice the difference in the scales [422]

measurements of cast films prepared from **34** established the surfactant alignment to be perpendicular to the substrate surface with 45° and 65° tilts (Fig. 64)
[445]. Thin films cast from **33** were found to be self supporting, stiff, and brittle;
whereas those prepared from **34** could be bent (Fig. 65). Three different types of
anionic copper(II) porphyrins (Fig. 66) have been incorporated into the cast
multibilayers of **33** and **34**. In Type I porphyrins, the anionic substituents were
localized on one side of the ring; in Type II, the three sulfonate groups were
oriented horizontally; and, in Type III porphyrins, the anionic groups were
symmetrically placed. The orientation of the porphyrin guests in the cast-
multibilayer hosts was evaluated by electron spin resonance spectroscopy and
by computer simulation of the observed EPR spectra (Figs. 67 and 68) [445].
Type III porphyrins were found to lay flat on the surfaces of the polar headgroups indiscriminately in cast multibilayers of both **33** and **34**. Intercalation of
Type I and Type II porphyrins into the cast multibilayers was, however,
structure-dependent. Thus, while Type I porphyrins aligned parallel with the
hydrocarbon chains in cast multibilayers of **34** (Fig. 67), they assumed a random
orientation in **33** (Fig. 68). Type II porphyrins could not penetrate into the
hydrocarbon regions of cast multibilayers of **33** (Fig. 68), but they partially
interdigitated into **34** (Fig. 67). Methodologies developed for the preparation
and characterization of porphyrin guests in cast-multibilayer hosts will pave the
way for preparation of the desired supramolecular assemblies [445].

Cast multibilayers prepared from an azobenzene-containing surfactant (**33**)
[426] and from mixtures of **33** and **5** display only short-term stabilities and low
solubilities in water. These disadvantages have been overcome by coating the
multibilayers with cellulose acetate and by using poly(vinyl alcohol) as a binder
[429]. The method of choice in preparing these composite cast multibilayers

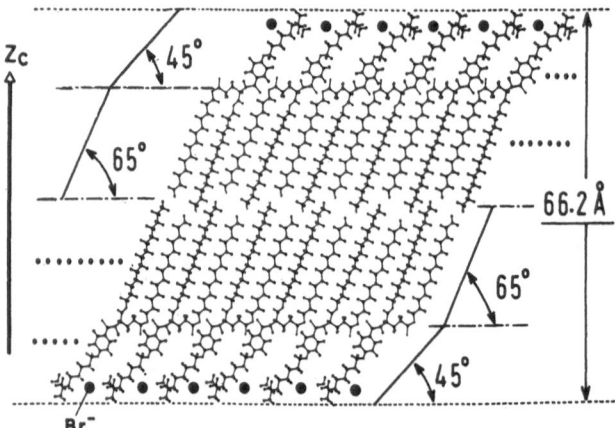

Fig. 64. Plausible model of the molecular packing of the cast multibilayer of **34** as produced by
electron-density matching of the X-ray diffraction data. The plane of the benzene ring is arbitrarily
set in the plane of the two alkyl chains. When the benzene plane is assumed to be in the
perpendicular disposition, the tilt angles for the spacer and tail chains become slightly greater: 50°
for the spacer and 75° for the tail [445]

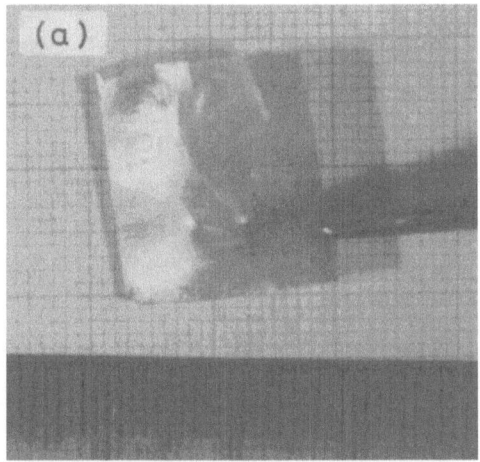

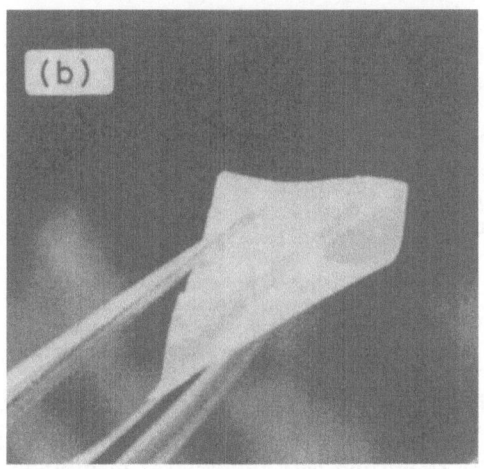

Fig. 65a, b. Photographs of cast multibilayer films: **a** stiff **33** film; **b** flexible **34** film [445]

involved the following steps: (i) preparation of SUVs from mixtures of **5** and **33**, typically in a 10:1 mole ratio; (ii) gently mixing an aqueous poly(vinly alcohol) solution (200 mg in 4.0 ml water) with 1.0 ml of the SUV solution; (iii) placing this mixture on a precast film of 300 mg cellulose acetate in 6.0 ml acetone; (iv) drying in vacuo; (v) coating the cast multibilayer with cellulose acetate (300 mg in 6.0 ml acetone); and (vi) drying slowly under saturated acetone vapor. An electron micrograph of the resultant composite cast multibilayer is shown in Fig. 69 [429]. In some cases, the poly(vinyl alcohol) matrix was cross-linked by γ-radiation [428].

Thin films have also been cast from fluorocarbon-surfactant-containing SUVs with bound poly(vinyl alcohol) [436, 437]. Wide angle X-ray diffraction measurements have been interpreted in terms of highly oriented multibilayers and complete substrate coverage by the fluorocarbon surfactants, even at low surfactant-to-poly(vinyl alcohol) ratios (Fig. 70) [437].

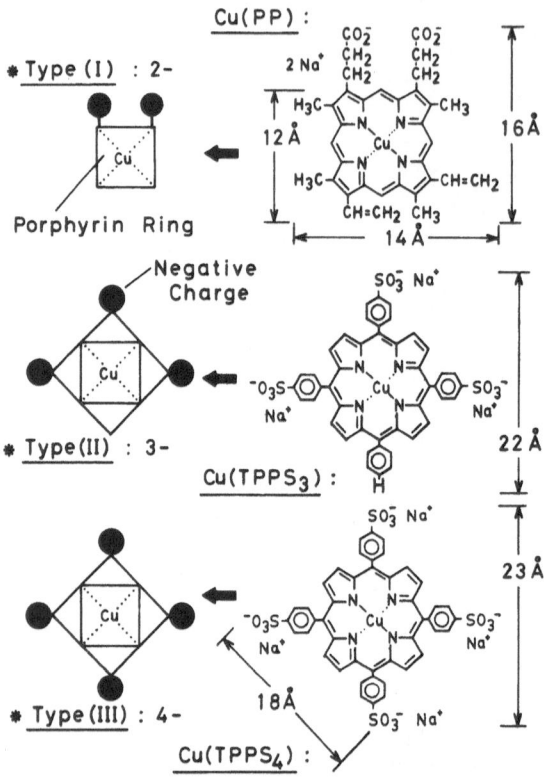

Fig. 66. Porphyrin derivatives classified into type I, type II, and type III [445]

Utilization of polymerizable surfactants [431] and surfactants containing polymerizable head groups ($CH_2=CH–COO^-$ instead of Br^- in **5**) [435] provided an alternative method for stabilizing cast-multibilayer membranes. Two alternative approaches have been taken; the first one involved the casting of polymerized vesicles, while prepolymerized vesicles were cast in the second method and the cast films were then polymerized. X-ray scattering provided evidence for the perpendicular orientation of the surfactants in films cast from SUVs prepared from **5** with the $CH_2=CH–COO^-$ headgroup and for a higher degree of order in films which were polymerized after casting. These results are not unexpected since polymerized SUVs are more likely to retain their structures in the drying process than are their non-polymerized counterparts (Fig. 71) [435]. The films which were cast from SUVs and subsequently polymerized also remained stable in electrolyte solutions. In contrast, those made from prepolymerized SUVs swelled in electrolytes [435].

Composite cast multibilayers provided a route to the formation of multilayer, two-dimensional polymer networks [443, 445]. This method utilized the following steps: (i) ultrasonic dispersal of 15 mM of the dialkylammonium surfactant, **35**, and 15 mM of the bisacrylate monomer. **36**; (ii) addition of

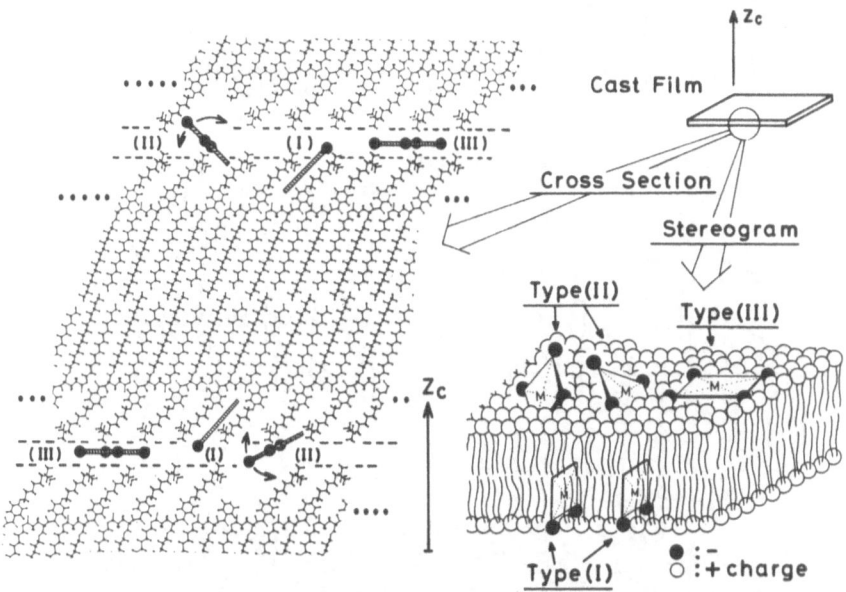

Fig. 67. Schematic representation of three types of anionic porphyrins in a cast multibilayer film of **34**. The overall bilayer organization is assumed to be the same as that of Fig. 64. For clarity, counterions are not shown and the bilayer units are separated from each other. The spacer portion is also not shown in "Stereogram". Type I porphyrins (Fig. 66) are inserted into the bilayer along the molecular axis of the spacer chain. Type II porphyrins are randomly placed on the bilayer surface. Type III porphyrins lie flat on the bilayer [445]

2 mol % of the initiator, $HOCH_2CH_2OC_6H_4COC(CH_3)_2OH$; (iii) casting the mixture on a fluorocarbon membrane filter; (iv) incubation at 25 °C and 60% relative humidity for 48 h; (v) photo-polymerization by a 500-W, ultrahigh-pressure Hg lamp; (vi) peeling the cast composite bilayer off of the fluorocarbon membrane support; and (vii) extraction of the surfactant component by repeated dipping into fresh methanol over a period of three hours. Complete removal of the surfactant was confirmed by IR spectroscopy and elemental analysis. X-ray diffraction and XPS measurements provided evidence for the presence of regular multibilayer parallel to the film plane, as indicated in the proposed schematics (Fig. 72) [444]. A photograph and a scanning electron micrograph of the multilayer, two-dimensional polymer network are shown in Fig. 73 [444].

A similar approach has been taken for preparing molecularly controlled siloxane networks [440, 446]. Composite bilayer films were cast from mixtures of alkoxysilane ($CH_3Si(OCH_3)_3$) and **37** or **38** on a fluorocarbon sheet and were kept at 25 °C and 60% relative humidity for three days. Exposure to gaseous ammonia in a closed vessel for ten days resulted in hydrolysis and subsequent condensation. The surfactants were then extracted by repeated immersion in methanol. Manipulation of the composition of the cast multibilayers allowed

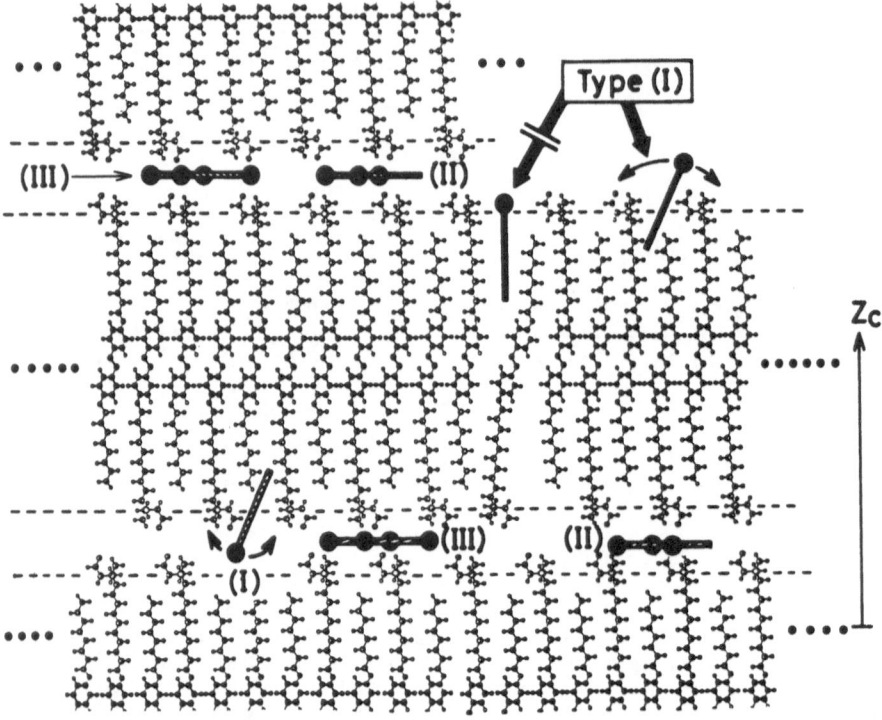

Fig. 68. Schematic representation of three types of anionic porphyrins in a cast multibilayer film of 33. For simplicity, counterions are not shown. The bilayer packing is based on the X-ray diffraction data. Type I porphyrins (Fig. 66) assume random orientations. Type II and III porphyrins stay horizontally on the bilayer surfaces [445]

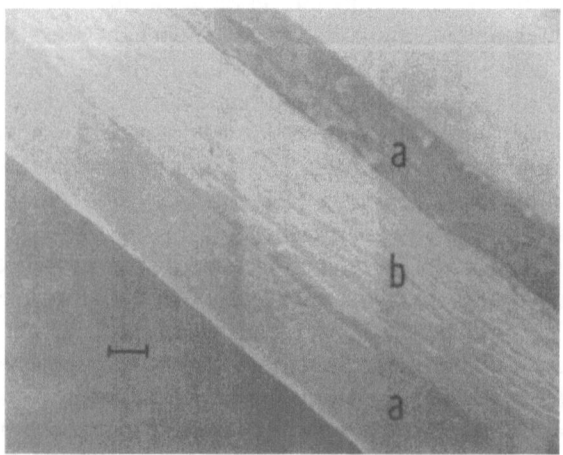

Fig. 69. SEM of a cross-section of a three-layered film (*bar* = 10 μm); CA layer (*a*); PVA-bilayer composite layer (*b*). The vertical shades are artifacts [429]

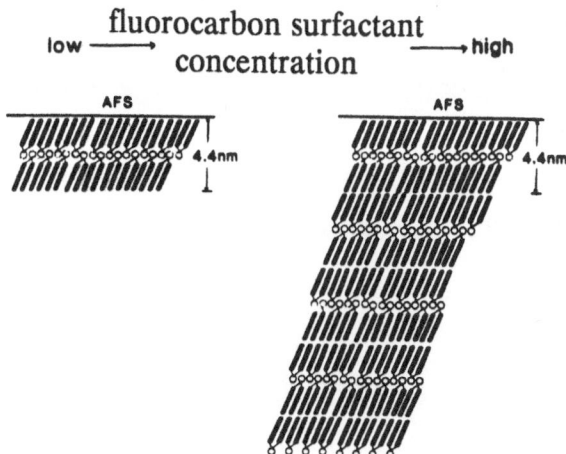

Fig. 70. Schematic representation of the structure of the outermost layer of the composite thin film with small and large weight percents of fluorocarbon amphiphile [437]

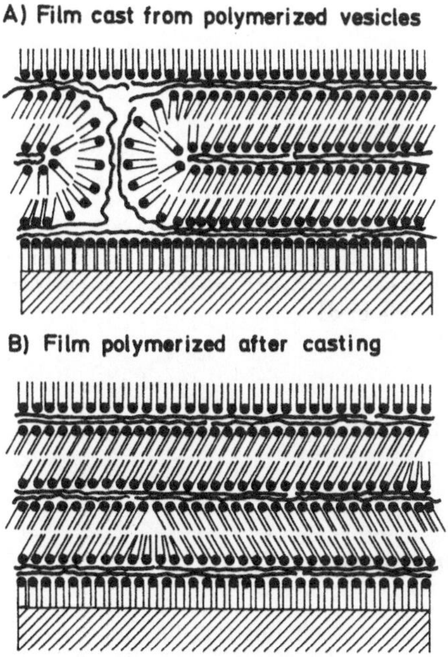

A) Film cast from polymerized vesicles

B) Film polymerized after casting

Fig. 71a,b. Schematic representation of possible structures in: **a** films cast from polymerized vesicles; **b** films polymerized after casting [435]

the formation of polysiloxane films having tube-like, spherical, and plate-like nanostructures [446].

Myoglobins [442] and iron-oxide particles [447] have also been organized in cast multibilayers. Flexibility and versatility render composite cast multibilayers to be suitable for the construction of functional, ultrathin, polymeric superstructures with molecularly defined morphologies. Cast-multibilayer

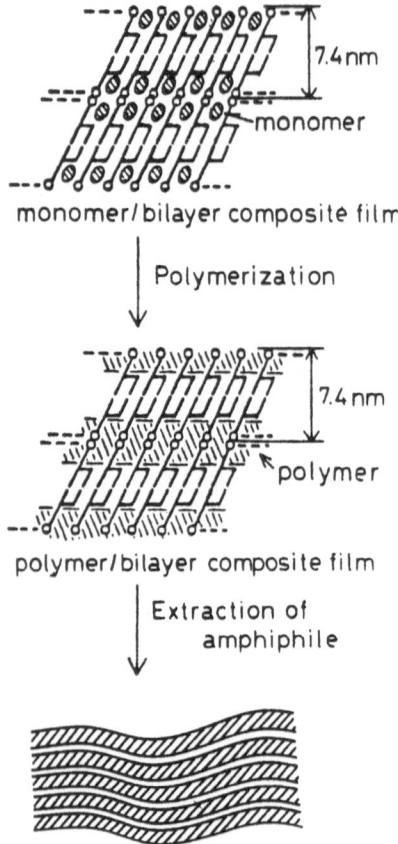

monomer/bilayer composite film

Polymerization

polymer/bilayer composite film

Extraction of amphiphile

multi-layered 2D network

Fig. 72. Schematic illustration of template synthesis of a multilayered two-dimensional polymer network [444]

membranes also serve as permeability controls in capsule membranes [343–345, 448].

2.9 Polymers and Polymeric Membranes

Creative interplay between colloid and polymer chemistries has increasingly contributed to the development of membrane-mimetic systems and advanced materials. On the one hand, the employment of polymer methodologies and/or the addition of polymers have favorably altered the properties of colloidal systems. On the other hand, the introduction of surfactants and surfactant assemblies prior, during, or subsequent to polymerization has resulted in distinctly different polymers.

Incorporation of polymerizable moieties into surfactants or into their counterions has permitted the formation of polymerized or polymer-coated

Fig. 73a,b. Multilayered film of two-dimensional polymer network: **a** photograph, diameter 25 mm; **b** cross-sectional view by scanning electron microscopy [444]

monolayers, LB films, BLMs, and surfactant vesicles. Polymeric monolayers, vesicles, and cast bilayers have also been prepared from prepolymerized surfactants. Such polymerized and polymeric membrane-mimetic systems have been shown, in previous sections of this monograph, to be more stable than their non-polymerized counterparts. Appropriate oligomers and polymers have also been used for stabilizing and destroying surfactant assemblies.

The removal of surfactants from polymer-coated vesicles and from composite cast bilayers to yield "ghost" vesicles and molecularly thick, two-dimensional polymer networks (detailed previously) illustrates a colloid-chemical approach to the construction of specific polymers.

The inter-relationship between colloid and polymer chemistries is completed by colloidal polymer particles. The formation of 50-nm-diameter, 100- to 200-nm-long polyaniline fibrils in a poly(acrylic acid)-template-guided polymerization, similar in many ways to those produced from polymerized SUVs (see above), provides a recent example of polymer colloids [449]. The use of poly(styenesulfonic acid) as a template yielded globular polyaniline particles which were found to be quite different morphologically from those observed in the regular chemical synthesis of polyaniline [449].

Many aspects of the chemical behavior of colloids are also exhibited by polymer gels [450, 451]. Certain polymers which possess the appropriate three-dimensional networks may increase their size several fold, as do colloids, by taking up solvent molecules. Such colloids are termed "sols" in their swollen state and "gels" in their compact state. Transition from gel to sol occurs at the so-called "critical gel point". If the solvent is organic, the gels are called "organogels". Organogel formation has been investigated in reversed micelles [283–285]. A useful property of polymeric hydrogels is that their critical gel points can be shifted by changes in temperature and pH, through the addition of solvents and electrolytes, or by the application of sound or mechanical stress (Fig. 74) [452]. A particularly interesting property of environmentally sensitive hydrogels in their ability to convert chemical energy into mechanical energy [453]. A 1-mm-thick, 5-mm-wide, and 20-mm-long strip of hydrogel, prepared from weakly cross-linked poly(2-acrylamido-2-methyl propane) sulfonic acid, was suspended by two plastic hooks in an aqueous 1.0×10^{-2} M solution of n-dodecylpyridinium chloride containing 3.0×10^{-2} M sodium sulfate. The polymeric hydrogel swelled by a factor of 45 in the solution. It was also shown that applying an oscillating, 20 V electric field (altered polarity; 2 s) across the hydrogel induced a worm-like motion upon it, resembling muscle flexing (Fig. 75) [453]. Polymeric gels will find applications ranging from the controlled release of entrapped drugs to electronic actuators and sensors.

Polymers themselves, cast into membranes, spinned into hollow fibers, pressed into sheets, or extruded into selected shapes and configurations, have been widely used as compartments for the generation and stabilization of

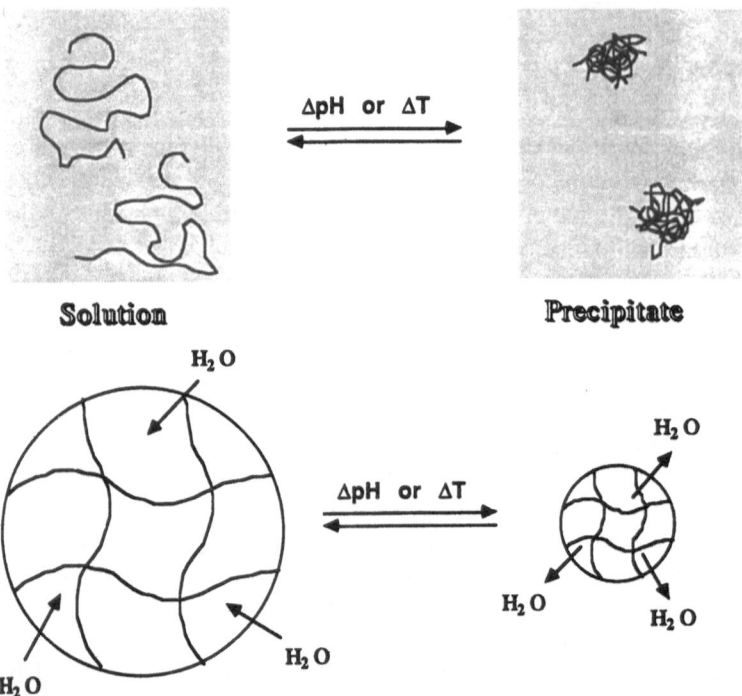

Fig. 74. Phase change behavior of pH and temperature-sensitive polymers and hydrogels [452]

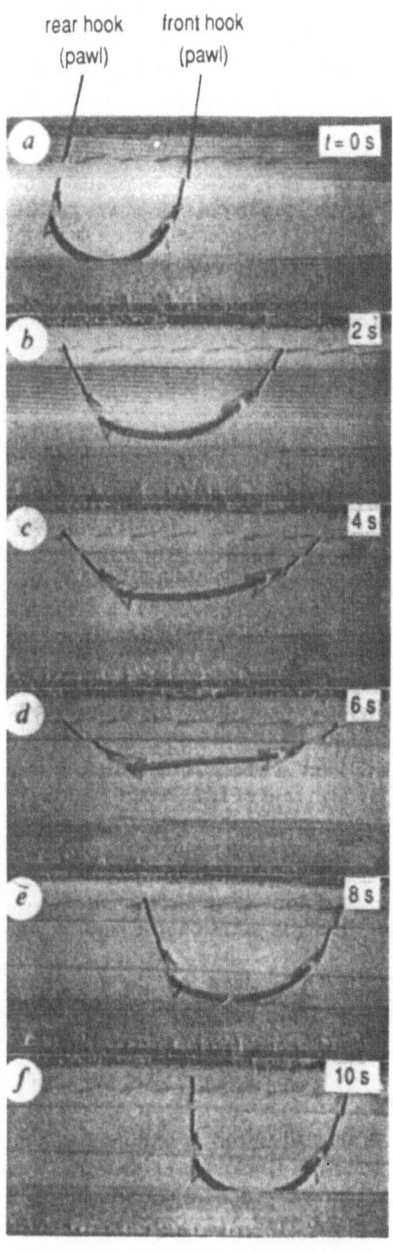

rear hook (pawl) front hook (pawl)

electrode

plastic ratchet bar

electrode

20 mm

Fig. 75a–f. Time profiles of the polymer gel in action: **a–d** 'stretching' process. The front hook attached to the gel can slide forward along the plastic ratchet bar, but the rear hook is prevented by the teeth of the ratchet from sliding backwards; **e, f** bending process. When the polarity of the electric field is reversed and the gel bends, the rear hook can move forward along the ratchet bar, but the front hook is prevented from sliding backwards. This action being repeated, the gel walks forward. The gel is poly(2-acrylamido-2-methyl propane) sulphonic acid (PAMPS); degree of crosslinking 5 mol%; degree of swelling in the solution, a factor of 45. Electric field, 20 V; current 15 mA cm^{-2}; electrode distance, 20 mm. Solution: 1×10^{-2} M n-dodecyl-pyridinium chloride, containing 3×10^{-2} M sodium sulphate. Total ionic strength: 0.1 [453]

metallic, catalytic, semiconducting, and magnetic particles. A description of polymer synthesis, characterization, and engineering [454, 455] is outside the scope of the present monograph. It is sufficient to draw attention to the ever increasing use of conductive [32, 455–458], liquid crystalline [459] and ferro-electric [460] polymers, as well as to those having non-linear optical properties [461].

2.10 Other Systems

The range of compartments available for hosting particles in controlled sizes and morphologies is limited only by the imagination and diligence of chemists. Many examples have been cited; thus, only brief references will be made here to some of the systems developed for the in situ generation and investigation of metallic, catalytic, semiconducting, and magnetic particles. Inevitably, some-body's favorite system will be omitted.

Channels provide by molecular-sieve type materials, such as zeolites, have been extensively utilized as hosts and templates, often for the oriented growth of particles [6]. Appreciable non-linear optical effects were observed, for example, in large needle-like zeolite crystals which contained p-nitroaniline in their channels upon alignment by a 3-kV/cm electric field [462]. Molecular sieves are constructed, typically, from tetrahedral TO_4 (where T = Si, Al or P) units. Classical zeolites contain 12 T-atom rings with 7- to 8-Å channel diameters [463, 464], which act as sites for the incorporation of small inorganic or organic species [465], or conjugated polymers [466-468]. Synthesis of zeolites with larger channel diameters has been particularly significant; indeed, zeolites having 14 to 18 T-atom rings, with 12- to 13-Å-[469, 470] and 29- to 30Å-[471, 472] diameter channels, have been prepared. A gallophosphate molecular-sieve (called "cloverite"), 20-T-atom-ring zeolite, having a 30-Å-diameter cavity, has also been synthesized [471]. It had a cubic unit cell of $\{Ga_{768}P_{768}O_{2976}(OH)_{192}(RF)_{192}\}192RF$, where RF is quinuclidinium fluoride (Fig. 76) [471]. Zeolites with even larger sized pores (up to 100 Å in diameter)

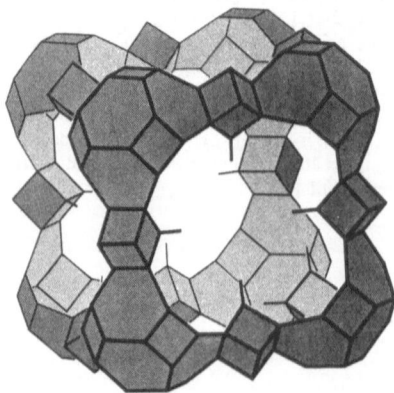

Fig. 76. A skeletal drawing of the large supercage of the framework structure showing the clover-leaf-shaped windows and the pockets at the corners of the cage [471]

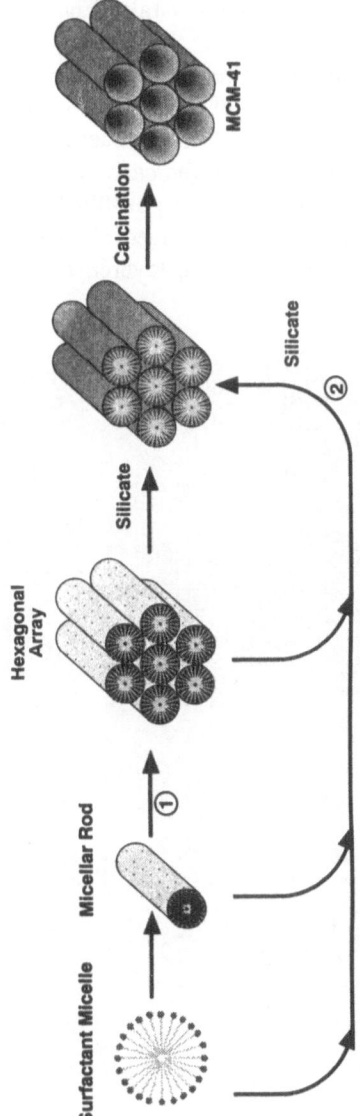

Fig. 77. Schematic drawing of the liquid-crystal templating mechanism. Hexagonal arrays of cylindrical micelles form (possibly mediated by the presence of silicate ions), with the polar groups of the surfactants (*light grey*) to the outside. Silicate species (*dark grey*) then occupy the spaces between the cylinders. The final calcination step burns off the original organic material, leaving hollow cylinders of inorganic material [473]

have been synthesized by using liquid crystals (Fig. 77) [473] or reversed micelles as templates [474].

The recently discovered and much researched buckminster fullerenes [475, 476] provide interesting opportunities for particle incorporation or attachment. Graphitic tubes, ranging from 4 to 30 nm in diameter and up to 1 μm in length, have been formed on the negative end of the carbon electrode used in the d.c. arc-discharge evaporation of carbon in an argon-filled vessel (Fig. 78) [477].

Advantage has also been taken of dispersed clays to host small particles [478]. Montmorillonite clays are colloidal, layered aluminosilicates with ex-

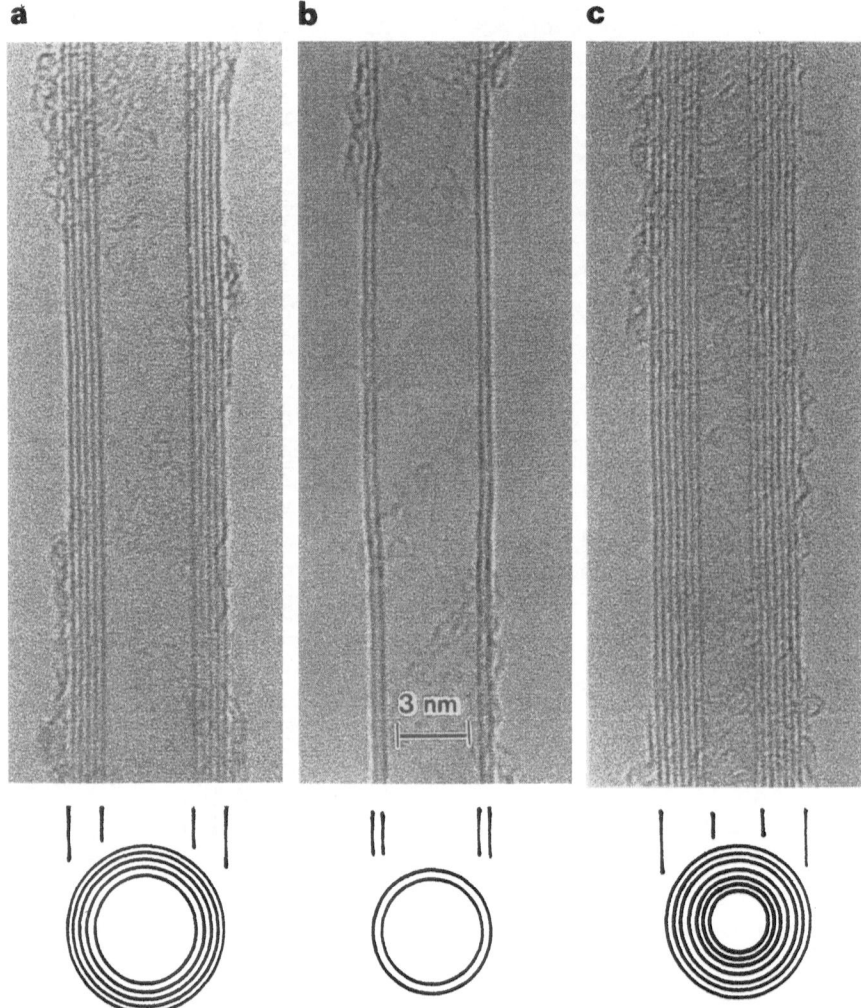

Fig. 78a–c. Electron micrographs of microtubules of graphite carbon. Parallel dark lines correspond to the (002) lattice images of graphite. A cross-section of each tubule is illustrated: **a** tube consisting of five graphitic sheets, diameter 6.7 nm; **b** two-sheet tube, diameter 5.5 nm; **c** seven-sheet tube, diameter 6.5 nm, which has the smallest hollow diameter (2.2 nm) [477]

changeable cations distributed over their external and internal surfaces
[479–481]. The layers, each consisting of a sheet of $AlO_4(OH)_2$ octahedra
sandwiched between two sheets of SiO_4 tetrahedra, are stacked parallel to one
another to form plates. The flat, composite, aluminosilicate-layer lattice carries
a net negative charge due to the isomorphous substitution of Al and Si atoms by
atoms of lower valence. The number of exchangeable cations in a given clay
sample are expressed in terms of the cation-exchange capacity. Montmorillonites typically have cation-exchange capacities of 70–80 milliequivalent grams per
100 g. The interlayer or basal spacing, d_L, in clay sheets depends upon the
solvation of the exchangeable cations [480, 481]. Ion exchange of the sodium
cations in montmorillonites with surfactants leads to organoclay complexes
which are often referred to as pillared clay minerals [482, 483]. The length of the
pillars (d_L) depends on the type of surfactant used and on the absence or
presence of solvents. Solvation of clays results in swelling and in a concomitant
increase in d_L (Fig. 79) [484]. X-ray diffraction techniques have been proven to
be important in providing precise information on the basal spacing of organoclay complexes. The basal spacing for octadecylammonium organoclay complexes in the dry state (14.5–18.9Å), for example, was found to increase to 34 Å
in ethanol and to 42 Å in benzene [482]. Clearly, organoclay complexes provide
versatile hosts for nanoparticles. They can be used, for example, in solvents
[484], in coating electrodes [478, 485], and, indeed, in the solid state [486].

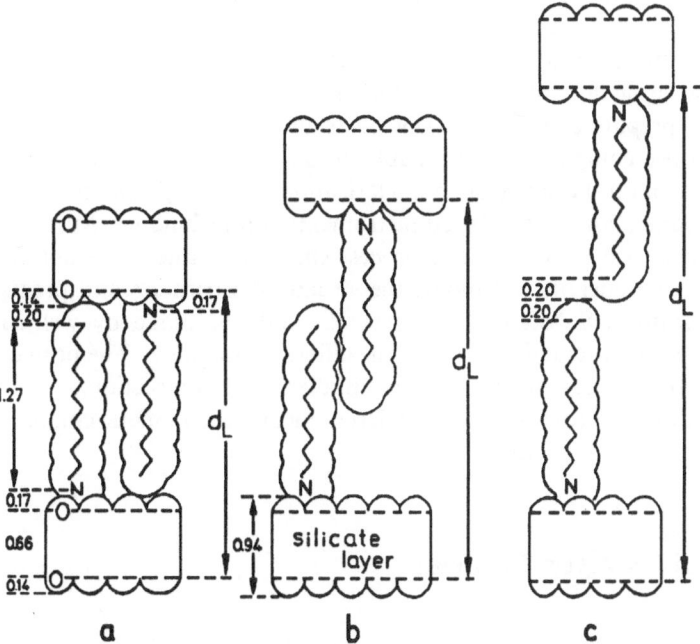

Fig. 79a–c. An oversimplified representation of the proposed mode of swelling of organoclay
complexes. Illustrated are: **a** the non-swollen; **b** the semi-swollen; **c** the completely swollen organoclay complexes. Distances are in nanometers [484]

Space available in porous glass [487], ultrafine Nafion [488, 489], and metallic membranes [490, 491] has also been utilized for the development of smal particles. Cylindrical micropores in alumina membranes have been used, for example, as templates for the electrodeposition of parallel arrays of gold particles (0.26 µm in diameter, 0.3 µm to 3 µm in length) which were infrared transparent [491] and could be used as chemical sensors [490].

Last, but by no means least, reference should be made to the use of proteins in nano-fabrication [492]. One approach is illustrated by the fabrication of a 1-nm-thick metal film with 15-nm-diameters holes, periodically arranged on a triangular protein lattice [493]. Advantage was taken of the 10-nm-thick, uniformly porous surface (or S) layer of the crystalline protein obtained from the thermophilic bacterium *Sulfolobus acidocaldarius*. The protein was adsorbed from a dilute solution onto a molecularly smooth carbon-film surface, metal coated by evaporation, and ion milled to give spatial ordering of holes with the same nanometer periodicity as the protein lattice [493].

2.11 Comparative Merits of Different Compartments

The variety of different compartments available for hosting metallic, catalytic, semiconducting, and magnetic particles, or for the construction of nanosized materials, are bewildering. Which is best suited to a given application? Unfortunately, there is no simple answer to this question. Selection of a membrane-mimetic system will be determined, by and large, by the background and experience of the investigator. Synthetically inclined chemists will be likely to functionalize the surfactants needed to assemble a compartment for a specific purpose and will prepare conceptually new hosts and templates. Biologists and biochemists will use compartments available in nature. Physical chemists will focus their attention on the superior characterization of existing compartments and on their modification by advanced instrumentation techniques. Materials scientists, together with engineers, will process the compartments to optimal configurations. Theoreticians will model the structural intricacies and predict the best organizations. Substantial progress can only be made by well-coordinated and creative interaction between the different disciplines. The primary purpose of the present chapter has been to stimulate such interaction.

The most important properties of the different membrane-mimetic compartments are summarized in Table 4.

3 Metallic and Catalytic Particles

Metallic particles have been extensively used as efficient and selective industrial catalysts [494–497]. Their performance strongly depends on the structure and area of the exposed surface; in general, the larger the catalytic area the more

Table 4. Properties of membrane-mimetic compartments

Compartment	Dimensionality; size of host	Method of preparation	Stability	Comment	Key references
Monolayers	2D, 3D; depends on trough	Spreading of water-insoluble surfactants, in an organic solvent, onto the surface of clean water	Days	Parties grown under monolayers could be transferred to solid substrates at any stage of their growth	104, 105
Langmuir-Blodgett (LB) films	2D, 3D; depends on substrate dimensions	Monolayers transferred to substrates	Weeks to months	Desired molecules could be intercalated into specified positions; small particles could grow in situ or be placed between the layers	106, 137
Self-assembled (SA) films	2D, 3D; depends on substrate dimensions	Spontaneous adsorption of surfactants from solution onto substrates	Weeks to months	Convenience of formation and versatility rendered them to be viable alternatives to LB films	68, 183
Aqueous micelles	3D; 20–40 Å diameter	Dissolution of surfactant in water at concentrations above the critical micelle concentration (CMC)	Dynamic equilibrium between monomers and micelles	Particle formation was only possible at the micellar surface or close to it	55, 101
Reversed micelles	3D; 20–40 Å diameter	Dissolution of surfactant in a non-polar solvent in the presence of a small amount of water	Concentration-, surfactant-, solubilizate-, and solvent-dependent dynamic equilibria	Particles could be generated and derivatized in aqueous pools and, subsequently, separated from surfactants	55, 101
Surfactant vesicles	3D; 300–10 000 Å diameter	Shaking, sonication, or alcohol injection	Days to weeks	Solubilizates and particles could be in situ generated or placed into the bilayer or aqueous interior; a high level of organization was possible	55, 72, 101
Polymerized-polymeric vesicles	3D; 300–10 000 Å diameter	Polymerization of surfactants prior (polymeric) or subsequent (polymerized) to vesicle formation	Months to a year	Highly stable systems with controllable morphologies could be generated	55, 71, 72

Table 4. Contd.

Compartment	Dimensionality; size of host	Method of preparation	Stability	Comment	Key references
Tubules	2D; 0.2–1.0 μm diameter, 10–1,000 μm long	Precipitation of diacetylenic phosphatidylcholines, dissolved in organic solvents, by water below the phase transition temperature	Weeks	Interior and exterior could be metallized or coated by substrates or particles	361
Fibers	2D, 3D; micron size or longer, depending on system	Cooling aqueous, previously heated (to 100 °C) solutions of *chiral* aldonamides	Weeks in the presence of surfactants	A large variety of morphologies was possible	372
Bilayer lipid membranes (BLMs)	2D, 3D; 30- to 50-Å-thick, 1- to 2-mm-diameter membrane, supported by a solvent surfactant reservoir the Plateau-Gibbs border or torus) and separating two aqueous solutions macroscopically	Painting of the surfactant (or lipid), dissolved in a hydrocarbon solvent, across a teflon pinhole which separates two compartments of aqueous solution	Hours	Convenient system for fundamental studies as simultaneous electrical and spectroscopic measurements were possible	385, 387
Cast multibilayers	2D,3D; depends on substrate dimensions	Controlled evaporation of surfactant deposited on substrates	Weeks	Viable alternative to LB films	427
Zeolites	2D for classical, 2D and 3D for superzeolites; varies	Chemical synthesis	Months to years	Particles could be grown in situ or incorporated into channels; complex supramolecular architecture was possible	463, 464
Clays-pillared clays or organoclays	2D, 3D; varies	Purification and ion exchange by surfactants of naturally occurring products or synthesis	Weeks to months	Interlayer distance could be varied and precisely determined X-ray diffraction measurements; particles could be grown in situ or incorporated between layers; complex supramolecular architecture was possible	478, 480, 482

efficient the catalyst is [498]. Indeed, a phenomenological relationship has been described between catalytic activity, a, and the radius of catalytic particles, R, in terms of the scaling law [499]:

$$a \propto R^{D_R} \tag{13}$$

where D_R, defined as the reaction dimension (the scale), relates the sensitivity of the catalytic performance to changes in geometric parameters and to the distribution of active sites. Equation (13) has been shown to describe many reactions successfully, including oxidation, reduction, isomerization, photolysis, and electron transfer [499]. Not surprisingly, the development of a homogeneous population of catalytic particles with controllable morphologies and surface structures, together with routes to their uniform deposition on solid supports, continues to be a major scientific endeavor.

Membrane-mimetic compartments have provided a viable means for generating monodispersed catalytic particles [500]. In particular, reversed micelles and microemulsions have been used extensively as hosts. A complete summary of work reported on the in situ generation of catalysts in membrane-mimetic media, including publications up to 1987, has been produced [500] and, therefore, will not be reiterated here. Attention will be focused on more recent research utilizing monolayers, bilayer lipid membranes (BLMs), Langmuir–Blodgett (LB) films, zeolites, and clay particles as membrane-mimetic templates.

Significant new insight has been gained into the formation of small clusters and nanosized metallic particles [501, 502]. This fundamental information is not only inherently fascinating, but it is vitally important for the construction of new generations of advanced nanostructured materials. Evolution of nanosized metallic particles from non-metallic clusters and the chemistries of these species will, therefore, be discussed in the following sections.

3.1 Non-metallic Clusters, Quasi-metallic Clusters, and Nanosized Metallic Particles – Evolution from Bonds to Bands in Aqueous Solution

Transition from non-metallic clusters consisting of only a few atoms to nanosized metallic particles consisting of thousands of atoms and the concomitant conversion from covalent bond to continuous band structures have been the subject of intense scrutiny in both the gas phase and the solid state during the last decade [503–505]. It is only recently that modern-day colloid chemists have launched investigations into the kinetics and mechanisms of cluster formation and cluster aggregation in aqueous solutions. Steady-state and pulse-radiolytic techniques have been used primarily to examine the evolution of nanosized metallic particles in metal-ion solutions [506–508].

The assembly of silver colloids has proven to be a favorite subject of investigations [47, 509, 510], due, in the main part, to the availability of detailed information [9, 511–515]. Aqueous silver colloids, in the millimolar concentration range and containing spherical 5- to 10-nm-diameter particles, are characteristically yellow ($\lambda_{max} \approx 380$ nm; see Fig. 80) [521]. The yellow color may arise from collective surface plasma oscillations or interband transitions [516–523]. The absorption coefficient of the colloidal solution, A, can be calculated by inserting dielectric constants from the Drude model into the Mie formula for absorption by colloid particles [517, 524]:

$$A = 9\pi N V n_0^3 c \lambda^2 / \sigma [(\lambda_m^2 - \lambda^2)^2 + \lambda^2 \lambda_m^4 / \lambda_a^2] \tag{14}$$

where N is the concentration of the particles, V is the volume per particle (related to the filling factor, f; f = volume fraction of the particles in the solution), c is the velocity of light, λ_m is the wavelength at which maximum absorption occurs, and λ_c is the plasma wavelength defined as:

$$\lambda_m^2 = \lambda_c^2 (\varepsilon_0 + 2n_0^2) \tag{15}$$

$$\lambda_c^2 = (2\pi c)^2 m / 4\pi N_e e^2 \tag{16}$$

where ε_0 is the high frequency contribution of the dielectric constant of the silver, m is the effective electron mass (0.995 m_0), and N_e is the electron density. For a

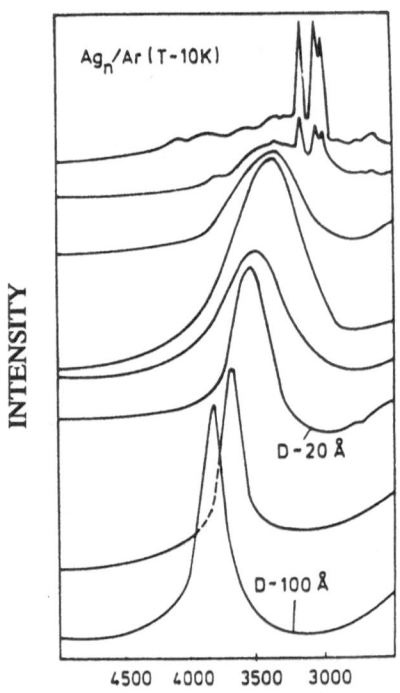

Fig. 80. Absorption spectra for Ag/Ar colloids. The mean diameter D of the spherical Ag particles decreases from *bottom* to *top* [521]

narrow Lorentzian band, the band-width at half maximum, w, is given by:

$$w = \lambda_m^2/\lambda_a = (\varepsilon_0 + 2n_0^2)c/2\sigma \tag{17}$$

where σ is the particle's d.c. electrical conductivity:

$$\sigma = N_e e^2 R/mu \tag{18}$$

and R is the mean free path of the electrons in the colloid, normally taken to be $1/R = (1/r + 1/r_\infty)$, where r_∞ is the electron mean free path in bulk silver (520 Å), u is the electron velocity at the Fermi level (1.4×10^8 cm s^{-1}), and λ_a is a constant related to the width of the absorption band ($\lambda_a = 2\lambda_c^2\sigma/c$).

Absorption spectra of well-defined silver clusters and a series of growing colloidal particles, prepared by the gas aggregation technique, are shown in Fig. 80 [521]. Silver monomers and small aggregates (D < 2 nm) appear to have sharp absorption bands in the 300–350 nm region (see upper two spectra in Fig. 80). Increasing the sizes of the silver clusters can be seen to manifest in the appearance of a single broad absorption band whose bandwidth sharpens and maximum shifts to the red occur as the diameter of the colloids increases. The spectrum of the 100-Å-diameter silver particles corresponds clearly to the surface plasmon absorption band. Further increase in the size of the silver colloids and their concentrations results in spectra which are dominated by light scattering.

The detection of sharp plasmon absorption signifies the onset of metallic character. This phenomenon occurs in the presence of a conduction band intersected by the Fermi level, which enables electron-hole pairs of all energies, no matter how small, to be excited. A metal, of course, conducts current electrically and its resistivity has a positive temperature coefficient. On the basis of these definitions, aqueous 5–10 nm colloidal silver particles, in the millimolar concentration range, can be considered to be metallic. Smaller particles in the 100-Å > D > 20-Å size domain, which exhibit absorption spectra blue-shifted from the plasmon band (Fig. 80), have been suggested to be quasi-metallic [513]; these particles are size-quantized [8–11]. Still smaller particles, having distinct absorption bands in the ultraviolet region, are non-metallic silver clusters.

Irradiation of a 1×10^{-4} M aqueous, degassed AgClO$_4$ solution in the presence of 0.1 M methanol by submicrosecond high energy (3.6 MeV) electrons led to the development of sequential transient absorbances (Fig. 81), ascribed to the formation of the following species [512, 513]:

$$Ag^+ + e_{aq}^- \rightarrow Ag^\circ \quad \lambda_{max} = 360 \text{ nm} \tag{19}$$

$$Ag^\circ + Ag^+ \rightarrow Ag_2^+ \quad \lambda_{max} = 310 \text{ nm} \tag{20}$$

$$2Ag_2^+ \rightarrow Ag_4^{2+} \quad \lambda_{max} = 275 \text{ nm} \tag{21}$$

Longer irradiation than that shown in Fig. 81 resulted in decreased absorption of Ag$_4^{2+}$ at 275 nm and in an increase at 380 nm, which are characteristic of a developing silver plasmon band [525, 526].

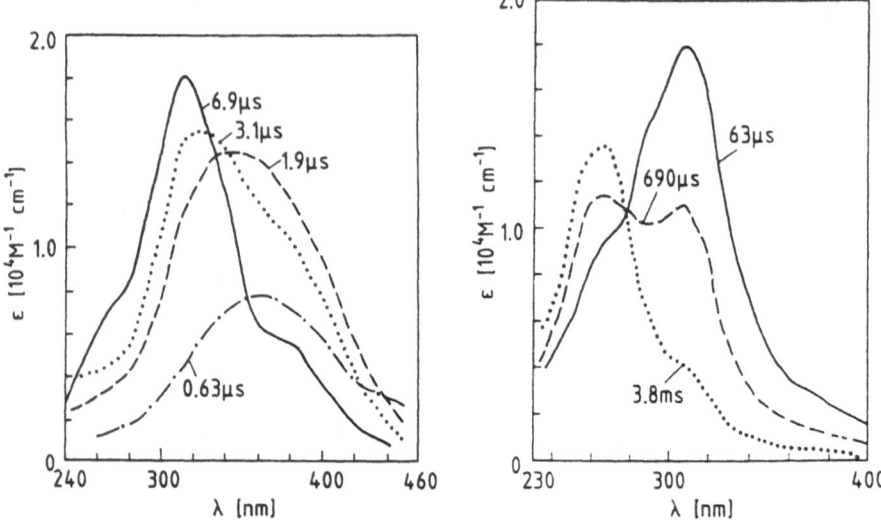

Fig. 81a. Absorption spectrum of a solution at various times after the pulse. 1×10^{-4} M AgClO$_4$ and 0.1 M methanol; absence of polyphosphate. **b** Absorption spectrum of the solution of (a) at longer times after the pulse [512]

Addition of sodium polyphosphate appreciably altered the rate constants for reactions (19)–(21) and stabilized the small non-metallic silver clusters [512, 513]. Advantages of the steady-state and pulse-radiolytic approaches to silver-cluster formation are manifold. Firstly, experimental conditions can be precisely adjusted such that the reactive species is exclusively e_{aq}^- or, alternatively, that it is a known alcohol radical. Secondly, the concentration of the reducing species (the number of reducing equivalents generated) is readily calculable. Thirdly, in time-resolved experiments, rate constants for the individual reaction steps can be determined by monitoring absorption and/or conductivity changes. These latter determinations permitted the assessment of agglomeration numbers [512, 513].

Interestingly, non-metallic silver clusters, depending on their sizes, may act either as electron donors or as electron acceptors. Using sulfonatopropyl-viologen, SPV (E^0 for {SPV/SPV$^-$} = − 0.41 V/NHE), pulse radiolysis established that small silver clusters (n ≤ 4) acted as electron donors (i.e. E^0 for {Ag$_n^+$/Ag$_n$} < E^0 for {SPV/SPV$^-$}) while, conversely, large silver clusters (n ≥ 4) were electron acceptors (i.e. E^0 for {Ag$_n^+$/Ag$_n$} > E^0 for {SPV/SPV$^-$}) [511]. Size-dependent electrochemical potentials of silver aggregates have been elucidated (Fig. 82) [506].

Non-metallic gold [527], copper [528], and lead [529] clusters have also been generated by pulse radiolysis. Extension of the irradiation period resulted in cluster growth and in the concomitant development of absorption maxima at 520 nm, 580 nm, and 220 nm, corresponding to the plasmon bands of metallic gold, copper, and lead colloids, respectively [527–529].

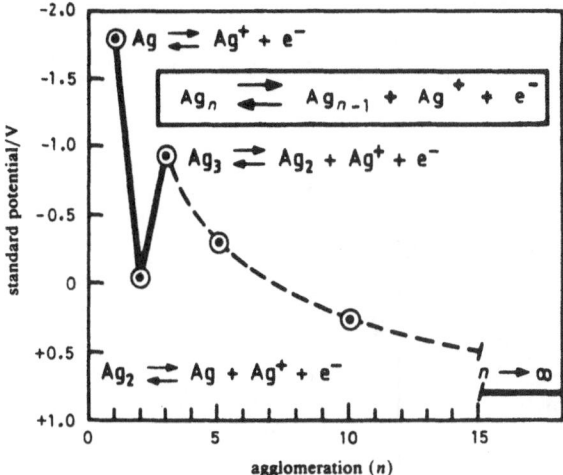

Fig. 82. Standard redox potential of the silver microelectrode as a function of the agglomeration number [506]

3.2 Chemistry of Nanosized Metallic Particles

Charge alteration on the surfaces of nanosized metallic silver particles has been investigated by simultaneously monitoring absorption and conductivity changes during pulse-radiolytic experiments [506]. Pulse radiolysis of a nitrous-oxide-(N_2O) saturated aqueous solution of 3.0 nm diameter metallic silver particles containing 0.2 M 2-propanol resulted in electron injection to the colloid. N_2O functions to double the yield of hydroxyl radicals (·OH) generated in water radiolysis:

$$H_2O \rightarrow e_{aq}^- + \cdot OH + H^+ + \cdot H + H_2 \tag{22}$$

$$N_2O + e_{aq}^- + H_2O \rightarrow N_2 + \cdot OH + OH^- \tag{23}$$

The hydroxyl radical reacts with 2-propanol to form 2-hydroxypropan-2-yl radicals, which transfers electrons, in turn, to the colloidal silver particles:

$$CH_3CH(OH)CH_3 + \cdot OH \rightarrow CH_3 \cdot CHOCH_3 \tag{24}$$

$$Ag_n + x(CH_3)_2 \cdot COH \rightarrow Ag_n^{x-} + x(CH_3)_2CO + xH^+ \tag{25}$$

The reaction described by Eq. (25) occurs in the millisecond time scale and results in an exponential increase in the conductivity of the solution and a parallel decrease of the absorbance at 440 nm. The net result of the reaction was both a blue shift and a narrowing of the silver plasmon absorption band (see $0 \rightarrow a$ change in Fig. 83) [506].

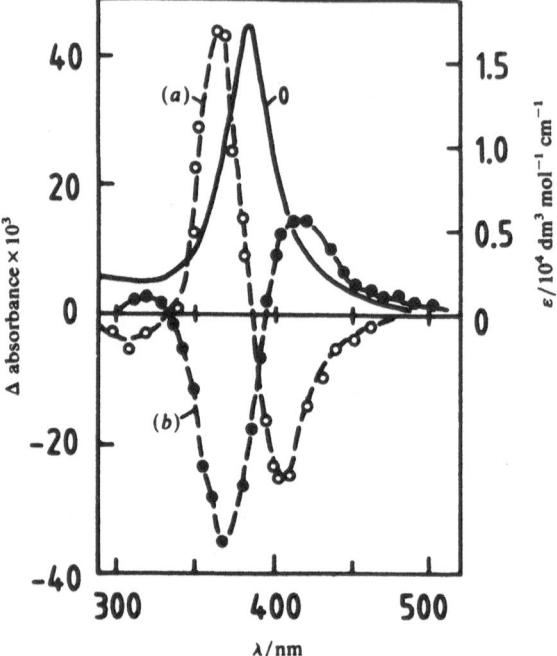

Fig. 83. Surface plasmon absorption band of a 1×10^{-4} mol dm^{-3} silver sol (O). Changes in absorption after electron donation (*a*) and positive hole injection (*b*) by free radicals [506]

Hole injection into the silver particles was accomplished by allowing ·OH (formed in the pulse radiolysis of N_2O-saturated, aqueous, 3.0-nm-diameter Ag particle solution (Eqs. 22, 23) in the absence of the ·OH scavenger, 2-propanol) to extract electrons from the surface of colloidal, metallic silver particles. The process resulted in a red shift, broadening, and a decrease in intensity of the silver plasmon absorption band (see $0 \rightarrow b$ change in Fig. 83) [506]. Addition of silver ions to metallic silver colloids elicited a similar change in the absorption spectrum [506].

Alteration of the silver plasmon band spectrum upon electron and hole injection has been rationalized in terms of changes in the density, N_e, and conductivity, σ, of the electron gas in the metal particles as described by Eqs. (16)–(18) [506]. Thus, a decrease in N_e by electron extraction from the metallic silver particles increases λ_c (Eq. 16) and λ_m, thereby shifting the absorption maximum (Eq. 15) of the plasmon band to a longer wavelength (Fig. 83). A decrease in N_e also decreases σ (Eq. 18), which leads, in turn, to an increase of w (Eq. 17); that is, to an increase in the bandwidth of the plasmon band absorption (Fig. 83). Similarly, the increase in N_e by electron transfer to the silver colloids is paralleled by a decrease in λ_c (Eq. 16) and, hence, by a decrease in λ_m (Eq. 15), as seen by the shift of the plasmon absorption band to a shorter wavelength (Fig. 83). Electron donation to the silver particles also causes an increase in σ (Eq. 18) and, thus, in a decrease in the bandwidth of the plasmon absorption band, w (Eq. 17), as seen in Fig. 83. Since the number of electrons transferred is known from conductivity measurements, N_e and, hence, the spectra could be calculated. Good agreement has been obtained between the theoretical and observed spectral changes accompanying electron and hole donation to metallic colloidal silver particles [506].

The surfaces of colloidal metallic particles are electron deficient. Partial charge donation from negatively charged ions such as polyphosphate, polyacrylate, and poly(vinyl sulfate) ions, thus, results in the formation of weak complexes. Charge repulsion between the polyanion-coated metallic particles ensures their stability. Simple nucleophiles, N (cyanide anions, ammonia, thiolate anions, and sulfide ions, for example), have also been shown to interact with nanosized metallic silver particles and, in the presence of oxygen, this interaction was found to lead to AgN formation [506, 507, 530]. The mechanism proposed involved two steps [506]. In the first step, donation of an electron pair by N into a vacant orbital on the particle surface was suggested to result in the development of a partial positive charge on a silver atom on the particle surface (designated by Ag) and a partial negative charge in the interior of the silver particles:

$$Ag_nAg + N \rightarrow Ag_n^{\delta-} Ag^{\delta+} N \tag{26}$$

In the second step, the excess negative charge was believed to be transferred to oxygen with the concomitant formation of AgN^+:

$$Ag_n^{\delta-} Ag^{\delta+} N + O_2 \rightarrow AgN + O_2^- \tag{27}$$

Equilibrium was reached in the addition of N^- (Eq. 26) when the accumulated negative charge in the interior of the silver particles precluded further electron transfer. Addition of N to metallic colloidal silver particles also shifted the Fermi potential to a more negative value (see top part of Fig. 84) [531].

The excess negative charge located in the interior of metallic silver colloids could also be transferred to other electron acceptors, including methylviologen, nitrobenzene, nitropyridinium oxide, anthracene quinone sulfonic acid, and potassium cyanohexaferrate(III)[506, 531]. The efficiency and, indeed, the direction of electron transfer were found to depend on the position of the Fermi level of the surface-modified silver particles. For example, chemisorption of AgN to a silver particle is shown to result in a shift of the Fermi level to a more positive potential, as shown in the lower line in Fig. 84.

Electron transfer was mediated by metallic silver colloids whose surfaces contained either a strong (SH^-) or a weak (CN^-) nucleophile [531]. The former case is illustrated by changes in the absorption spectrum of a 1.0×10^{-4} M, deaerated solution of metallic silver particles, subsequent to the consecutive addition of 2.0×10^{-4} M NaSH and 3.0×10^{-4} M anthracene quinone sulfonic acid, AQS (Fig. 85) [506]. The origin of the intensity decrease and the broadening of the silver plasmon absorption band upon the addition of nucleophilic SH^- is incompletely understood. However, that an absorption

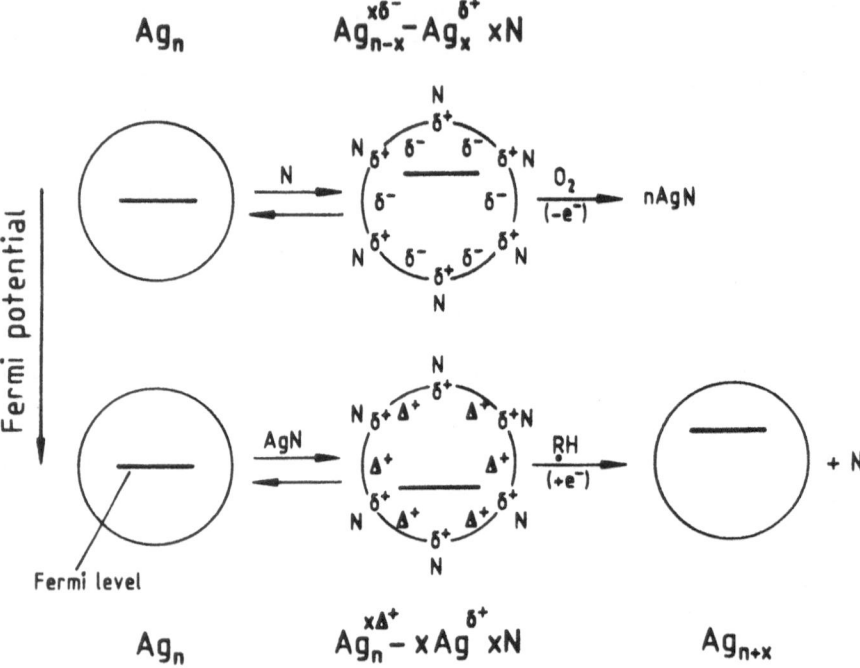

Fig. 84. (*Upper*) Schematic presentation of surface complexation (preoxidation) by a nucleophilic reagent N and final oxidation of the silver particle by oxygen. (*Lower*) Chemisorption of a complex AgN and final reduction of the silver ions in chemisorbed AgN by excess electrons deposited by reducing radicals. In both parts of the figure, the changes in the position of the Fermi level in the colloidal silver particles are indicated [531]

spectrum characteristic of hydroquinone develops concomitantly with broadening of the broad silver plasmon band upon addition of AQS clearly indicates that electron transfer occurs to the quinone from the silver atoms. Furthermore, the amount of hydroquinone formed corresponded to the concentration of silver in the colloidal particles. Exposure of the solution to air resulted in the usual hydroquinone-to-quinone transition (Fig. 85) [506]. Electron transfer mediated by a weak nucleophile is demonstrated in the simultaneous addition of KCN and methylviologen (MV^{2+}) to a solution containing 1.0×10^{-4} M metallic silver particles, which resulted in the formation of $MV^{+\cdot}$ whose concentration increased with additional amounts of added MV^{2+} or CN^- (Fig. 86) [531].

Metal ions have also been reduced at the surfaces of nanosized colloidal silver particles [532, 533]. Irradiation of an aqueous, degassed, 0.9×10^{-4} M metallic silver colloid solution (stabilized by 4.1×10^{-4} M sodium polyphosphate) in the presence of 1.9×10^{-4} M $Cd(ClO_4)_2$, 9.5×10^{-4} M 2-propanol, and 3.0×10^{-4} M acetone for increasing amounts of time resulted in a sequential blue shift of the silver plasmon absorption band from 390 nm to 260 nm (Fig. 87) [532]. These spectral changes are explicable in terms of cadmium ion reduction at the silver particle surface. The reduction is mediated by alcohol radicals

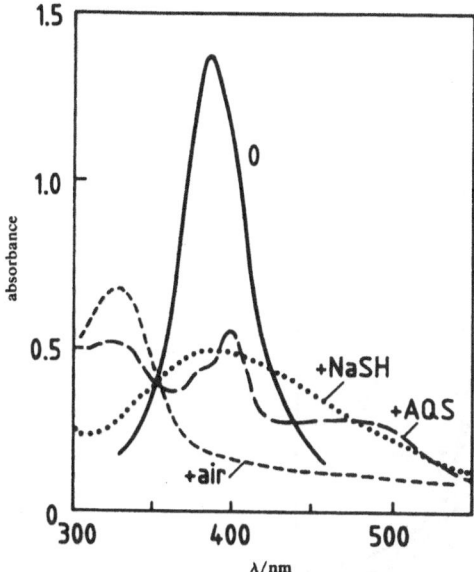

Fig. 85. Plasmon absorption band before and after addition of NaSH, and after subsequent addition of AQS. The solution was finally exposed to air [506]

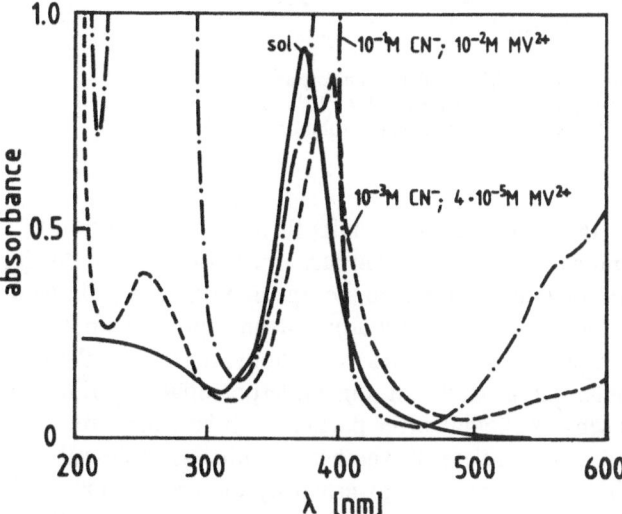

Fig. 86. Spectrum of a 1×10^{-4} M silver sol before and after the addition of KCN plus methylviologen. The sol was prepared by γ-irradiation of a 1×10^{-4} M AgClO$_4$ solution containing 0.5 M propanol-2, 1×10^{-5} M polyphosphate, and 0.01 M acetone (pH 6.5) [531]

(formed in an analogous reaction to that shown in Eq. 25) via the colloidal silver particles. This interpretation is in accord with the inability of the alcohol radicals to reduce cadmium ions in homogeneous aqueous solutions (i.e. in the absence of silver particles) [532]. Similar experiments have been reported for the deposition of lead on colloidal gold particles [533].

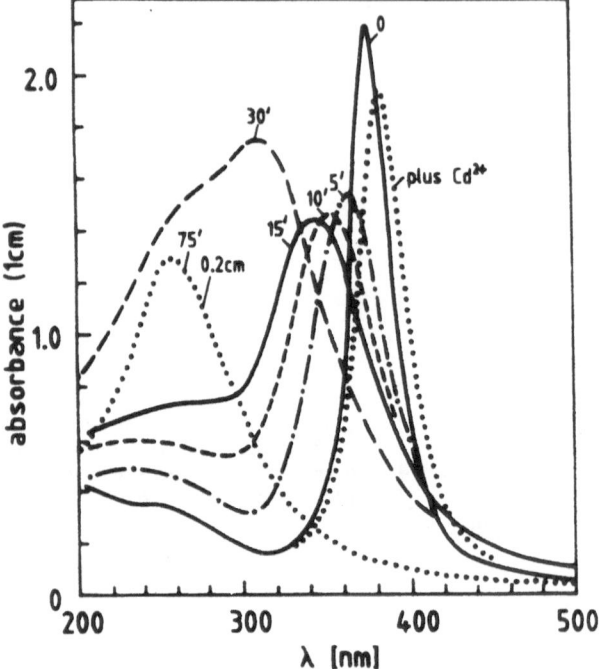

Fig. 87. Absorption spectrum of the 1×10^{-4} M silver sol before and after addition of 1.9×10^{-4} M Cd^{2+}, and after various times of γ-irradiation. Dose rate 8.7×10^4 rad/h. The thickness of the optical cuvette was generally 1.0 cm, but 0.2 cm at 75 min. pH of solution, 6.5 [532]

Electron transfer to the colloid particles shifts their Fermi levels to a level which is suitably more negative to induce cadmium-ion reduction ($E = -1.8$ V). Silver particles act as electron storage pools and, depending on the extent of their charging, their Fermi levels adjust to an appropriate potential. To state it differently, the mechanism involves the cathodic charging of the silver colloids until their potential is sufficient for underpotential deposition of cadmium ions. In this sense, colloidal silver particles can be considered to be dispersed microelectrodes. Regardless of whether the process is described in solid-state or electrochemical terms, it is governed by the need to match the silver-particle oxidation potential with the cadmium-ion reduction potential. In electrochemistry, this is accomplished by changing the electrode potential (hence, the Fermi level). In experiments with nanosized particles, this is realized by adsorbing (or desorbing) electrons, ions, or molecules onto (or from) the colloids.

Controlled reduction of cadmium (or lead) ions on surfaces of nanosized silver (or gold) metallic particles results in the formation of double-layer colloids [532–534]. Depending on the coverage, the second layer can vary from being non-metallic clusters to quasi-metallic and metallic colloids. Growth of the second-layer particles can be monitored by absorption spectrophotometry. For example, three monolayers of cadmium have to build up on the silver colloid prior to the observation of the cadmium plasmon band (Fig. 87) [532]. Absorption spectra of double-layer colloids of different thicknesses can be calculated from the Mie equation [510, 535] by using the optical constants of the pure metals [536] and by correcting for the mean free path in the colloidal particles [537]. Such a calculation for gold-coated silver particles is shown in Fig. 88 [534]. The calculated plasmon band absorption spectra for pure silver and pure gold metallic particles agree well with those found experimentally. Coating the silver particles with just a monolayer of gold (Au atom diameter = 2 Å) is seen to decrease and broaden the silver spectrum; the gold plasmon absorption band gradually appears at the expense of the silver band. Subtle differences remain, however, between the calculated and observed spectra for mixed colloidal particles which must await interpretation [534]. Preparation of triple-decker, colloidal particles possessing a gold interior, lead middle layer, and cadmium exterior layer has also been reported [533].

Mixing equal concentrations of two separately prepared solutions of colloidal lead and silver particles (each degassed by repeated freeze-pump-thaw cycles on a high-vacuum line and mixed under vacuum) resulted in a slow blue shift of the silver plasmon band from 390 nm to 337 nm and a concomitant broadening

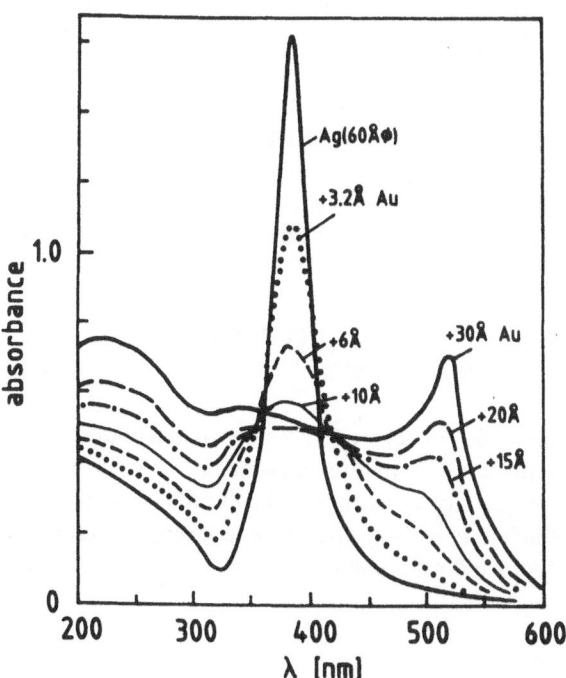

Fig. 88. Theoretical spectra of gold-coated silver colloids. The optical constants were taken from Johnston and Christy [536]. The dielectric constants were corrected for the effect of the mean free path using the method proposed [537] by Kreibig [534]

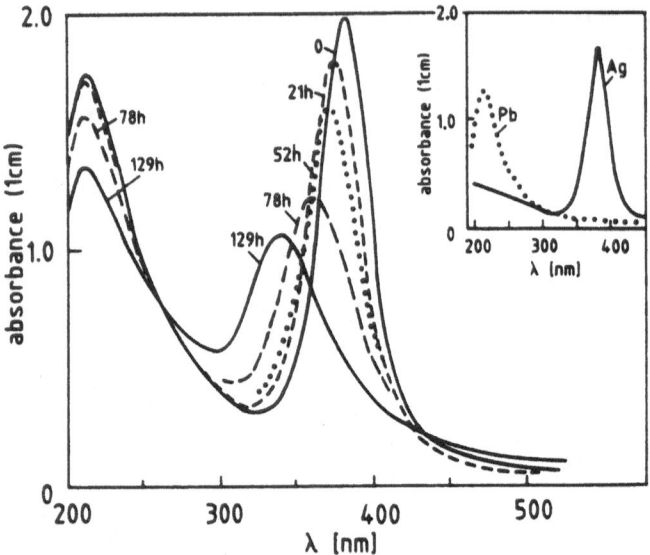

Fig. 89. Absorption spectrum of a solution containing 1×10^{-4} M Ag and 5×10^{-5} M Pb at various times after mixing. pH = 4.5. *Inset*: absorption spectra of the separate lead and silver sols [538]

and parallel decrease in the intensity of the lead plasmon absorption band (Fig. 89) [535]. These results clearly indicated the equilibration of the Fermi levels between the two separate colloid particles via the transfer of lead atoms to the silver colloid. Addition of small amounts of methylviologen increased the rate of equilibration, probably by shuttling electrons between the particles [335].

Fine tuning of the Fermi levels of nanosized metallic and size-quantized quasi-metallic particles by adsorbing (or desorbing) charges, ions, or molecules opens the door to the construction of tailor-made advanced materials [538].

3.3 Metallic and Catalytic Particles and Particulate Films in Membrane-Mimetic Compartments

It is important to state the difference between particles and particulate films at the onset of this section. Particles are separate nanometer- to micron-sized colloids dispersed in solution. Physically interconnected colloidal metal particles constitute a particulate film which may be supported by a monolayer floating on an aqueous subphase or be deposited on a solid substrate.

3.3.1 Metallic and Catalytic Particles

Reversed micelles and microemulsions continue to be used as hosts for the in situ generation of nanosized platinum [539, 540], palladium [539], rhodium

[539], copper [541], and silver [542] particles [500, 543, 544]. Monodispersed nickel boride [543, 545–547], cobalt boride [543, 545, 548], nickel-cobalt boride [543, 545, 548], rhenium oxide [540], platinum-rhenium oxide [540], and iron boride-iron oxide [543] catalysts have also been prepared in these systems. Well-characterized reversed micelles and microemulsions were selected such that their structures and chemical compositions were not affected by the introduction of the particle precursors (metal ions and reducing agents). Generally, micro-emulsion-entrapped catalysts remained stable for several months. An important advantage of generating metal and catalytic particles in microemulsions is that they can be transferred to solid supports without appreciable structural changes.

Nanosized cobalt, copper, gold, nickel, rhodium, and silver particles have been stabilized by polyions and polymers [514, 549–553]. Particularly signific-ant has been the simultaneous reduction of $HAuCl_4$ and $PdCl_2$ in the presence of poly(N-vinyl-2-pyrrolidine) to give relatively uniform, 1.6-nm-diameter, palla-dium-coated gold bimetallic clusters [554].

Zeolite channels have provided sites for silver [555, 556], silicon [557], and selenium [558, 559] clusters; copper clusters have been generated between the layers of montmorillonite [560]; and copper, platinum, and palladium clusters were formed in silicon dioxide matrices [561].

The available information on generating metallic and catalytic particles in the different membrane-mimetic compartments is summarized in Table 5 [539–568].

3.3.2 Metallic Particulate Films

Formation of silver and gold particulate films under monolayers, floating on aqueous solutions, will be highlighted in this section.

There are several advantages to using monolayer matrices for nanoparticula-te-film generation. Firstly, stable, well-characterized, and long-lasting mono-layers can be formed from a large variety of surfactants. Secondly, monolayer surface areas and charges are two-dimensionally controllable and the composi-tion of the aqueous subphase is readily varied. In the third place, monolayers, along with the particulate films grown under them, can be conveniently transferred to solid supports (i.e. to substrates). Differences between inorganic particles generated between the headgroups of LB films and particulate films formed under monolayers floating on an aqueous subphase should be recog-nized [68]. Packing of the headgroups in LB films limits the growth of the particles to 40–60 Å. Conversely, particulate films up to several thousand Å can be grown under monolayers. Furthermore, the surfactant monolayer can, if desired, be removed from the particles after transfer of the film to a substrate.

Experimental systems used in the chemical generation and in situ optical monitoring of nanosized metallic particulate films are illustrated in Fig. 90. A precursor gas (CO, for example) may be injected in the arrangement shown in the upper part of Fig. 90, while that shown in the lower part of Fig. 90 permits the

Table 5. Metallic and catalytic particles and particulate films in membrane-mimetic compartments

Membrane-mimetic compartment	Incorporated particles	Comments	Reference
W/o microemulsions prepared from: (i) pentaethyleneglycolddecylether (PEDGE)-hexadecane or hexane-water; (ii) hexa-decyltrimethylammonium chloride (CTACl)-octanol-water; and (iii) polyethylene sorbitanate tristearate (Tween)-hexane-decane-water	Monodispersed, 2- to 5-nm-diameter Pt, Pd, Rh, Ir, Ni, Co, and Au	Particles, formed from their metal salts by reducing agents, were used as catalysts. Micellar core size controlled particle sizes	539–544
W/o microemulsions prepared from: (i) pentaethyleneglycolddodecylether (PEDGE)-hexadecane or hexane-water; and (ii) hexa-decyltrimethylammonium chloride (CTACl)-octanol-water	Monodispersed, 2- to 6-nm-diameter nickel boride, cobalt boride, nickel-cobalt boride, and iron boride-iron oxide	Particles, formed from their metal salts by $NaBH_4$ reduction, were used as catalysts. Micellar core size controlled particle sizes	540, 543, 545, 548
W/o microemulsions prepared from: (i) pentaethyleneglycolddodecylether (PEDGE)-hexane-water; and (ii) cetyltrimethyl-ammonium bromide-hexane water	Pt, ReO_2, and Pt-ReO_2	Particles were formed from K_2PtCl_4 and $NaReO_4$ by hydrazine reduction. Surface charge and surfactant adsorption on particles governed their aggregation	540
W/o microemulsions prepared from pentaethyleneglycolddodecyl-ether (PEDGE)-cyclohexane-water	Ag	Particles were formed from silver salts by $NaBH_4$ reduction. Water: surfactant ratio determined silver particle diameters	542
W/o microemulsions	Cu	Precursor concentration determined particle size	541
W/o microemulsions, prepared from pentaethyleneglycolddodecyl-ether (PEDGE)-hexane-water	Au	Monodispersed particles were prepared by in situ photolysis of $HAuCl_4$. The mechanism of aggregation was examined by laser flash photolysis	562
1430-Å-diameter single unilamellar vesicles (SUVs) prepared from mixtures of **2** and **18**	Pt	Pt particles were in situ generated by photolysis of K_2PtCl_4. Vesicle-incorporated methylene blue (MB) or 10-methyl-5-deazaisoalloxazine-3-propanesulfonic acid (MAPS) were catalytically reduced by bubbling H_2 through Pt-containing vesicles; subsequent extravesicular addition of Fe^{3+} led to Fe^{2+} and MB or MAPS reformation; these steps could be repeated several times	563

1430-Å-diameter SUVs prepared from mixtures of **2** and **18**	Pt	Vesicle-entrapped or isolated Pt were used as catalysts for ethylene and cyclohexane hydrogenation	564
SUVs prepared from mixtures of hexadecylphosphate and hexadecylviologen ($C_{16}MV^{2+}$) and containing size-quantized CdS particles in their interiors	Ag	Band-gap irradiation of CdS, led to sacrificial photoreduction of Ag^+ at the outer surfaces of SUVs via electron exchange between $C_{16}MV^{2+}$ and $C_{15}MV^+$, located at the inner and at the outer surfaces of the vesicles, at the expense of benzyl alcohol as a sacrificial electron donor	565
Polyacrylic acid	Ag	Gamma irradiation of aqueous $AgNO_3$ containing polyacrylic acid led to the formation of silver clusters in solutions or in films	514
Poly(vinyl alcohol)	Rh	Reduction of rhodium(III) chloride in methanol-water, in the presence of the polymer, led to 30–70 Å Rh particles	549
Poly(diphenylbutadiyne) and poly[5,7-dodecadiyne-11,12-diol-co-bis[((n-butoxycarbonyl)methyl)urethane]	Au	Non-linear optical effects were observed for the polymer-encased gold particles	551
Poly(vinyl alcohol), polyacrylic acid	Cu	Gamma irradiation of aqueous $CuSO_4$ in the presence of polymers led to the formation of copper clusters	553
Polyaniline	Co, Ni, and Cu	Suspension of the metal oxide (CuO, $Co(OH)_2$, or $Ni(OH)_2$) in a mixture of ethylene glycol, aniline, and sulfuric acid led to simultaneous metal cluster formation and polymerization	552
Poly(N-vinyl-2-pyrrolidone)	Palladium-coated gold particles	Simultaneous reduction of $HAuCl_4$ and $PdCl_2$ in the presence of the polymer led to uniform, 1.6-nm-diameter, Pd-coated Au particles	554
Polystyrene	Si	Nanometer to micrometer silicon particles, exhibiting size-quantization effects and luminescence, were cast in polystyrene films	557
Zeolite rho	Ag_4^{n+}	Silver clusters, formed by H_2 reduction of silver-exchange zeolite rho, were characterized by electron spin resonance spectroscopy	555

Table 5. Contd.

Membrane-mimetic compartment	Incorporated particles	Comments	Reference
Ag-NaA zeolites	Silver clusters	Silver clusters, formed in X-ray irradiated samples at 77 K and annealed at 280 K, were characterized by electron spin resonance spectroscopy	556
Synthetic mordenite (a zeolite having 7-Å-diameter, one-dimensional channels	Se	Se particle domains were incorporated into zeolite channels	558
A-, X-, Y-, and AIPO-5-molecular sieves and synthetic mordenite	Se	Predominantly trigonal helical chains of selenium were incorporated into the zeolite channels	559
Montmorillonite	Cu	4- to 5-Å-diameter Cu clusters were in situ generated in the reduction of Cu^{2+} in ethylene glycol containing dispersed montmorillonite	560
Negatively charged monolayers floating on an aqueous silver nitrate solution	Ag	Infusion of formaldehyde across the monolayer resulted in the formation of silver particulate films which could be transferred to solid substrates at any time during their growth	110
Negatively charged monolayers floating on an aqueous silver nitrate solution	Ag	Metallic silver particulate films generated electrochemically at monolayer surfaces could be transferred at any stage of their development to solid substrates	566, 567
Glyceryl monooleate (GMO) (22) bilayer lipid membranes (BLMs)	Ag	Silver particulate films were generated photochemically in situ on the BLM surface and used in surface enhanced Raman spectroscopy	568
Tubules prepared from diacetylenic phospholipids (21)	Copper and nickel films	Electron microscopy and X-ray fluorescence measurements indicated 20- to 30-nm metallic coatings on the interiors and exteriors of the tubules	356
Porous aluminium oxide membranes	Au	0.26 μm diameter and 0.3-μm- to 3.0-μm-long gold particles were formed in the pores of aluminium oxide membranes	491

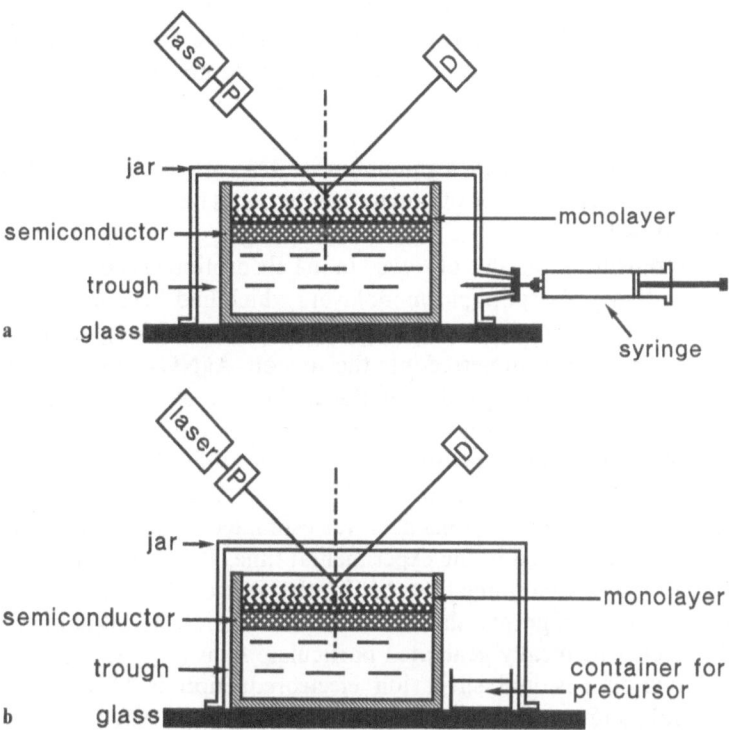

Fig. 90a, b. Schematics of the experimental arrangements used for the generation of semiconductor particles at the negatively charged, surfactant-headgroup-aqueous subphase interface showing the arrangements used for the in situ monitoring of reflectivities. P = polarizer and D = detector

generation of the desired precursor (HCHO vapor, for example). Silver particulate films were also generated by electrochemical reduction of silver ions at monolayer-solution interface in a cell shown in Fig. 91 [566, 567].

In the chemical generation, the time of exposure to formaldehyde had a dramatic effect on the structure of the silver particulate films formed under

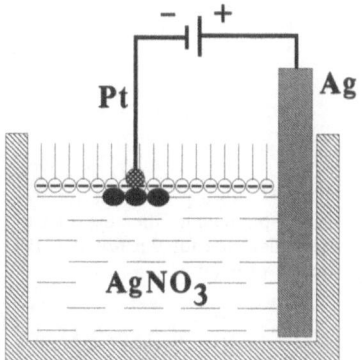

Fig. 91. Schematics of the circular trough (not drawn to scale) used for the electrochemical generation of silver particulate films at monolayer interfaces

monolayers. This is illustrated in Fig. 92 [110]. At relatively short exposure times, only small silver clusters formed. Longer exposure to formaldehyde resulted in the development of larger aggregates whose absorption maxima broadened and appeared at increasingly longer wavelengths. After HCHO infusion for a very long period of time (typically days to weeks), an apparent negative absorption appeared with a minimum at 320 nm (see Fig. 92). This interband transition is characteristic of rough silver particles having large surface area to volume ratios [569].

An interband transition was also observed in the absorption spectra of silver particles under synthetic phospholipid monolayers which had been formed by electrochemical generation (4, for example) [566, 567]. The working electrode (Pt, Pt/Ir, Ag tip, 0.5 mm) was immersed into the aqueous $AgNO_3$ subphase and a mirror-like silver layer was observed upon the application of a potential. The rate of growth of the silver film depended on the applied potential. A circular film with a diameter of about 1.0 cm was formed upon application of -700 mV (vs SCE) for 30 s. Increasing the potential to -1700 mV (vs SCE) produced a 5.0-cm-diameter silver ring in 30 s; the thickness was estimated as 60–80 nm using the amount of electricity passed in the experimental time. The resistivity of the silver film, in situ, was determined to be in the range of 5–20 ohm/cm. Transmission electron micrographs showed interconnected, 50- to 80-nm silver particles in the electrochemically generated particulate films (Fig. 93) [570].

Preferential two-dimensional silver ion electroreduction is likely to be mediated by lateral conductivity at the monolayer headgroups. The first silver particles formed extended the cathode and, thus, continued the reduction of silver counterions at the monolayer surface [566, 567].

Importantly, no silver particulate film formation could be observed at the air/liquid interface, either in the chemical or in the electrochemical generation, in the absence of surfactants or under monolayers prepared from positively charged surfactants.

Gold particulate films have been formed under thiol monolayers spread on aqueous $HAuCl_4$ solutions by exposure to carbon monoxide exposure [110].

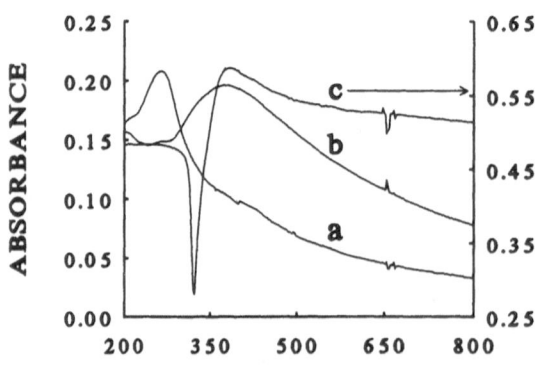

Fig. 92. Absorption spectra of silver particulate films exposed to formaldehyde for 2 h (*a*), 12 h (*b*), and 1 week (*c*)

WAVELENGTH, nm

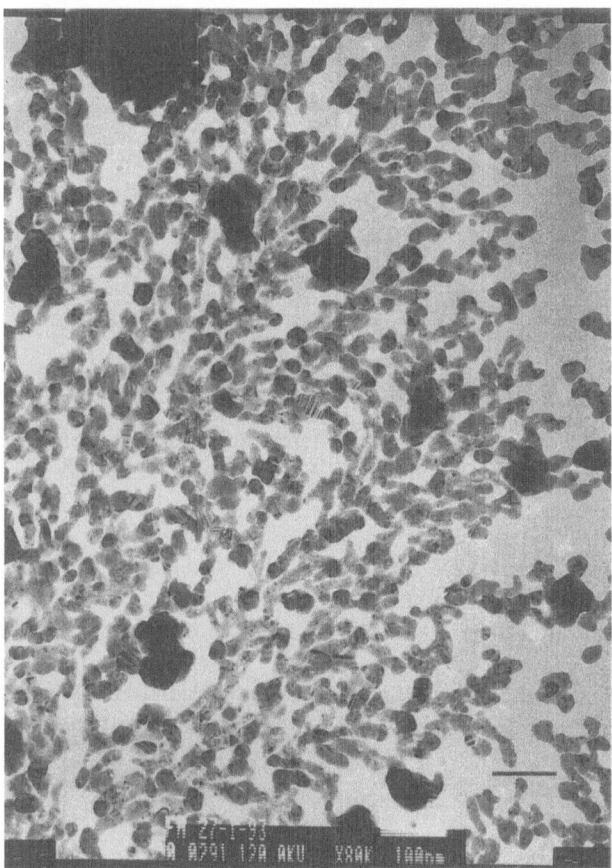

Fig. 93. Transmission electron micrograph of electrochemically generated silver particles under dihexadecyl phosphate monolayers kept at a surface area of 45 Å^2/molecule. The monolayer was floated on a 1.0×10^{-2} M AgNO$_3$ solution

Annealing the particulate film (transferred to a quartz substrate) at 140 °C for 10 min led to the development of colloidal gold particles showing a characteristic plasmon absorption band with a maximum at 560 nm (Fig. 94) [110].

Exteriors and interiors of tubules, prepared from diacetylenic phospholipids, have been coated with a 20 to 30-nm-thick layer of nickel and copper [356]. Metallization greatly improved the thermal, mechanical, and electrical properties of the tubules. Indeed, tubules coated with magnetic nickel could be aligned with a magnetic field (10^3 or less) and tracked the field rotating over 1000 rev/min [356]. It would be interesting to investigate size and dimensionality reductions of these metallic particulate films. Gold plating of the interiors of porous, aluminum-oxide membranes represents an alternative approach to metallization [491].

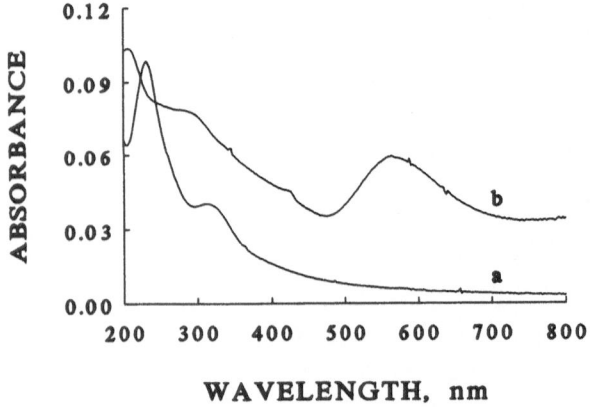

Fig. 94. Absorption spectra of gold particulate films formed under thiol monolayers spread on aqueous $HAuCl_4$ solutions exposed to CO prior (*a*) and subsequent (*b*) to annealing at 140 °C for 10 min (unpublished results)

Ultrathin metal films, prepared either chemically or electrochemically, contain interconnected, roughened metal particles which can be exploited for catalysis, electron transfer, non-linear and surface-enhanced optics, and related applications.

Available information on the generation of metallic particulate films in the different membrane-mimetic compartments is summarized in Table 5.

4 Semiconductor Particles and Particulate Films

To most scientists, the term "semiconductor" implies a crystalline or amorphous solid-state material whose conductivity is intermediate between that of metals and insulators and whose resistance decreases with increasing temperature. Orbital overlap in bulk semiconductors generates electronic energy continua or energy bands. A semiconductor is characterized by the filled conduction and empty valence bands; the energetic separation of these bonds is termed the "band-gap" (see Fig. 95), the magnitude of which defines the position of the absorption edge. Band-gap excitation results in the promotion of an electron from the conduction band to the valence band with the concomitant creation of a positive charge, or hole, in the latter. Movements of the electron or the hole in the semiconductor crystal are responsible for electrical conductivity. The motions of the electrons and holes are correlated; electrostatic attraction results in the formation of a bound electron-hole pair, termed "exciton". The binding energy of excitons in semiconductors is small (of the order of 0.15 eV) and, consequently, their diameters are rather large (in the range of 50–150 Å). Sharp exciton absorption and emission bands are normally observed, therefore, only at low

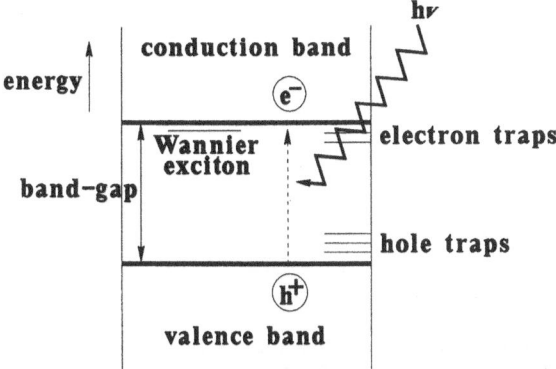

Fig. 95. Energy scheme of a semiconductor

temperatures. Real semiconductors also contain electron and hole traps (Fig. 95). Depending on the depths of these traps, charge carriers may recombine therein or undergo electron transfer to external donors or acceptors.

Much less is known about colloidal semiconductor particles and particulate films than about their solid-state counterparts. Physical chemists have launched investigations into the properties of dispersed colloidal semiconductors during the last decade [8, 9, 63, 64, 571]. Dispersed colloidal semiconductors offer a number of advantages. Firstly, they exhibit a size- and dimensionality-dependent behavior which is of fundamental importance to, and is exploitable in, electro-optical devices. They have broad absorption spectra and high extinction coefficients at appropriate band energies and may also be sensitized by doping or via physical or chemical modifications. Additionally, dispersed colloidal semiconductors can be prepared with sufficiently small dimensions and low concentrations to allow spectroscopic measurements with minimal interference from light scattering; they possess high surface areas and are, therefore, efficient light harvesters. Photoexcitation generates electrons and holes which are rapidly surface bound and can readily transfer from the small particle to an acceptor species (Fig. 96) [571]. Finally, dispersed colloidal semiconductors are relatively inexpensive to produce.

However, there are a number of difficulties associated with the synthesis of colloidal semiconductor particles. The preparation of stable, monodispersed, well-characterized populations of nanosized, colloidal semiconductor particles is experimentally demanding and intellectually challenging. Small and uniform particles are needed to diminish non-productive electron-hole recombinations; the mean distance by which the charge carriers need to diffuse to reach the particle surface from which they are released is necessarily reduced in small particles. Monodispersity is a requirement for the observation of many of the spectroscopic and electro-optical manifestations of size quantization in semiconductor particles. Small semiconductor particles are difficult to maintain in solution in the absence of stabilizers; flocculations and Ostwald ripening

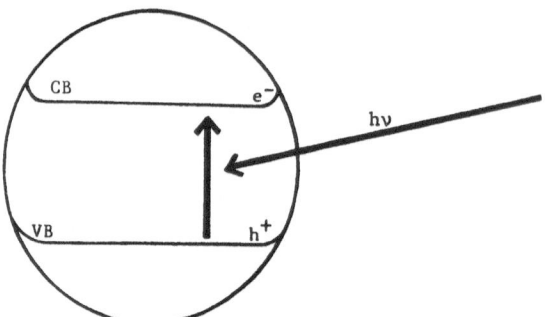

Fig. 96. Schematic illustration of a colloidal semiconductor. Band-gap excitation promotes electrons from the valence band (*VB*) to the conduction band (*CB*). In the absence of electron donors and/or acceptors of appropriate potential at the semiconductor surface or close to it, most of the charge-separated, conduction-band electrons (e^-_{CB}) and valence-band holes (h^+_{VB}) non-productively recombine. Notice the band bending at the semiconductor interface [500]

ultimately lead to precipitation. However, such stabilizers modify, of course, the photoelectric behavior of the semiconductor particles. Some of these difficulties have been overcome by ingenious chemical and physical methodologies, including the utilization of membrane-mimetic compartments. Following brief summaries of our current understanding of size quantization (Sect. 4.1), preparation of monodispersed semiconductor particles (Sect. 4.2), and the manipulation of their surface properties (Sect. 4.3), emphasis will be placed on the in situ preparation of semiconductor particles and particulate films in membrane-mimetic compartments and on their characterization and utilization therein (Sect. 4.4).

4.1 Nanosized Semiconductor Particles – Size Quantization

Size quantization is a fundamental property of semiconductor particles whose diameters are equal to or smaller than the exciton length in the bulk material [9, 11, 572–576]. For example, an exciton diameter of 50–60 Å in bulk cadmium sulfide suggests that size quantization effects would be observed in CdS particles of diameters less than or equivalent to 60 Å (equivalent to 3000–4000 atoms) [573]. In size-quantized semiconductors, charge carriers increase their kinetic energy in order to be accommodated in the particle. This manifests itself in the conversion of bands to bonds (i.e. in the splitting of the bands into discrete quantized levels) and in an increase of the band-gap. Importantly, electron microscopy and X-ray scattering measurements established the structures and unit cells of size-quantized semiconductor particles to be identical to those of their bulk counterparts. There is a progressive blue shift of the absorption edge with decreasing semiconductor-particle size. For example, bulk CdS has an absorption edge at 515 ± 5 nm, corresponding to the onset of absorption of 80-Å diameter particles, while the absorption edges of 40-Å- and 20-Å-diameter CdS particles are at 480 ± 5 nm and 400 ± 10 nm, respectively. Additionally, there are pronounced absorption maxima for CdS particles whose diameters are 25 Å or smaller which are believed to result from the optical transition of the first excitonic state of CdS [575]. Color changes from black to red, orange, yellow, and finally white, which accompany progressive decreases in the sizes of Cd_3P_2 particles, spectacularly demonstrate size quantization [577].

Fluorescence spectra of semiconductor particles originate from the recombination of charge carriers in either a trapped or excitonic state [578, 579]. The former manifests itself in the appearance of a broad and Stokes-shifted band. In contrast, the spectrum, due to excitonic fluorescence, appears as a sharp band near the absorption onset and is considered to arise from the detrapping of the trapped electrons [579]. The highly environmentally dependent position of the semiconductor emission maxima has been related to semiconductor sizes and size quantization [579].

Measurement of non-linear optical properties [580] also provides a means for characterizing size-quantized semiconductor particles. Third-order optical non-linearity of size-quantized semiconductor particles has been discussed in terms of resonant and non-resonant contributions [11]. Resonant non-linearity is expected to increase with decreasing particle size and increasing absorption coefficients.

Size quantization has been treated theoretically by several methods [581–590]. The simplest model utilized the particle-in-the-box approach [581, 582]. The exciton was considered as a particle with reduced mass, $\mu = (1/m_{e-} + 1/m_{h+})^{-1}$, orbiting around a fixed h^+. Remarkably good agreement was obtained between the calculated and observed semiconductor absorption edges for CdS and ZnS. Surprisingly, more rigorous calculations, which considered either the potential energy difference between the lower edge of the conduction band and the vacuum level or which used a two body-in-the-box model, did not bring about significant improvements. Similarly, no amelioration was achieved by performing molecular orbital calculations [585]. Theoretical treatments of size quantization have been reviewed and compared [589]. Clearly, a better theoretical understanding of semiconductor size quantization is required.

4.2 Preparation of Monodispersed, Nanosized Semiconductor Particles

Colloidal sulfide, selenide, telluride, phosphide, and arsenide semiconductor particles are prepared by the controlled precipitation of appropriate aqueous metal ions by H_2S, H_2Se, H_2Te, PH_3, and AsH_3, respectively. Colloids are stabilized, typically, by sodium poly-phosphate. A large number of experimental parameters determine the size, size distribution, morphology, and chemical composition of a semiconductor particles in a given preparation. Concentrations, rates, and the order of addition of the reagents; the counterions selected;

the pH of the solution; temperature; the extent of stirring; and incubation times all influence semiconductor-particle formation. In spite of the extensive work carried out, the synthesis of colloidal semiconductor particles is more of an art than a science. Newcomers to the field are well advised to rigorously follow the published recipe for a given preparation. Designing new methodologies for the preparation of long-lasting, uniform, nanosized semiconductor particles requires an innate sense of intuition and the ability to observe and interpret subtle, and often unexpected, chemical behavior.

Highly monodispersed, size-quantized semiconductor particles have been prepared by size fractionation of the colloidal samples by gel electrophoresis [591]. The method is illustrated by quoting the published recipe for CdS particle formation [591]:

"Cadmium sulfide colloid was prepared by injecting 4.5 ml of H_2S into 1000 ml of an aqueous solution containing 2×10^{-4} M $Cd(ClO_4)_2$ (Alpha) and 2×10^{-4} M sodium polyphosphate (Riedel de Haën) at pH 9.99. During the precipitation of CdS, the pH of the solution fell below 5. After stirring for 10 min, the resulting colloidal solution was concentrated to 100 ml by using a rotary evaporator at 35 °C. Of the 2×10^{-3} M CdS colloid (remained stable for weeks) containing 10% w/v sucrose, 4.0 ml was layered on top of the gel. A voltage of 100 V was applied, and the electrophoresis was run for 6 h in the anodic direction. After 6 h on the preparative gel, the blue fluorescing CdS (smallest particles) had moved down 14 cm on the gel whereas the green fluorescing CdS had only moved 9.5 cm. The gel was removed from the glass tube and cut into 10 slices approximately 4 mm thick in the region containing the CdS particles. Each slice was placed into 10 ml of 1×10^{-4} M NaOH solution and left standing overnight in the dark. During this time, some of the CdS particles leached out into the alkali solution. Approximately 5 ml of solution was decanted from the gel and filtered through a 0.65 μm membrane filter (Sartorius 113).

The gel was prepared by mixing 66 ml of gel buffer (0.01 M $Cd(ClO_4)_2$, 0.01 M sodium polyphosphate, pH 11) 7.4 ml acrylamide solution (44.4 g of acrylamide, 1.2 g of methylene-bisacrylamide in 100 ml water) and 62.4 ml water in round bottom flask and deaerated for 5 min with a water pump. Then 1.2 ml of freshly prepared ammonium persulfate solution (150 mg in 10 ml water) and 0.1 ml of N,N,N',N'-tetramethylethylenediamine (TEMED) were added and mixed carefully to avoid the introduction of air. Finally, 100 ml of the resulting solution was poured carefully, to minimize the introduction of air, into a 25 mm-diameter tube. The monomer solution was overlayered with water to remove the meniscus and reduce the entry of air. The polymerization was allowed to proceed overnight at room temperature. The resulting gel, having a polymer concentration of 2.5% and a cross-link density of 2.6%, was used in a home made apparatus."

Table 6. Summary of the measured physical and optical properties of unfractionated and fractionated colloids [591]

Fraction number	d, cm	Fluorescence maximum, nm	Fluorescence width (fwhm), nm	$\bar{r}$, Å	Standard deviation, Å	$\bar{r}_w$, Å
Unfractionated		498.0	26.1	29.14	5.2	33.04
1	13.3	475.0	19.1	20.38	3.6	22.69
2	12.9	480.0	19.5			
3	12.5	483.8	19.1	24.86	2.8	26.03
4	12.1	486.8	19.5			
5	11.7	489.7	20.6	26.23	3.2	27.75
6	11.3	491.5	19.9	30.32	2.8	31.39
7	10.9	494.5	19.1	33.88	3.2	34.97
8	10.5	497.8	19.5			
9	10.1	501.5	18.8	35.98	4.0	37.62
10	9.7	503.3	19.5			

Some properties of unfractionated and fractionated CdS particles are given in Table 6 [591].

Arresting the growth of CdS particles by capping them with thiophenol provided an ingenious approach to the control of semiconductor particle sizes

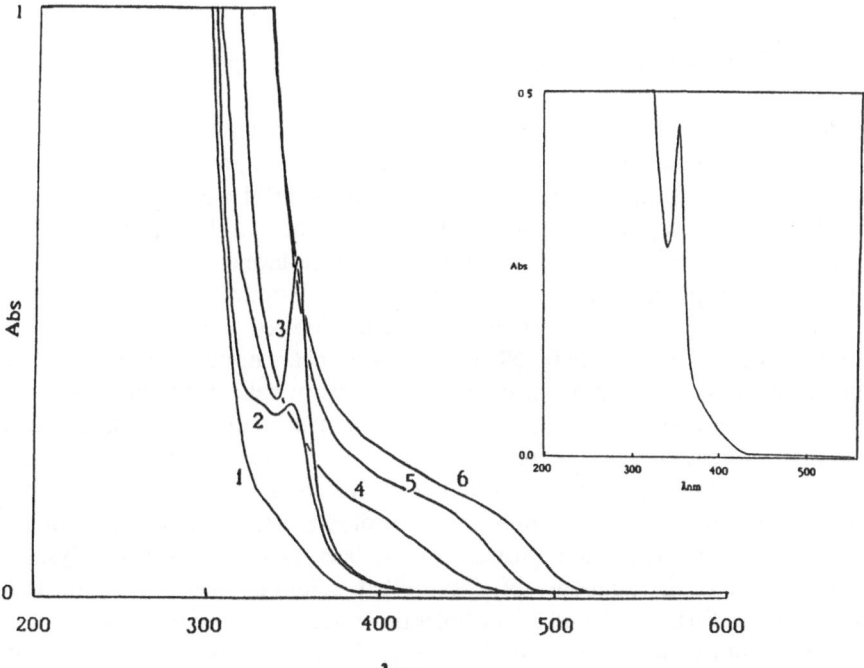

Fig. 97. Electronic spectra showing the titration of sulfide ion into a DMF solution of $(CD_{10}S_4(SPh)_{16})^{4-}$ $(1.25 \times 10^{-4}$ M) in 1 mm pathlength cells. Ratios of cluster to added sulfide are: (1) 1 : 0; (2) 1 : 1; (3) 1 : 2.5; (4) 1 : 4; (5) 1 : 10; (6) 1 : 20. The *inset* shows the spectrum of the titration ratio of 1 : 2.5 [594]

and size distributions [592–594]. The method involved the initial mixing of 0.1 M sodium sulfide (in MeOH:H$_2$O = 1:1, v/v) with 0.1 M thiophenol (in MeOH) followed by the addition of 0.1 M cadmium acetate (in methanol) under continuous stirring. Subsequent to stirring for 15 min the solution was filtered and, finally, suction dried by drawing dry nitrogen through the filtrate for 10 min [592]. Control of the particle size was achieved by variation of the thiophenol-sodium sulfide ratios. Competition between thiophenol and sulfide ions for Cd^{2+} governed the concentration of cadmium available for growth of the incipient CdS colloid. The process was considered to be analogous to an inorganic polymerization. The nucleation and growth of CdS molecules corresponded to the initiation and propagation steps, while sulfide and thiophenolate ions constituted the propagation and termination reagents [573]. Indeed, the mechanism was shown to be a type of living polymerization; CdS particles could be grown further simply by adding extra sulfide ions to the solution containing the thiophenol-capped colloid [593, 594]. Careful manipulation of the experimental conditions facilitated the preparation of $\{Cd_{10}S_4(SPh)_{16}\}^{4-}$, a species which contained about 55 atoms in a sphalerite arrangement, had a 10 Å diameter, and showed an excitonic absorption at 351 nm (Fig. 97) [594].

The analogous capping of cadmium selenide particles by organo(silyl)-selenides in reversed micelles will be discussed in Sect. 4.4.

4.3 Surfaces of Nanosized Semiconductor Particles

Particles in the nanometer-size regime necessarily have large surface-to-volume ratios; approximately one-third of the atoms are located on the surfaces of 40 Å CdS particles, for example. Furthermore, colloid chemical preparations typically result in the development of surface imperfections and in the incorporation of adventitious or deliberately added dopants. Such surface defects act as electron and/or hole traps and, thus, substantially modify the optical and electro-optical properties of nanosized semiconductor particles. Altered photostabilities [595], fluorescence [579, 594, 596, 597], and non-linear optical properties [11, 598–600] are manifestations of the surface effects in colloidal semiconductors.

Separation of bulk and surface properties in macroscopic semiconductors is less than straight forward and requires highly sensitive experimental techniques. In contrast, the large surface-to-volume ratios in nanosized semiconductor particles render the examination of surface processes in and/or on these colloids to be experimentally feasible. Advantage has been taken of pulse radiolysis to inject electrons (in aqueous, N$_2$O-saturated solutions which contained 2-propanol; see Eqs. 22, 23, and 25) or holes (in aqueous, N$_2$O-saturated solutions which did not contain 2-propanol; see Eqs. 22 and 23) into nanosized semiconductor particles [601, 602]. Electron injection into CdS particles, for example, decreased the extinction coefficient at 470 nm (the absorption onset) by -5×10^4 M^{-1} cm^{-1} (Fig. 98) [576]. Hole injection resulted in the appearance of a transient absorption band in the long-wavelength region and in much less

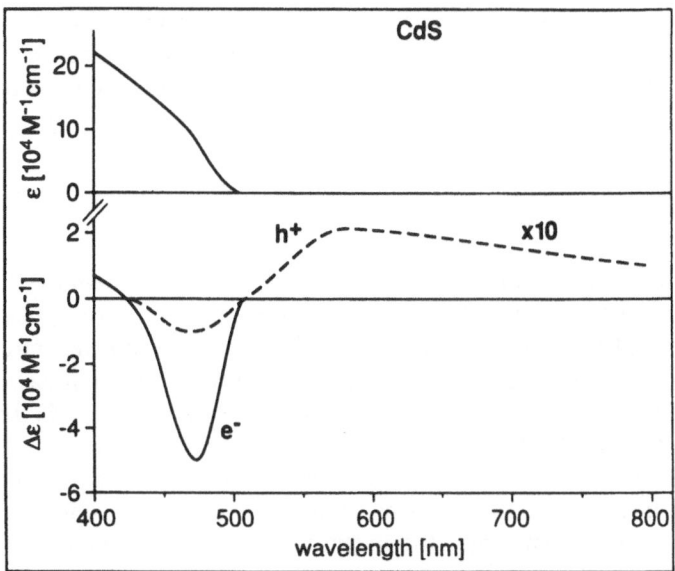

Fig. 98. *Lower part*: difference spectra of Q-CdS after the reaction with hydrated electrons (*full line*) and after the reaction with OH radicals (*dashed line*, 10x magnified). *Upper part*: absorption spectrum of the investigated sample [576]

bleaching of the ground-state absorption of CdS at 470 nm (Fig. 98). A similar behavior was observed in laser flash photolysis [500, 600, 603]. Both laser flash photolysis and pulse radiolysis experiments indicated that the presence of a single trapped electron-hole pair per CdS particle was sufficient to saturate the ground-state absorption. Indeed, it was calculated that the addition of an electron-hole pair to a 25 Å diameter CdS particle would result in a red shift of the exciton energy by some 50 eV and in a reduction of the oscillator strength by 90% [600]. These results were rationalized in terms of a theoretical model based on the electron and hole wave function overlap [573]. The hole wave function (smaller dark circle in Fig. 99) was considered to be located at the center of the CdS particle and to have a good overlap with the electron wave function (shaded circle; see drawing on the left-hand side of Fig. 99). Perturbation of the system by

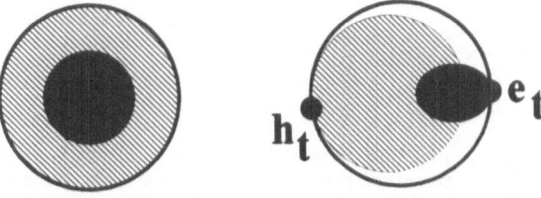

Fig. 99. Schematic illustration of the overlap of the electron and hole wave function in the absence (*left-hand side*) and in the presence (*right-hand side*) of the trapped electron-hole pair

trapping of the charge carriers at surface defects was rationalized in terms of the higher effective mass of the hole than the electron; the hole wave function is localized around the defect site while the electron wave function remains delocalized over the entire particle. Consequently, the spatial overlap of the electron and hole wave functions is reduced in the presence of surface defects (drawing on the right-hand side of Fig. 99). The model is also in accord with the more efficient bleaching of CdS-particle absorption by a trapped e^- than by a trapped h^+ [573].

The supporting medium (aqueous or organic solvents; membrane-mimetic compartments) also has a profound influence on the optical and electro-optical properties of nanosized semiconductor particles. This dielectric confinement (or local field effect) originates, primarily, in the difference between the refractive indices of semiconductor particles and the surrounding medium [573, 604]. In general, the refractive index of the medium is lower than that of the semi-conductor particle, which enhances the local electric field adjacent to the semiconductor particle surface as compared with the incident field intensity. Dielectric confinement of semiconductor particles also manifests in altered optical and electro-optical behavior.

Formation of composite and sandwich-type semiconductors constitutes an interesting development (Fig. 100) [576, 605–609]. A composite semiconductor, like a cherry and its stone, contains one material as its core and a second material as its shell. CdS particles coated by $Cd(OH)_2$, CdSe coated by ZnS, and HgS coated by CdS are examples of recently prepared composite semiconductors

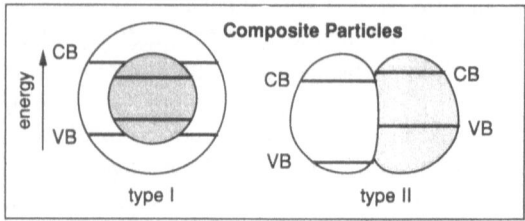

a

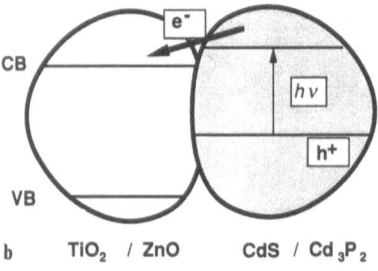

b TiO_2 / ZnO   CdS / Cd_3P_2

Fig. 100a. Schematic structure of different types of composite particles [576]. **b** Schematic representation of a sandwich colloid. VB = valence band edge, CB = conduction band edge [575]

[576, 605], the layered structures of which were demonstrated by high-resolution electron microscopy [606]. Composite semiconductor particles showed strong excitonic fluorescence. Parts of both semiconductors are exposed to the media in sandwich structures (Fig. 100). Excitation by visible light results in an initial charge separation in the smaller band-gap semiconductor. If the conduction band of the second semiconductor particle lies at a lower energy than that in the first material, the electrons are promptly transferred into it. Thus, efficient charge separation is accomplished in nanosized sandwich semiconductors. Electron transfer has been investigated in TiO_2/ZnO and CdS/Cd_3P_2 sandwich semiconductor particles [608–610].

4.4 Semiconductor Particles and Particulate Films in Membrane-Mimetic Compartments

Membrane-mimetic compartments are being increasingly utilized as hosts for semiconductor particles and particulate films. Available information is summarized in Table 7 [611–709].

Aqueous pools of reversed micelles have been fruitfully employed for the in situ generation of semiconductor particles. The first publication in this area described the formation of CdS in sodium bis(2-ethylhexyl)sulfosuccinate (AOT) aggregates in isooctane [611]. The preparation involved the addition of aqueous $CdCl_2$ or $Cd(NO_3)_2$ to isooctane solutions of AOT. Exposure to controlled ammeters of the CdS particles formed. Irradiation of degassed, AOT-reversed-micelle-entrapped , platinized CdS by visible light (450-W Xenon lamp; $\lambda > 350$ nm) in the presence of thiophenol (PhSH) resulted in sustained hydrogen formation. Sacrificial electron transfer occurred from thiophenol to positive holes in the colloidal CdS and, consequently, diminished undesirable electron-hole recombinations (Fig. 101) [611].

A higher level of size and morphology control in the incipient semiconductors has been accomplished in reversed micelles prepared from cadmium AOT [614] and from mixtures of cadmium AOT and sodium AOT [615] or, alternatively, by arresting particle growth by surface derivatization [592, 621, 622]. Indeed, surface derivatization of semiconductor clusters was first reported for particles in reversed micelles [621]; the reversed micelles act to confine precursor ions and to control the growth of the semiconductor particles. Conditions are typically arranged so that, initially, there is no more than one metal ion (say Cd^{2+}) per water pool. Addition of a heptane solution of bis(trimethylsilyl) selenium resulted in the formation of size-quantized metal selenide particles (say CdSe) in the reversed micelles. This solution could be evaporated to dryness and the resultant particles could be reconstituted in a hydrocarbon solvent. Alternatively, addition of metal (say Cd^{2+}) ions to the reversed-micelle-entrapped metal selenide particles, followed by the addition of alkyl(trimethylsilyl)selenium, $RMSiMe_3$, led to the formation of alkyl-capped

Table 7. Semiconductor particles and particulate films in membrane-mimetic compartments

Membrane-mimetic compartment	Incorporated semiconductor(s)	Comments	Reference
AOT-isooctane-H_2O reversed micelles	50-Å-diameter CdS particles generated in situ in reversed micelles from $CdCl_2$ or $Cd(NO_3)_2$ by H_2S	Reversed-micelle-entrapped CdS was fluorescence quenched by methylviologen; band-gap excitation in the presence of Rh as catalyst and PhSH as sacrificial electron donor resulted in water photoreduction	611
AOT-heptane-H_2O reversed micelles	Small (10 Å diameter) CdS particles prepared in situ in reversed micelles from $Cd(ClO_4)_2$ and Na_2S	Agglomeration number grew discontinuously; CdS was fluorescence quenched by MV^{2+}	612
AOT-isooctane-H_2O reversed micelles	Monodispersed CdS particles generated in situ in reversed micelles	Size of CdS depended on the ratio of Cd^{2+} to S^{2-}; strong fluorescence was observed in the presence of excess Cd^{2+}; high yield of CdS-mediated MV^{2+} photoreduction	612
AOT-isooctane-gelatin-water gels	50-Å-diameter CdS particles generated in situ	High yields of CdS-mediated MV^{2+} and $Fe(CN)_6^{4-}$ photoreduction	613
Cadmium lauryl sulfate-, $Cd(LS)_2$, and cadmium dioctylsulfosuccinate-, $Cd(AOT)_2$, isooctane-H_2O reversed micelles	Size-quantized CdS particles generated in situ in reversed micelles	Immobilized $Cd(LS)_2$ or $Cd(AOT)_2$ controlled the sizes of incipient CdS	614
NaAOT + $Cd(AOT)_2$ mixtures, isooctane-H_2O reversed micelles	Size-quantized CdS particles generated in situ in reversed micelles	Mixed surfactants favored the formation of monodispersed CdS; low $[H_2O]$ minimized CdS photocorrosion	615
AOT-isooctane-H_2O reversed micelles	TiO_2 particles prepared by mixing anhydrous isooctane solutions of titanium tetraisopropoxide with AOT-isooctane reversed micelles	$[H_2O]$ to $[AOT]$ ratio determined TiO_2 formation	616
NaAOT-heptane and toluene-H_2O reversed micelles	Size-quantized CdS particles generated in situ in reversed micelles	Low $[H_2O]$ and reversed-micelle interface were important in controlling the size and polydispersity of CdS	617

NaAOT-n-heptane-H_2O reversed micelles	Size-quantized CdS particles generated in situ in reversed micelles	[H_2O]/[AOT] ratios influenced the equilibrium sizes of CdS particles generated; rates of CdS growth, determined in a stopped-flow spectrophotometer, were consistent with the rate-determining intermicellar exchange of solubilizates	618
NaAOT-n-heptane-H_2O reversed micelles	Size-quantized CdS and Cd_3As_2 generated in situ in reversed micelles	Addition of alkylamines dramatically increased exciton luminescence of the semiconductor particles	619
NaAOT-n-heptane-H_2O reversed micelles	Size-quantized CdS particles generated in situ in reversed micelles	Transient bleaching and recovery rates of CdS excitonic absorption, determined by picosecond pump-probe spectroscopy, depended on [H_2O]/[AOT] ratio and micellar surface. Fluorescence spectra and lifetimes depended on [Cd^{2+}]/[S^{2-}] ratios	620
NaAOT-n-heptane-H_2O reversed micelles	Size-quantized CdSe and CdS generated in situ in reversed micelles	CdSe or CdS growth were arrested by capping with (organo)(silyl)selenide or thiophenolate ions; the process was a living polymerization	592, 621, 622
NaAOT-n-heptane-H_2O reversed micelles	Thiophenolate-ion-capped, size-quantized CdS generated in situ in reversed micelles	Capping coverage and dynamics were shown to be dependent on the CdS size by ^{1}H nmr	623
Single-compartment, 800-to 1000-Å-diameter dihexadecylphosphate (DHP) surfactant vesicles	CdS particles prepared in situ in vesicles from Cd^{2+} by controlled exposure to H_2S	CdS particle sizes were controlled by the amounts of Cd^{2+} and H_2S added	624
Single-compartment dioctadecyldimethylammonium chloride (DODAC) (6) vesicles	CdS particles generated in situ in DODAC vesicles by several methods	Quantum-size effects for vesicle-incorporated CdS were rationalized	625
Single-compartment DHP vesicles	CdS particles prepared in situ in vesicles from Cd^{2+} and H_2S at inner, at outer, and at both of these surfaces	Visible-light irradiation of DHP-vesicle-incorporated, Rh-coated CdS in the presence of PhSH as a sacrificial electron donor resulted in water photoreduction	626, 627
Single-compartment vesicles prepared from DHP, DODAC (6), and 2	CdS particles prepared in situ in vesicles from Cd^{2+} and H_2S at inner, at outer, and at both of these surfaces	Sacrificial water photoreduction was optimized	628

Table 7. Contd.

Membrane-mimetic compartment	Incorporated semiconductor(s)	Comments	Reference
Single-compartment vesicles prepared from mixtures of DODAC (6) and $(n\text{-}C_{18}H_{37})_2N^+(CH_3)CH_2\text{-}CH_2SH, Br^-$	CdS particles prepared in situ at both the outer and inner surfaces of vesicles from Cd^{2+} and H_2S	Visible-light irradiation of mixed-vesicle-incorporated, Rh-coated CdS resulted in sacrificial water photoreduction; a thiol-functionalized surfactant acted as the sacrificial electron donor; system could be recycled	629
Single-compartment DODAC (6) vesicles	CdS particles prepared in situ in vesicles from Cd^{2+} and H_2S	Sacrificial water photoreduction was optimized	630
Single-compartment DHP vesicles	CdS particles prepared in situ in vesicles from Cd^{2+} and H_2S	Conduction-band electron transfer to Rh and methylviologen was examined	631
Single-compartment DHP vesicles	Mixed Cd-ZnS particles prepared in situ in vesicles from mixtures of Cd^{2+} and Zn^{2+} by H_2S	Mixed crystals (solid solutions) or ZnS-coated CdS could be produced depending on the method of preparation	632
Single-compartment DHP vesicles	Size-quantized CdS particles generated in situ in vesicles	Photocurrent onset, pH = 10.1, and quasi-Fermi level, $nF_f^* = -0.85$ V vs SCE, were determined from pH-dependent photocurrent measurements in the presence of methylviologen and glucose	633
Single-compartment DHP and DODAB (5) vesicles	Size-quantized CdSe particles generated in situ in vesicles	Band-gap irradiation, in the presence of sacrificial electron donors (glucose or cysteine), led to charge separation and methylviologen reduction	634
Single-compartment DHP vesicles	Size-quantized CdS particles generated in situ in vesicles	Transmembrane photoelectron transfer was demonstrated in DHP vesicles which contained methylviologen in their interiors and CdS and sacrificial benzyl alcohol in their exteriors; photochemical events were followed by simultaneous in situ optical and electrochemical monitoring	635, 636
Single-compartment DHP vesicles	Size-quantized CdS particles generated in situ in vesicles	Band-gap irradiation of CdS, contained exclusively in the vesicle interior, led to sacrificial photoreduction of Ag^+ at the outer vesicle surfaces via electron exchange between $C_{16}MV^{2+}$ and $C_{16}MV^+$, located at the inner and at the outer surfaces of vesicles, at the expense of benzyl alcohol as a sacrificial electron donor	565

System	Semiconductor	Description	Ref.
DL-β,γ-dipalmitoyl-α-lecithin surfactant vesicles	Cds and CuS particles	Band-gap irradiation of CdS, contained exclusively in the vesicle interior, led to photoreduction of Co^{III}-EDTA at the outer vesicles surfaces via electron exchange between $C_{16}MV^{2+}$ and $C_{16}MV^{+}$, located at the inner and at the outer surfaces of vesicles	637
Monolayers prepared from arachidic acid, bovine-brain phosphatidylserine (16), and n-hexadecyl-4-vinylbenzamido)undecyl hydrogen phosphate (4)	Size-quantized CdS, ZnS, PbS, and CuS semiconductor particulate films generated in situ under monolayers	The thickness of the film was controlled by the amount and rate of H_2S (or H_2Se) infusion; particulate films could be transferred to solid supports at any stage of their growth	638, 639, 640, 641
Monolayers prepared from poly(vinylbenzyl)-phosphonate	Size-quantized CdS, ZnS, PbS, and CuS semiconductor particulate films generated in situ under monolayers	The thickness of the film was controlled by the amount and rate of H_2S(or H_2Se) infusion; particulate films could be transferred to solid supports at any stage of their growth	642
Monolayers prepared from n-$C_{16}H_{33}C(H)$ [$CON(H)(CH_2)_2NH_2$]$_2$	Size-quantized CdS particulate films generated in situ under monolayers	pH of the subphase determined headgroup surface areas and CdS sizes; particulate films could be transferred to solid supports at any stage of their growth	643
Monolayers prepared from arachidic acid	Size-quantified CdS particulate films generated in situ under monolayers	CdS particulate films were characterized in situ by scanning tunneling microscopy under potentiostatic control; particulate films could be transferred to solid supports at any stage of their growth; scanning tunneling spectroscopy of CdS on a conducting substrate indicated an n-type metal-insulator-semiconductor (MIS) junction behavior	644
Monolayers prepared from DODAB (5)	Size-quantized CdSe particulate films	CdSe particulate films could be transferred to solid supports at any stage of their growth; scanning tunneling spectroscopy on a conducting substrate indicated an n-type metal-insulator-semiconductor (MIS) junction behavior	645

Table 7. Contd.

Membrane-mimetic compartment	Incorporated semiconductor(s)	Comments	Reference
Monolayers prepared from arachidic acid	PbS size-quantized particulate films prepared by the infusion of NH_3 through the monolayer floating on 1.0×10^{-3} M $Pb(NO_3)_2 + 2.0 \times 10^{-2}$ $CS(NH_2)_2$ (Method I) or by the infusion of H_2S through the monolayer floating on 1.0×10^{-3} M $Pb(NO_3)_2$ (Method II)	PbS particulate films could be transferred to solid supports at any stage of their growth; PbS prepared by Method I showed p-type, and that prepared by Method II showed n-type, rectification behavior	646
Monolayers prepared from arachidic acid	Size-quantized PbS particulate films	Epitaxial growth of equilateral triangular crystalline PbS was demonstrated	647
Monolayers prepared from arachidic acid	Size-quantized PbSe particulate films	Epitaxial growth of equilateral triangular, plus a few rod-like, crystalline PbSe was demonstrated	648
Monolayers prepared from $(n\text{-}C_{18}H_{37})_2$ N^+ $(CH_2CH_2SH)_2$, Br^-	Size-quantized CdS, ZnS, PbS, and PbSe particulate films	Extremely small, well-separated, stable semiconductor particles were formed	110
Bilayer lipid membranes (BLMs) prepared from glyceryl monooleate (GMO) (22), phosphatidylserine (16), and $[n\text{-}C_{15}H_{31}CO_2(CH_2)_2]_2N^+$ $(CH_3)\text{-}CH_2C_6H_4CH{=}CH_2$, Cl^-	CdS particles generated in situ on the BLM surface or incorporated from the bathing solution	CdS formation was monitored by simultaneous electrical and spectroscopic techniques; photovoltage, photocurrent, and semiconductor-sensitized polymerization were observed on and across BLMs	650
BLMs prepared from GMO, phosphatidylserine, and $[n\text{-}C_{15}H_{31}CO_2(CH_2)_2]_2N^+$ $(CH_3)\text{-}CH_2C_6H_4CH{=}CH_2$, Cl^- (1)	ZnS, PbS, CdS, CuS, Cu_2S, HgS, and In_2S_3 generated in situ on the BLM surface or incorporated from the bathing solution	Semiconductor formation was monitored by simultaneous electrical and spectroscopic techniques	649, 650, 651
BLMs prepared from GMO	CdS, ZnS, PbS, CuS, and In_2S_3; System A = single composition of particulate semiconductor on one side of the BLM; System B = two different compositions of particulate films on the same side of the BLM; System C = two different compositions of particulate films on opposite sides of the BLM	Systems A, B, and C were electrochemically characterized; CdS, ZnS, and In_2S_3 = n-type while CuS = p-type ES junction; System C = SS' heterojunction. Prolonged cyclic-voltammetric cycling converted n-type to p-type behavior	652

BLMs prepared from phosphatidylcholine (14) and phosphatidylethanolamine (15) mixtures (PC:PE = 1:1, w/w) and from lecithin	CuS, FeS, CdS, and AgBr generated in situ on the BLM surface	Electrical and photoelectrical properties were investigated	653
BLMs prepared from GMO	CdS generated in situ on the BLM surface	A quantitative model is provided for the rationalization of photopotential kinetics	654, 655
Cast multibilayers prepared from cyclam-containing double-chain surfactants	Size-quantized CdS particles generated in situ in cast bilayers	Physical state of cast films determined CdS sizes	656
Pt, Ti, Ni, and brass foils	CdSe-coated metal foils placed between two aqueous compartments (BLM analog = septum electrode)	Photoelectron transfer and hydrogen generation were demonstrated in a GC|3-S|CdSe|Ni|Sea water|GC cell where GC = glassy carbon electrode, 3-S = a solution of 1 M NaOH + 1 M sodium sulfide + 1 M sulfur, and CdSe|Ni = the septum electrode	657
Langmuir-Blodgett (LB) films prepared from cadmium arachidate	Size-quantized CdS particles formed in situ between the polar headgroups of the LB film	Ellipsometry indicated an increase of about 3 Å upon Cds formation	658
LB films formed from dioctadecyldimethylammonium chloride (6)	Hexamethylphosphate-stabilized, size-quantized CdS	Preformed, hexamethylphosphate-stabilized, size-quantized CdS was incorporated between the polar headgroups of LB films	659
LB films prepared from behemic acid	Cu₂S particles in situ formed between the polar headgroups of LB films	Temperature-dependent resistivity and photoconductivity measurements indicated the presence of (i) thin, continuous semiconducting layers and (ii) insulated aggregates	660
Polymerized LB films prepared from nonacose-10,12-diynoic acid and LB films prepared from mixed (dihexadecylphosphate and diynoic acid; dimethyl-dioctadecylammonium bromide and diynoic acid) surfactants)	Size-quantized CdS, CdSe < CdTe, CdS$_x$Se$_{1-x}$, CdS$_x$Te$_{1-x}$, and CdSe$_x$Te$_{1-x}$ particles formed in situ between the polar headgroups of LB films	Using mixtures of metal ions (Cd²⁺ and Ca²⁺) and mixed surfactants controlled the particle sizes	661

Table 7. Contd.

Membrane-mimetic compartment	Incorporated semiconductor(s)	Comments	Reference
LB films prepared from mixtures of behemic acid and octadecylamine	Size-quantized CdS particles	Preformed CdS particles were incorporated between LB films and characterized therein by TEM and XPS	662
LB films prepared from lead stearate	PbS particles formed in situ between polar head-groups of LB films	In air, PbS decomposed to lead stearate, elementary sulfur, and other volatile sulfur compounds	663
LB films prepared from stearic acid	α-Fe_2O_3 particles	Preformed α-Fe_2O_3 particles were incorporated between the polar headgroups of LB films	664
LB films prepared from lead stearate	Size-quantized PbS particles generated in situ between the polar headgroups of LB films	Heating the LB-containing particles in N_2 at 120 °C for 1 h removed the surfactants and produced PbS particles	665
LB films prepared from lead stearate	Size-quantized PbS particles generated in situ between the polar headgroups of LB films at 1 Torr pressure	PbS particulate films consisted of two-dimensional domain lines, suggesting one- or two-dimensional size quantization	666
Self-assembled (SA) alkanethiols on gold and aluminium	Size-quantized CdS particles	Size-quantized CdS particles were deposited onto metal substrates via SA monolayer bridges	667
Fluoroplastic MF-4SK cation exchange polymer (analogous to Nafion)	CdS, ZnS, and SnS_2 particles immobilized in the polymer	Photocatalytic H_2S on polymer-immobilized semiconductor particles was investigated	668
Nafion films	CdS and ZnS-CdS particles supported in Nafion films	Luminescence was analyzed	669
Nafion films	CdS and CdSe supported in Nafion films	Size-quantized particles were formed by varying Cd^{2+}-Ca^{2+} ratios in the Nafion film prior to exposure to H_2S or H_2Se; transition of molecular to bulk semiconductors was observed	670
Nafion films	Hydrous iron(III) oxide generated in Nafion films	Nitrogen fixation was demonstrated	671
Nafion films	Fe_2O_3 microcrystals prepared in Nafion films	Size quantization was demonstrated by TEM, absorption spectroscopy, and photocurrent measurements	672

Nafion films	PbS microcrystals prepared in Nafion films	Onset of anodic photocurrent at electrodes, coated by PbS-containing Nafion showed a cathodic shift with increasing band-gap	673
Nafion films	TiO$_2$ microcrystals prepared in Nafion films	Photocatalytic properties were investigated	674
Nafion films	Size-quantized CdS in Nafion films	Size quantization was investigated by picosecond spectroscopy	599, 600
Nafion films	CdSe prepared in Nafion films	Position of emission maximum depended on semiconductor size	675
Poly(methyl methacrylate) films	Pyridine-soluble, phenyl-capped Cd incorporated in polymers	Morphology of CdS in films was characterized by TEM and absorption spectrophotometry; a large third order non-linearity was observed by degenerate four-wave mixing	676
Ethylene-15% methacrylic acid copolymer	CdS and PbS incorporated into polymer films	Linear and non-linear optical properties were investigated	677
Polyvinylcarbazole	Size-quantized CdS particles incorporated into polymers	Wavelength and field dependence of charge generation were determined	678
Polypyrrole (conducting) polymer	Polymer is doped by α-CdS	Increasing CdS concentration decreased polymer conductivity and increased its p-semiconductor character	679
Polyacrylonitrile	NiS, PdS, MnS, Zn$_x$Cd$_{1-x}$S, Hg$_x$Cd$_{1-x}$S, Hg$_x$Cu$_{1-x}$S, Cd$_x$Mn$_{1-x}$S, Hg$_x$Cu$_{1-x}$S, and Mn$_x$Zn$_{1-x}$S organosols incorporated into polymer films	This method offered a route to metal-sulfide-polymer composites	680
Ultrathin (about 400 Å thick), polymer-blend membrane prepared from poly(styrenephosphonate diethyl ester) and cellulose acetate	Size-quantized CdS particles generated in membranes	Efficient CdS size dependent photoelectron transfer was demonstrated	681, 682

Table 7. Contd.

Membrane-mimetic compartment	Incorporated semiconductor(s)	Comments	Reference
β-cyclodextrin	CdS and TiO_2 particles stabilized by β-cyclodextrin	Selective photoreduction of CO_2/HCO_3^- to formate ion was demonstrated by the use of Pd-β-cyclodextrin-TiO_2	683, 684
Silicate glasses	Size-quantized HgSe, PbS, In_2Se_3, CdSe, CdS, Bi_2S_3, and AgI incorporated into transparent silicate glasses	Enhanced luminescence was observed in glasses	685
Silicate glasses	Size-quantized CdS incorporated into transparent silicate glasses	Non-linear optical and luminescence properties were modeled	685
Non-porous SnO_2 on glass	TiO_2 spin coated on conducting glass	Electron injection from excited ruthenium-2,2'-bipyridine-4,4'-dicarboxylic acid into the TiO_2 conduction band was examined under potentiostatic control of the electric field within the space charge layer of the membrane	687
GG 495 Schott filter precursors	CdS_xSe_{1-x} crystallites incorporated into transparent glasses	Low frequency inelastic scattering was observed	688
Glass films	Size-quantized CdS prepared in 0.1–0.2 μm glass films	X-ray diffraction, TEM, and non-linear optical properties were measured	689
Glass	TiO_2-coated glass	Photodegradation of 3-chlorosalicylic acid was investigated	690
Hollow glass microbeads	Microbeads coated by TiO_2 particles	Sunlight collection optic was analyzed	691
Porous ceramic membranes	Porous ceramic TiO_2 and ZnO membranes prepared by sol-gel technique	Semiconductor membranes were characterized	692, 693
Porous membrane	Colloidal CdS particles converted to Xerogels and dried at about 10^{-3} Torr at 30–40 °C to produce crack-free, optically transparent membranes	Mechanical and spectroscopic properties of CdS membranes were determined	694

System	Particles	Description	Ref.
Ceramic membranes	Size-quantized ZnO clusters concentrated to form syrups, gels, and crystals	Crystalline ZnO could be cast on porous or non-porous supports and fired to obtain ceramic ZnO membranes and thin films	695
Zeolite A, sodalite, zeolite X, and chabazite	Size-quantized CdS formed in zeolite channels	Absorption and emission spectra were determined	696
Zeolite Y	GaP	GaP was synthesized in zeolite channels	697
Zeolite A, zeolite Y, and zeolite mordenite	Size-quantized CdS and PbS formed in zeolite channels	Absorption and emission spectra were determined	698, 699
Zeolite mordenite	Size-quantized indium phosphide prepared in zeolite cages	Band-gap, flat-band potential, and photocurrent were investigated	700
Laponite clays	CdS particles	CdS interaction into clays was investigated	701, 702
Layered zirconium phosphate and zirconium phosphonates	Size-quantized ZnSe, PbS, CdS, and CdSe grown in the interlamellar region of layered zirconium hosts	X-ray diffraction, absorption spectra, and TEM were used to characterize systems	703
Sodium montmorillonite clay particles	α-Fe_2O_3 prepared in clay particles	Catalytic properties of clay-particle-incorporated α-Fe_2O_3 were investigated in the photodecomposition of saturated carboxylic acids	704
Ogranomontmorillonite complexes	Size-quantized CdS particles incorporated into the organomontmorillonite complexes	CdS formation was examined as a function of the surfactants present in the clay complexes, the composition of the solvents, and the concentration of the Cd-salts used	705
Yeasts (Candida glabrata and Schizosaccharomyces pombe)	Size-quantized CdS is biosynthesized	(γ-Glu-Cys)$_n$-Gly units controlled the nucleation and growth of CdS crystallites to peptide-capped, intracellular, 20-Å-diameter particles	706
Glutathione related isopeptides of general structure (γ-Glu-Cys)$_n$-Gly	Size-quantized CdS formed by sulfide-ion titration	Absorption spectra indicated nanometer sized particles	707
Ferritin	Nanometer-sized iron sulfide generated in ferritin	Biomineralization approach to nanophase engineering was proposed	708
Poly{A} and E. Coli DNA polynucleotides	CdS particles	CdS luminescence was quenched by polynucleotides	709

semiconductor particles [621]:

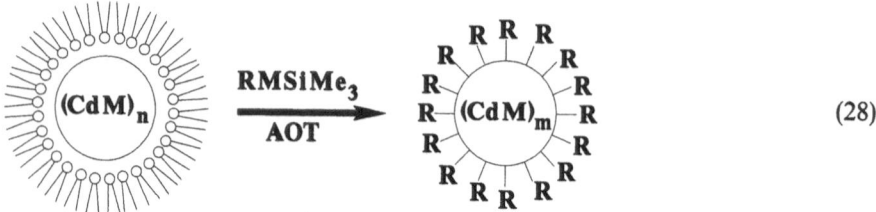

$$(28)$$

Capped semiconductor particles could be isolated as powders and redispersed in an organic solvent. This permitted convenient purification. Importantly, capped semiconductor particles could be grown, like living polymers, to desired sizes by the addition of appropriate amounts of metal ions and RMSiMe$_3$ [621].

Surfactant vesicles constitute a very flexible medium for the support of semiconductors. Semiconductor particles can be localized at the outer, the inner, or at both surfaces of single-bilayer vesicles (Fig. 102). Each of these arrangements has certain advantages. Semiconductor particles on outer vesicle surfaces are more accessible to reagents and can, therefore, undergo photosensitized electron transfer more rapidly. Smaller and more monodispersed CdS particles can be prepared and maintained for longer periods of time in the interior of vesicles than in any other arrangement.

Sizes of semiconductors in vesicles can be controlled by adjusting the number of precursor ions (Cd^{2+} ions, for example) either encapsulated in, or bound to the external surfaces of, vesicles and/or by regulating the amount and rate of addition of H$_2$S. Vesicles are also efficient in stabilizing colloidal particles for extended periods of time (several months) since spatial confinement in the

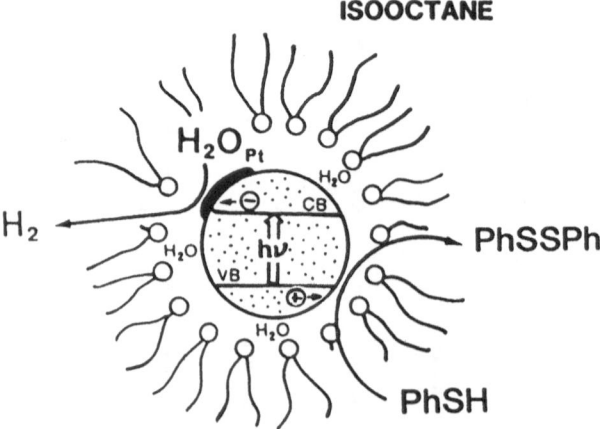

Fig. 101. Idealized model for the CdS-sensitized water photoreduction by PhSH in AOT reversed micelles in isooctane. VB = valence band, CB = conduction band [611]

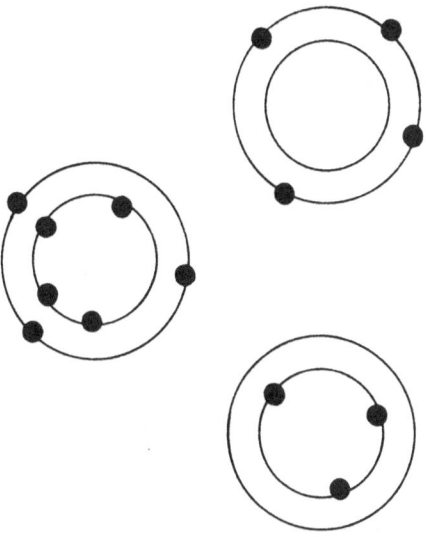

Fig. 102. Schematics of available sites for organizing colloidal semiconductors in single-bilayer surfactant vesicles [500]

bilayers precludes particle growth by either agglomeration or Ostwald ripening. Different size populations of semiconductors were shown to be generated at the inner and at the outer surfaces of DHP vesicles [624]. More precise control of CdS-particle sizes was obtained in DODAC vesicles [626]. Four different methods of preparation were investigated. In preparation A, CdS particles were generated from $CdEDTA^{2-}$ distributed at both the inner and outer surfaces of DODAC vesicles. CdS particles were localized only in the inner, or at the outer, surface of DODAC vesicles in preparations B and C, respectively. Preparation D involved the generation of CdS from Cd^{2+} ions electrostatically repelled from the DODAC-vesicle surfaces. CdS particles were formed in all cases by the careful addition of controlled amounts of H_2S. Absorption spectra of CdS on different vesicle preparations are shown in Fig. 103.

A considerable amount of basic information has been obtained by investigating the formation and properties of semiconductor particles produced on bilayer lipid membranes (BLMs) [649–652]. Simultaneous steady-state and ultrafast, time-resolved electrical and spectroscopic measurements provided a wealth of information on BLMs stabilized by polymerization and polymer coating. CdS particles were, for example, generated in situ on bovine brain phosphatidylserine (PS) membranes. BLMs were made by painting freshly prepared decane solutions of PS across a 0.80-mm Teflon hole separating two compartments which contained 0.10 M aqueous KCl buffered at pH 7.4 (Tris) at ambient temperature. Thinning of the film to a 50 ± 5-Å-thick BLM was monitored by observation of the reflected light and by capacitance measurements. Aqueous $CdCl_2$ was added to the left hand side of the BLM. Subsequent to incubation for 10–15 min, H_2S gas was slowly injected into the right hand side of the BLM. Attachment of Cd^{2+} ions onto the PS BLM surface and subsequent CdS formation were monitored by voltage-dependent capacitance

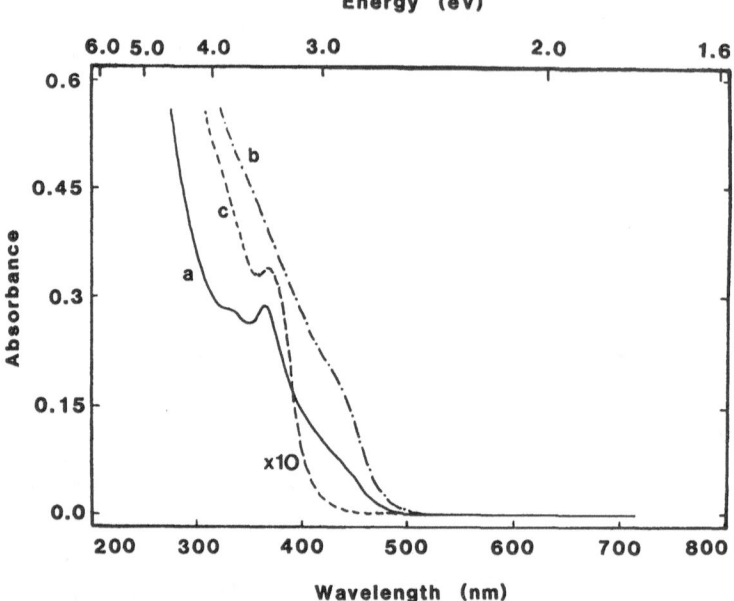

Fig. 103. CdS absorption spectra formed on outer and inner (a), outer (b), and inner surface (c) alone; [DODAC] = 2.0×10^{-3} M in all systems; [CdEDTA^{2-}] = 2.0×10^{-4} M (a), 1.0×10^{-4} M (b) and (c); pH 9.5 (a) and (b), 10.0 (c); the spectrum of (c) is enlarged 10 times [625]

measurements (Fig. 104) [650]. The minimum in curve a is at $V = 0 \pm 3$ mV, as expected for symmetrically charged PS membranes. Displacement of the curve upon the addition of Cd^{2+} (curve b) allowed the calculation of the charge density to be 360 Å^2 and the surface potential difference ($\Delta\psi$) to be 66 mV. It was estimated that [Cd^{2+}]:[PS] = 1:12 H$_2$S addition decreased $\Delta\psi$ to almost zero (curve c) and resulted in the development of Cds particles.

CdS, ZnS, Cu$_2$S, In$_2$S$_3$, and PbS semiconductor clusters have also been incorporated and generated in situ on a variety of BLMs. Optical microscopy of the process of semiconductor formation on BLMs provided a wealth of morphological information. Subsequent to injection of H$_2$S, the first observable change was the appearance of fairly uniform white dots of 1 μm or less on the black film (see photograph A in Fig. 105) [650]. The initially formed white dots moved rapidly about the BLM surface, encountering both the Plateau-Gibbs border and each other with resultant coalescence, particle growth, and island formation (see photograph B in Fig. 105). The islands themselves merged and, concurrently, a second generation of white dots began to appear on the surfaces of the black channels that surrounded the enlarged white islands. The behavior of the second generation of dots was similar to that of their predecessors. They moved rapidly about, combined with each other, and ultimately coalesced into the surrounding islands (see photograph C in Fig. 105). Eventually, a continuous film formed whose appearance was either that of a shiny, cloud-like layer (D) or that of a

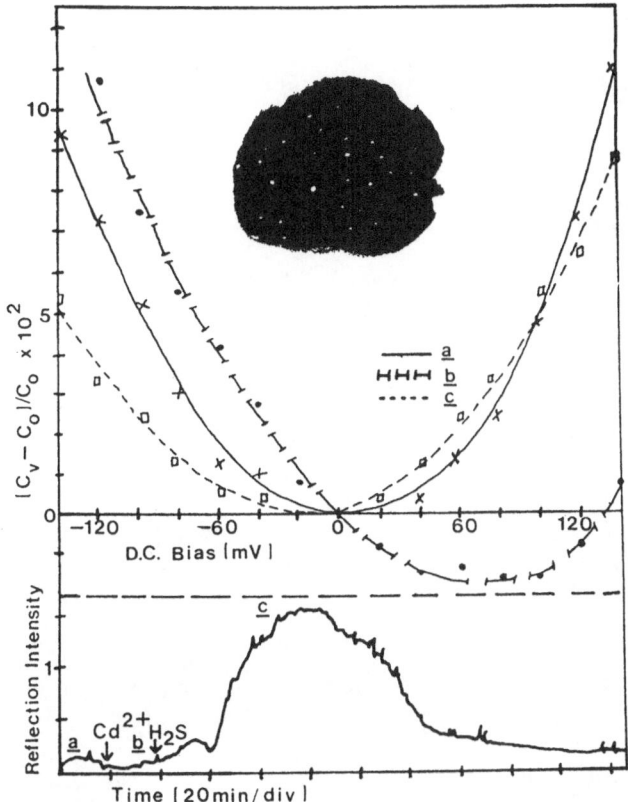

Fig. 104. *Upper part*: voltage-dependent capacitances of the BLMs prepared from PS prior (a) and subsequent (b) to the addition of $CdCl_2$. Cure (c) was obtained after H_2S addition. Grains of CdS particles are clearly seen in the photograph (*inset*) of the reflected light taken through a microscope. *Lower part*: reflected light intensity changes on CdS formation on PS BLM. Following the addition of $CdCl_2$ (b) and H_2S, formation of CdS results in a substantial increase in the intensity of the reflected light (c). Further illumination photocorrodes the CdS particles and, hence, decreases the reflected light intensity [500]

light, fog-like grain layer (E). In some instances, the film formed and pressed against the Plateau-Gibbs border, and a newly developing film overlapped it. Repetition(s) of this process led to films of several overlapping layers, which manifested in color development (F). Color formation, albeit with less hue, could also accompany the coalescence of islands (G, H). In general, a single layer of film which had formed quickly proved to be thinner than the arrested islands of particles. It needs to be emphasized that semiconductor particles, at all stages of their growth, were supported by bimolecular (black) membranes.

Band-gap excitation of PS-BLM-incorporated CdS by a laser pulse generated an electron-hole pair which, in turn, resulted in the reduction of oxygen on one side of the BLM and in the oxidation of HS^- on the other side. These processes produced a photovoltage across the membrane [651]. Three different

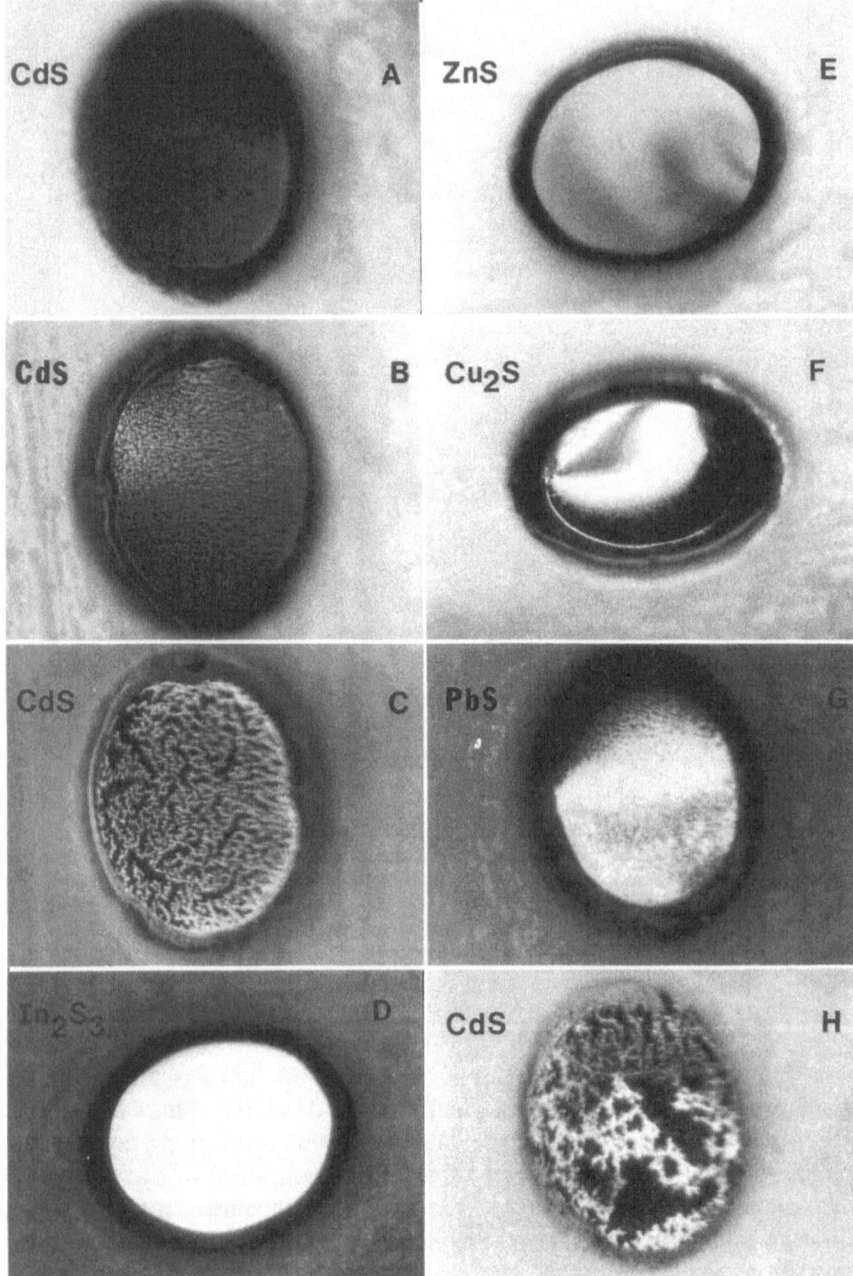

Fig. 105a–h. Micrographs of semiconductors in different stages of their growth on GMO BLMs [611]

systems have been investigated (Fig. 106). A single type of particulate semi-conductor deposited only on one side of the BLM constituted *System A*. Two different types of particulate semiconductors sequentially deposited either on the same or on opposite sides of the BLM represented *Systems B* and *C*, respectively.

In the absence of additives or adventitious impurities, the BLM is an electrical insulator. Current flow, in the order of only 10^{-9} A, was detected upon application of potential differences in the -0.10 to $+0.10$ V range (Fig. 107a). The determined resistance and capacitance of a 1.00-mm-diameter glyceryl monooleate (GMO) BLM bathed in 0.10 M KCl, $(3-5)\ 10^8$ ohm and 2.0–2.2 nF, agreed well with those reported previously (3×10^8 ohm; 0.380 $\mu F/mm^2$) [388]. In situ semiconductor formation on the BLM surface resulted in marked changes in the electrical response. Depending on the system, the current flow was found to increase asymmetrically and the BLM became very much stable and longer lived.

Single-phase microcrystalline semiconductor particles incorporated onto only one side of the BLM represent the most straightforward system. The composition of the electrochemical cell used in investigating *System A* is explicitly shown in Eq. (29):

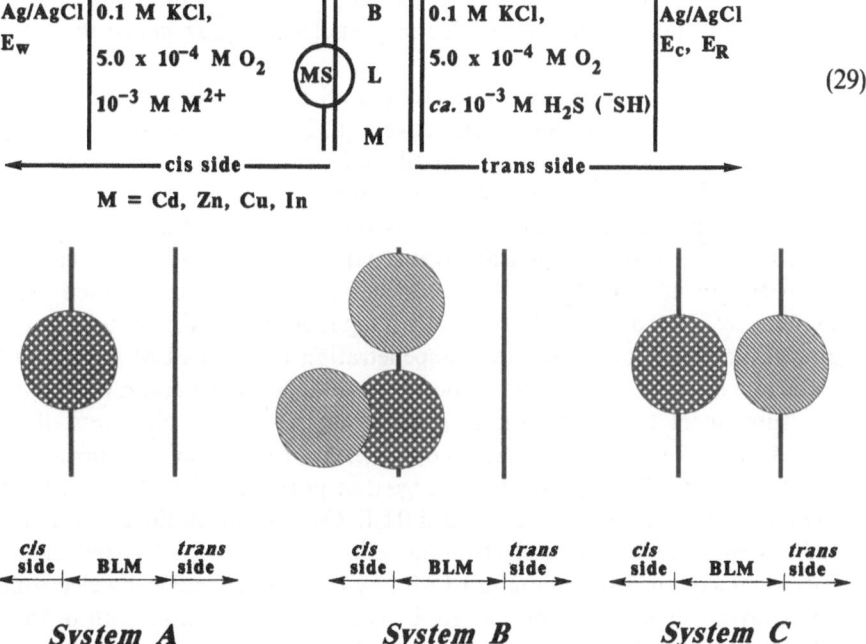

$$\begin{array}{c|c|c|c}
\text{Ag/AgCl} & \text{0.1 M KCl,} & \text{B} & \text{0.1 M KCl,} \quad \text{Ag/AgCl} \\
E_W & & & E_C,\ E_R \\
& 5.0 \times 10^{-4}\ \text{M O}_2 \quad \text{MS} \quad \text{L} & & 5.0 \times 10^{-4}\ \text{M O}_2 \\
& 10^{-3}\ \text{M M}^{2+} & \text{M} & \textit{ca.}\ 10^{-3}\ \text{M H}_2\text{S (}^-\text{SH)}
\end{array}$$

$$\qquad\qquad (29)$$

cis side ——————— ————trans side————

M = Cd, Zn, Cu, In

cis side   BLM   trans side   cis side   BLM   trans side   cis side   BLM   trans side

System A   **System B**   **System C**

Fig. 106. Schematic representation of the different semiconductor-coated BLMs. A single composition of particulate semiconductor deposited only on one side of the BLM constituted *System A*. Two different compositions of particulate semiconductors sequentially deposited on the same side of the BLM represent *System B*. Finally, two different compositions of particulates deposited on the opposite sides of the BLM made up *System C* [652]

where E_w, E_c, and E_R indicate the working, the counter, and the reference electrodes, respectively. Presumably, *Systems B and C* are also porous structures, explicitly shown in Eqs. (30) and (31), respectively.

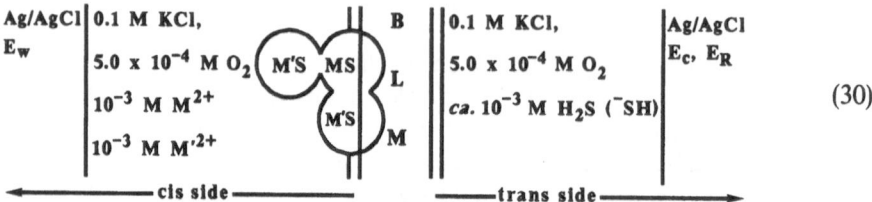

$$(30)$$

$$(31)$$

Equations (29), (30), and (31), as well as Fig. 106, are gross oversimplifications and should in no way be considered to convey any structural information beyond the location of the semiconductor particles and their penetration into the BLM.

Typical current–voltage, or I–V, curves of ZnS, CdS, and In_2S_3 on BLMs are illustrated in Figs. 107b, c, and d, respectively [652]. The shapes and characteristics of the I–V curves remained essentially independent of the rate of scanning in the range of 10 mV/s to 10^6 mV/s. Independence of the cyclic voltammetric behavior with respect to the frequency of scanning indicates that electronic (as opposed to electrolytic) charge transfer mechanisms are operative. Uncertainty combined with non-linearities of the resistances of BLM-incorporated semiconductor systems do not allow an unambiguous use of this criterion. For example, the extent of semiconductor penetration into the BLM varied with such factors as particle size and composition. Although the membrane remains intact subsequent to semiconductor growth (as seen by the blackness of the reflected light), formation of microscopic defects cannot be excluded. Such pinholes would facilitate electrolytic charge transport. Similarly, the presence of either adventitious or deliberately added (H_2S, O_2) dopants in the BLM matrix could well mediate resonance electron tunneling. Accordingly, electron transfer across a semiconductor-containing BLM may be governed either by electronic or electrolytic conductance, or, indeed, by some combination of both of these mechanisms. Regardless of the mechanism(s) involved, the role of the semiconductor particles is crucial. Different semiconductor particles have elicited substantially different current–voltage behavior; the observed photovoltage action spectra corresponded to the absorption spectra of semiconductor particles deposited on BLMs.

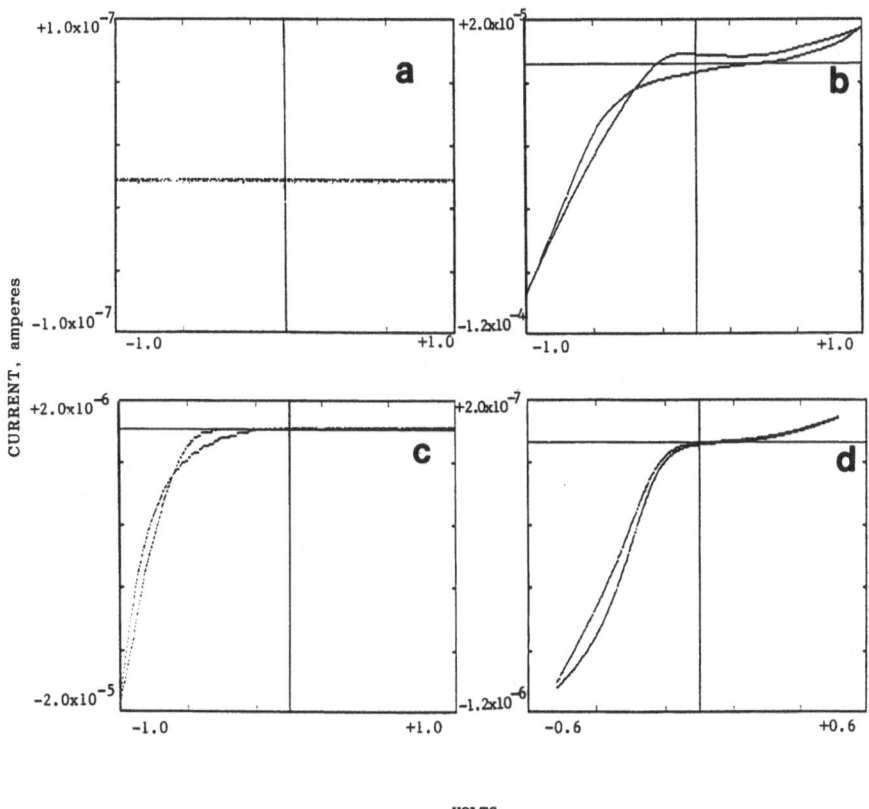

Fig. 107a–d. Cyclic voltammograms of a GMO BLM: **a** in the absence and; **b** in the presence of ZnS; **c** CdS and; **d** In_2S_3 particles on its surface (*System A*). Scan rate: 100 mV/s [652]

To a first approximation, the BLM can be considered to behave like a parallel plate capacitor immersed in a conducting electrolyte solution. In reality, even such a thin insulator as the modified BLM (designated by C'_m and R'_m in Fig. 108) could block the specific adsorption of some species from solution and/or modify the electrochemical behavior of the system. Similarly, *System C* may turn out to be a semiconductor(1)-insulator-semiconductor(2) (SIS') rather than a semiconductor(1)-semiconductor(2) (SS') junction. The obtained data, however, did not allow for an unambiguous distinction between these two alternative junctions; we have chosen the simpler of the two [652]. The equivalent circuit describing the working (E_w), the reference (E_R), and the counter (E_c) electrodes; the resistance (R_m) and the capacitance (C_m) of the BLM; the resistance (R_H) and capacitance (C_H) of the Helmholtz electrical double layer surrounding the BLM; as well as the resistance of the electrolyte solution (R_{sol}) is shown in Fig. 108a [652].

Deposition of a particulate semiconductor on the cis side of the BLM (*System A*) alters the equivalent circuit to that shown in Fig. 108b, where R_f and

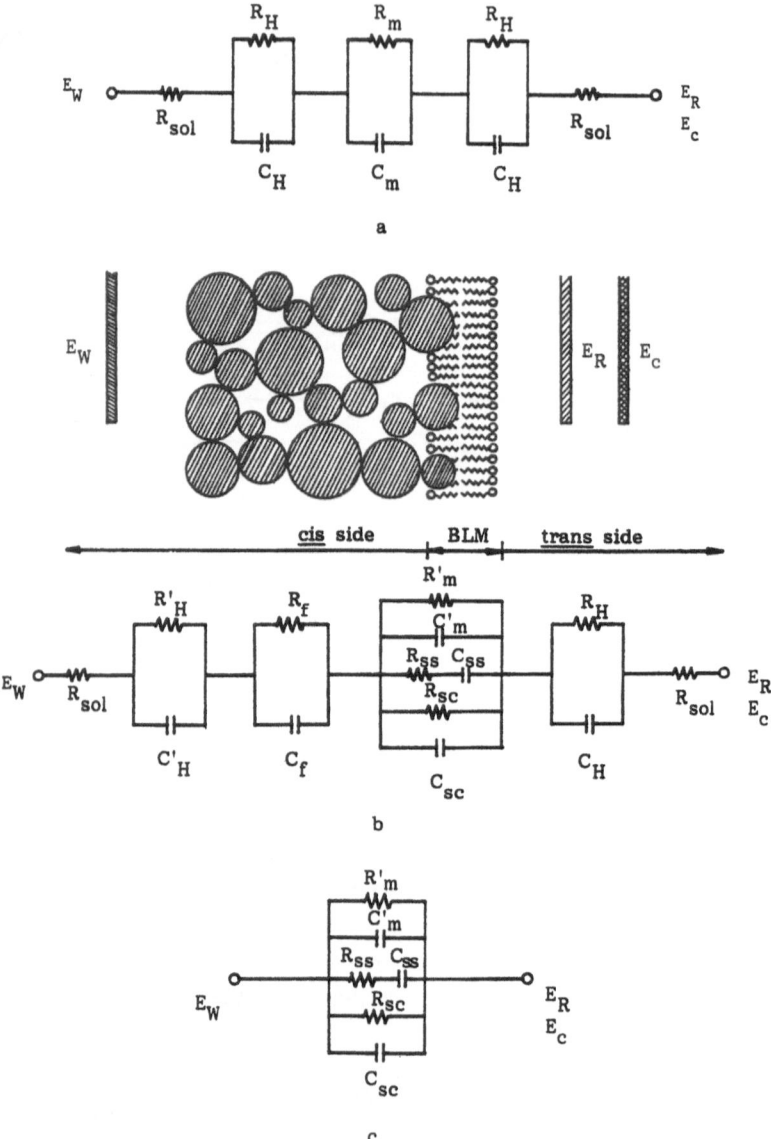

Fig. 108a–c. Proposed equivalent circuits for: **a** an "empty" and; **b** a semiconductor-particle-coated BLM. Porous structure of the semiconductor particles allowed; **c** the simplification of the equivalent circuit. R_m, R_H, and R_{sol} are resistances due to the membrane, to the Helmholtz electrical double layer, and to the electrolyte solutions, while C_m and C_H are the corresponding capacitances; R_f and C_f are the resistance and capacitance due to the particulate semiconductor film; R'_m and C'_m are the resistance and capacitance of the parts of the BLM which remained unaltered by the incorporation of the semiconductor particles; R_{sc} and C_{sc} are the space charge resistance and capacitance at the semiconductor particle-BLM interface; and R_{ss} and C_{ss} are the resistance and capacitance due to surface-state on the semiconductor particles in the BLM [652]

C_f are the resistance and capacitance due to the particulate semiconductor film; R'_m and C'_m are the resistance and capacitance of the parts of the BLM which remained unaltered by the incorporation of the semiconductor particles; R_{sc} and C_{sc} are the space charge resistance and capacitance at the semiconductor particle-BLM interface; and R_{ss} and C_{ss} are the resistance and capacitance due to surface-state on the semiconductor particles in the BLM. Electrolytes short circuit the porous semiconductor particles ($R_f = R_{sol} = 1.4\,k\Omega$) such that their contribution, along with that due to the Helmholtz layer, can be neglected. This allows the simplification of the equivalent circuit to that shown in Fig. 108c. As seen, the working electrode is connected (via ions) to the semiconductor particulate film.

Band models of n- and p-type electrolyte semiconductor-containing BLM (*System A*), ES, junctions are drawn in Fig. 109 [652]. Charge injection into the conduction band of the n-type semiconductor by a sufficiently active surface donor or by an applied voltage (making the trans side positive relative to the cis side) results in an accumulation of the majority carriers at the space-charge region of the semiconductor particles (Fig. 109a). Downward bending of the conduction-band energy (E_{cs}) provides for a favorable overlap with the energy level of the reducing agent (E_{red}) at the cis electrolyte surface; the semiconductor particle provides an ohmic contact with the cis electrolyte. A depletion layer is formed at the particle surface which is immersed in the BLM. Hence, the conduction band energy, E_c, is bent upwards to the energy of the conduction band edge at the surface, E_{cs}. Current flow is observed under cathodic (trans side positive with respect to the cis side) potential since E_{red} is located in the band-gap region. Electrons are assumed to tunnel through the modified, very thin BLM (or to be transported by electrolytes) to the overlapping unoccupied energy levels of the oxidizing agent (E_{ox}). The exponential increase of cathodic current with applied voltage is the expected consequence of decreased band bending and increased surface electron density. The behavior of p-type semiconductors can be analogously rationalized [652].

Methodologies developed for the in situ formation of semiconductor particulate films on BLMs have been applied to monolayer systems [638–641]. There are several advantages to the use of monolayer matrices for semiconductor particle generation. Monolayers are considerably more stable than BLMs and they also possess surface areas and charges which are two-dimensionally controllable. They may also, in association with the semiconductor particulate films grown in their matrices, by conveniently transferred to solid supports.

The experimental set-up used in the generation and in situ monitoring of semiconductor particulate films is identical to that used for nanosized, metallic, particulate films. Evolution of a nanocrystalline particulate film, illustrated by the formation of sulfide semiconductor particulate films (Fig. 110), has been discussed in terms of the following steps [639]:

a) formation of metal-sulfide bonds at a large number of sites at the monolayer-aqueous interface;

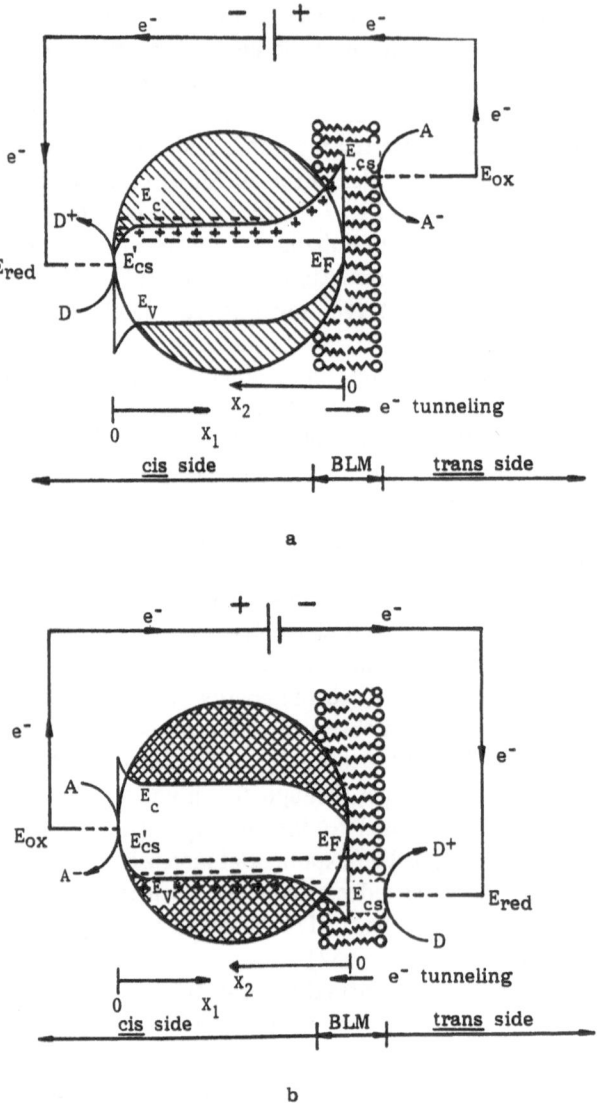

Fig. 109a, b. Band models for: **a** an n-; **b** a p-type semiconductor-containing BLM – (*System A*) ES junction. E_{cs} and E'_{cs} represent the conduction-band edge at the semiconductor-BLM interface and at the semiconductor cis surface. E_F is the Fermi energy level and E_{red} and E_{ox} are energy levels of the reducing and oxidizing agents, respectively. A and D stand for an electron acceptor and an electron donor, respectively [652]

   b) downward growth of well-separated nanocrystalline metal sulfide particles;
   c) coalescence of clusters into interconnected arrays of semiconductor particles;

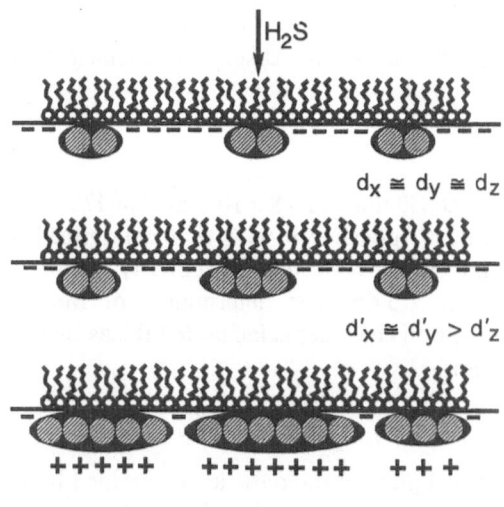

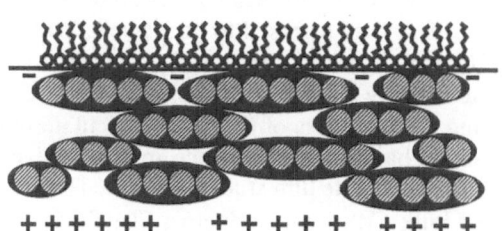

Fig. 110. Proposed schematics for the initial and subsequent growth of a monolayer-supported, porous, SQSPF. The d_x and d_y dimensions are in the plane and the d_z dimension is normal to the plane; they refer to the earliest observable particles. d'_x, d'_y, and d'_z are dimensions in the plane and are normal to the plane; they refer to particles observed at later stages of their growth [639]

d) formation of the "first layer" of a porous sulfide semiconductor particulate film composed of 20- to 40-Å-thick, 30- to 80-Å-diameter particles;

e) diffusion of fresh metal ions to the monolayer head group area;

f) formation of a "second layer" of the porous sulfide semiconductor particulate film (by using steps a, b, and c); and

g) build-up of "subsequent layers" of the sulfide semiconductor particulate film (by using steps a, b, and c) up to a plateau thickness (about 300 Å for CdS and ca. 3500 Å for ZnS) beyond which the film cannot grow.

The presence of a monolayer with an appropriate surface charge is an essential prerequisite for sulfide semiconductor particulate film formation. Infusion of H_2S over an aqueous metal-ion solution, in the absence of a monolayer, resulted in the formation of large, irregular, and polydispersed metal sulfide particles which precipitated in the bulk solution before settling to the bottom of the trough. Furthermore, no sulfide semiconductor particulate film formation could be observed upon the infusion of H_2S to a positively charged monolayer (dioctadecyldimethylammonium bromide, for example) floating on an aqueous metal-ion ($CdCl_2$, for example) subphase.

To date, cadmium sulfide, zinc sulfide, lead sulfide, cadmium selenide, and lead selenide semiconductor particulate films have been grown, in situ, under monolayers [639–648]. Absorbances (A) increased linearly with increasing thicknesses of the CdS and ZnS particulate films. Absorption coefficients, σ-values, were calculated from

$$\sigma = A/d_s' \tag{32}$$

and determined to be 2.4×10^5 cm^{-1} at 239 nm and 5.8×10^4 cm^{-1} at 475 nm for the CdS particulate film. These values agreed well with that determined for electrodeposited CdS films ($\sigma(435$ nm$) = 2.06 \times 10^4$ cm^{-1}). Similarly, an absorption coefficient of 5.8×10^4 cm^{-1} at 315 nm was determined for the ZnS particulate film. Knowledge of absorption coefficients facilitated the assessment of direct band-gap energies, E_g, from

$$(\sigma h\omega)^2 = (h\omega - E_g)C \tag{33}$$

where $h\omega$ is the photon energy. Typical plots of the data, as determined by Eq. (33), are shown in Fig. 111 [640]. Values of E_g for CdS particulate films of $d_s' = 63$ Å, 125 Å, 163 Å, 204 Å, and 298 Å were assessed to be 2.54 eV, 2.48 eV, 2.46 eV, 2.44 eV, and 2.43 eV (Fig. 111). Henglein's published E_g vs particle size curve [8] was used to estimate the average diameter of the 63-Å-thick CdS particles to be about 50 Å. Increasing the thickness of the CdS particulate film resulted in progressively decreased direct band-gaps and, hence, in progressively larger CdS particles. The thickest CdS particulate film studied exhibited a direct band-gap equal to that reported for bulk CdS semiconductors (2.4 eV) [8]. A direct band-gap of 2.75 eV was assessed for the 359-Å-thick ZnS particulate film.

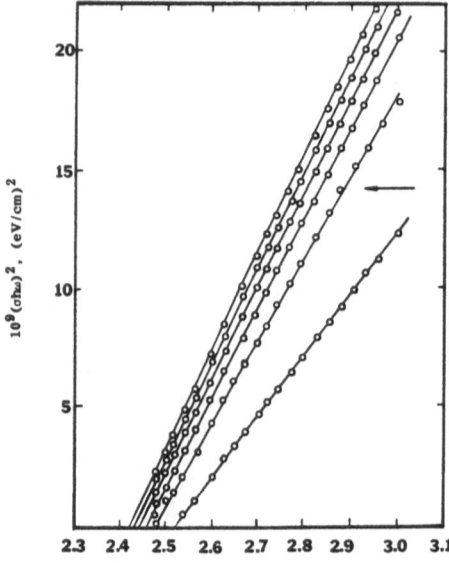

Fig. 111. Plots of $(\sigma h\omega)^2$ against energy for 63 Å, 125 Å, 163 Å, 204 Å, 263 Å, and 298 Å thick (in the order shown by the *arrow*) CdS particulate films [640]

Prolonged heating of the semiconductor particulate films at high temperatures resulted in pronounced changes in their absorption spectra. The absorbance of a 192 Å thick CdS particulate film (vacuum dried at 10^{-3} torr for three days) decreased upon heating at 490 °C for 5, 15, and 25 min [640–642]. Five min of heating shifted the direct band-gap from 2.47 eV to 2.40 eV (or to an absorption edge of 515 nm) equal to that of bulk CdS. A similar behavior was noted for 359-Å-thick ZnS particulate films; heating at 300 °C for 15 min shifted the direct band-gap from 3.75 eV to 3.64 eV (or to an absorption edge of 340 nm). Prolonged heating of semiconductor particulate films have, therefore, two important consequences. First, their properties become similar to those found for bulk semiconductors. Second, they are annealed to the substrate. Annealed semiconductor particulate films could not be washed or wiped away from their substrates. In contrast, vertical dipping of untreated semiconductor particulate films into water resulted in a partial loss of material from the subphase. Interestingly, annealing of semiconductor particulate films, prepared under thiol surfactants, resulted in band-gap shifts to higher energy [110].

Transmission electron micrographs of 30- to 50-Å-thick CdS particulate films indicated the presence of 20- to 80-Å-diameter particles possessing a relatively narrow size distribution and average diameters of 47 Å [638, 640]. HOPG was established, using scanning tunneling microscopy (STM), to provide an atomically flat surface with periodic roughnesses in the order of 1 Å. In two-dimensional STM images of HOPG-supported ZnS and CdS particulate films, the presence of 10- to 20-Å-thick, 30- to 40-Å-diameter ZnS and 20- to 30-Å-thick, 40- to 50-Å-diameter CdS particles is clearly discernable. The widths of the semiconductor particles observed by STM agree well with the corresponding diameters determined by transmission electron microscopy [638, 640].

Electrical and photoelectrical measurements were carried out on CdS particulate films deposited on glass substrates or teflon sheets [640]. The resistivity (ρ) of a semiconductor particulate film, measured between two parallel copper electrodes, is given by

$$\rho = R \frac{Ld_s'}{a} \tag{34}$$

where R is the measured resistivity in Ω, L is the length of the copper electrodes, a is the distance between them, and d_s' is the thickness of the semiconductor particulate film. Resistivities of 200- to 300-Å-thick CdS particulate films were determined to fall in the range of $(3–6) \ 10^7 \ \Omega$. These values represent the measurement of 10 samples of different thicknesses and may be attributed, in part, to the presence of different amounts of water in the films. The ρ values determined for CdS particulate films are some six orders of magnitude higher than those observed for materials having intrinsic conductivity.

The dark resistance of CdS particulate films was found to decrease exponentially with increasing temperature [640]. Illumination decreased the resistivity (i.e. increased the conductivity) of CdS particulate films by some two orders of magnitude and matched the absorption spectrum of the corresponding CdS particulate film nicely. Photoconductivity originates, therefore, in the production of conduction band electrons, e_{CB}^-, and valence band holes, h_{VB}^+, during bandgap irradiation of CdS:

$$CdS \xrightarrow{hv} e_{CB}^- + h_{VB}^+ \ . \tag{35}$$

Steady-state irradiation of CdS particulate films also resulted in the development of photovoltage. Irradiation by a 10-n, 343-nm laser pulse gave rise to a transient photovoltage. The magnitude of the photovoltage (1–8 mV) was found to increase linearly with the energy of the laser pulse (0.1–1.0 mJ). The rise time of the transient signal, corresponding to Eq. (35), was faster than the response time of the instrument used (10 n). The decay time of the signal was on the order of 3×10^{-4} s. This decay corresponds to charge recombination.

Cadmium sulfide particulate films, generated in thicknesses of 300 ± 50 Å at arachidic acid (AA) monolayer interfaces, have been characterized in situ by STM under potentiostatic control [644]. Electrical contact was made between the tip of the STM, acting as the working electrode (WE), which was in contact with the CdS particulate film floating on aqueous 0.30 M NaCl, and the reference (RE) and counter (CE) electrodes, placed in the subphase (Fig. 112) [644]. A well-defined single-reduction wave at about -1.15 V was observed. Prolonged exposure to room light shifted the reduction peak to -0.85 V. Electrical and photoelectrical characterizations have also been performed on Ti-foil-supported, 5000-Å-thick CdS particulate films in an electrochemical cell (Fig. 113) [644]. The Ti foil was used as the WE, while the RE and CE were placed into 0.50 M

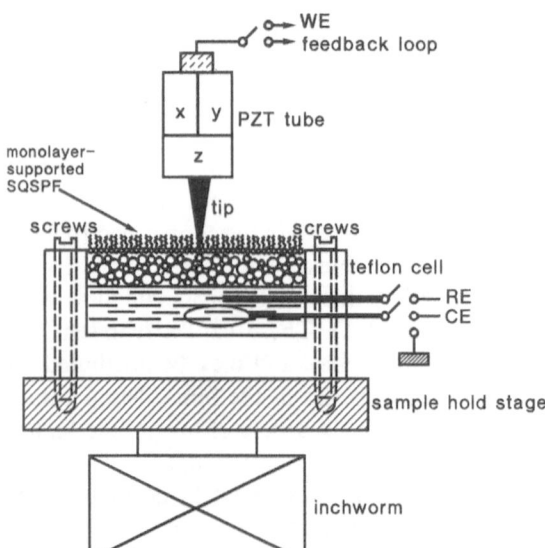

Fig. 112. Schematics of the cell used for the electrochemical characterization of monolayer-supported CdS particulate films. WE = working electrode, RE = reference electrode, and CE = counter electrode [644]

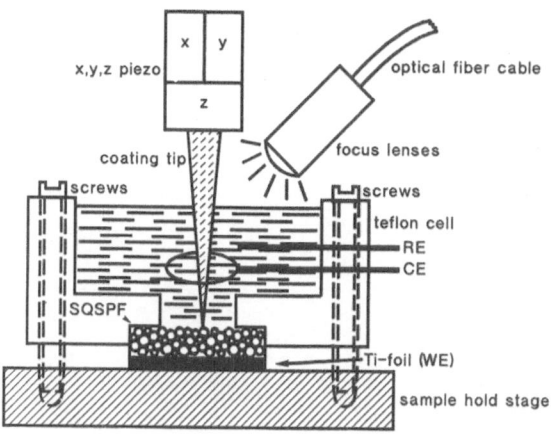

Fig. 113. Schematics of the photoelectrochemical cell used for characterization at Ti-foil-supported SQSPF-electrolyte junctions. WE = working electrode, RE = reference electrode, and CE = counter electrode [644]

KCl, which bathed the CdS film. This system was also investigated by cyclic voltammetry both in the dark and under illumination. Starting at about − 0.9 V, the dark cathodic current exhibited a peak at − 1.15 V due to Cd^{2+} reduction and then rose to − 1.4 V as a result of hydrogen production. The observed anodic peak at − 0.85 V was attributed to the stripping of cadmium deposits in the lattice (Cd_L^0). Cyclic voltammetry subsequent to illumination resulted in the appearance of cathodic waves at − 1.0 V and − 1.3 V, at the expense of that at − 1.15 V. The anodic peak broadened, as is indicative of photocorrosion.

Molecular recognition between the monolayer headgroups and incipient semiconductor nanocrystallites can, in many ways, be regarded as mimicking biomineralization [15–18] and represents an important milestone in the realization of the potential of a colloid-chemical approach to band-gap engineering. Lead sulfide (PbS) particulate films composed of highly oriented, equilateral-triangular crystals have been generated in situ by the exposure of AA-mono-layer-coated, aqueous, lead nitrate [$Pb(NO_3)_2$] solutions to H_2S (Fig. 114) [647]. AA monolayers, in their solid states, consist of $CH_3(CH_2)_{18}COOH$ molecules two-dimensionally arrayed at the air-water interface. Spread over the aqueous subphase, the carboxyl or the carboxylate groups of AA are aligned perpendicularly to the water surface. The alkyl chains of AA, fully extended in the air in a planar zig-zag conformation, are oriented approximately normal to the surface in a triangular lattice of hexagonal close packing with a lattice constant of $a = 4.85$ Å [96, 109]. Combined synchrotron X-ray reflection and diffraction data established a structural model for AA monolayers at air-water interfaces. The model required the hydrocarbon chains to be well packed in a pseudo-hexagonal lattice and tilted toward their nearest neighbor [96].

Rationalization of the packing of the AA headgroups at the water-air interface is, unfortunately, less than straightforward [97]. The absence of information concerning the extent of headgroup ionization (at a bulk pH of 5.5), counterion binding, and the degree of hydration hinders the interpretation of

Fig. 114. Transmission electron micrograph (TEM) at limiting aperture coverage of PbS crystals formed by the slow (30 min) infusion of H_2S to an AA monolayer in the Lauda film balance (kept at $\pi = 26 \, mNm^{-1}$ surface pressure) floating on an aqueous 5.0×10^{-4} M $Pb(NO_3)_2$ solution. The PbS particulate film was deposited on a formvar-coated, 200-mesh copper grid [647]

experimental results and the development of a reliable theoretical approach for predicting headgroup organization at the monolayer-subphase interface. Using the experimentally determined value for the surface area of one AA molecule (20.0 $Å^2$/molecule) permitted assessment of the lattice constant to be 4.81 Å (a) and the $d_{(100)}$ spacing to be 4.16 Å ($d_{(100)} = a \cdot \sin 60°$). These values are in good agreement with those determined for AA monolayers by synchrotron X-ray scattering (a = 4.85 Å and $d_{(100)} = 4.13$ Å) [96]. They are also similar to those determined for cadmium stearate (a = 4.89 Å and $d_{(100)} = 4.20 \pm 0.10$ Å) and other fatty acid monolayers [96, 109].

Reliable assessment of the arrangement and crystallinity of AA monolayers supported by a Pb^{2+} subphase is equally elusive. Our data is best accommodated in terms of an AA : Pb^{2+} = 3:4 ratio (Fig. 115). Grazing incidence X-ray diffraction measurements of lead arachidate monolayers demonstrated the existence of long-range ordering (250 Å) of Pb^{2+} [109].

PbS is known to crystallize in a cubic crystalline lattice with a lattice constant of a = 5.9458 Å. Atomic coordinates are (0, 0, 0) and (1/2, 1/2, 0) for Pb and (1/2, 1/2, 1/2) and (0, 0, 1/2) for S. The nearest-neighbor separation of Pb-Pb and S-S atoms of 4.20 Å matches the $d_{(100)}$ network spacing of the AA monolayer. This fit implies the alignment of PbS along its (1 1 1) plane to the (1 0 0) plane of the AA monolayer (Fig. 115). A comparison of the interatomic Pb to Pb distance of the (1 1 1) plane of the PbS crystal (4.20 Å) with that of the $d_{(100)}$ spacing of the AA monolayers (4.16 Å) revealed a mismatch of only 1% between these two crystals.

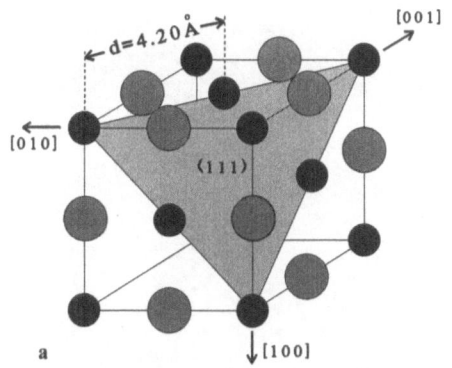

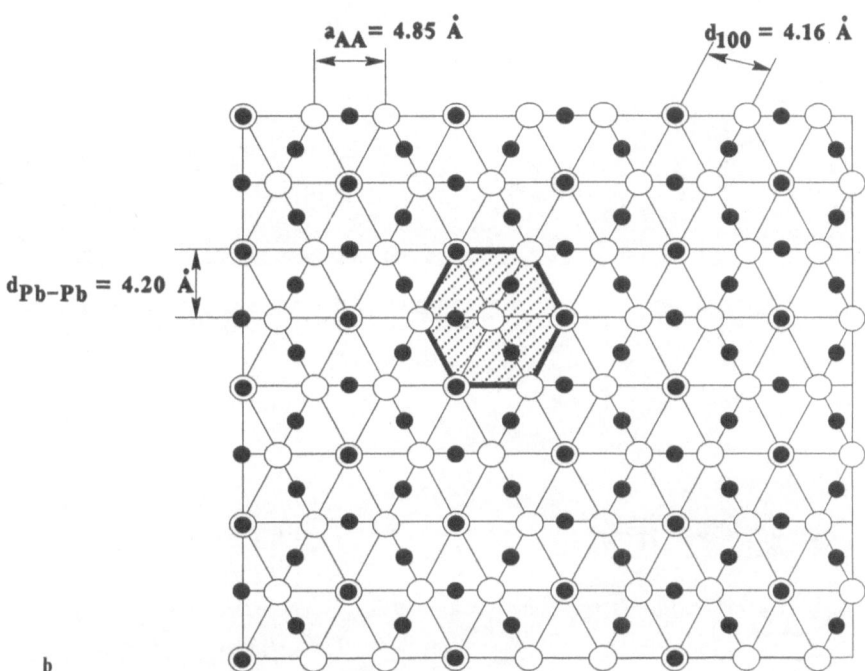

Fig. 115a. Three-dimensional representation of the PbS crystal lattice and (111) plane. **b** Schematic two-dimensional representation of the proposed overlap between Pb^{2+} ions and AA headgroups; $\bigcirc$ = AA headgroup, $\bullet$ = Pb^{2+}, and $\circledcirc$ = Pb^{2+} and AA headgroups. A unit cell is highlighted by the *dotted area* which is enclosed by *heavy lines* [647]

Epitaxial growth of PbS under well-compressed AA monolayers is explicable in terms of the geometrical complementarity between PbS and the AA head-groups. The strong intrinsic electrostatic interaction results in a very high Pb^{2+} concentration at the monolayer interface. The extremely low solubility of PbS in water ($K_{SP} = 8.81 \times 10^{-29}$ at 25 °C) favors its rapid and random nucleation. However, the presence of the monolayer acts to drastically diminish the reaction

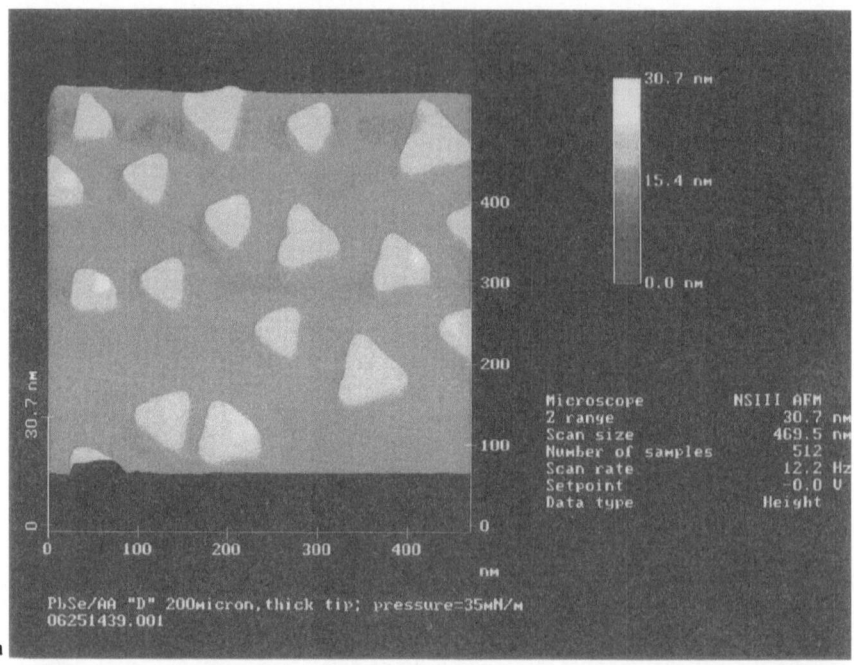

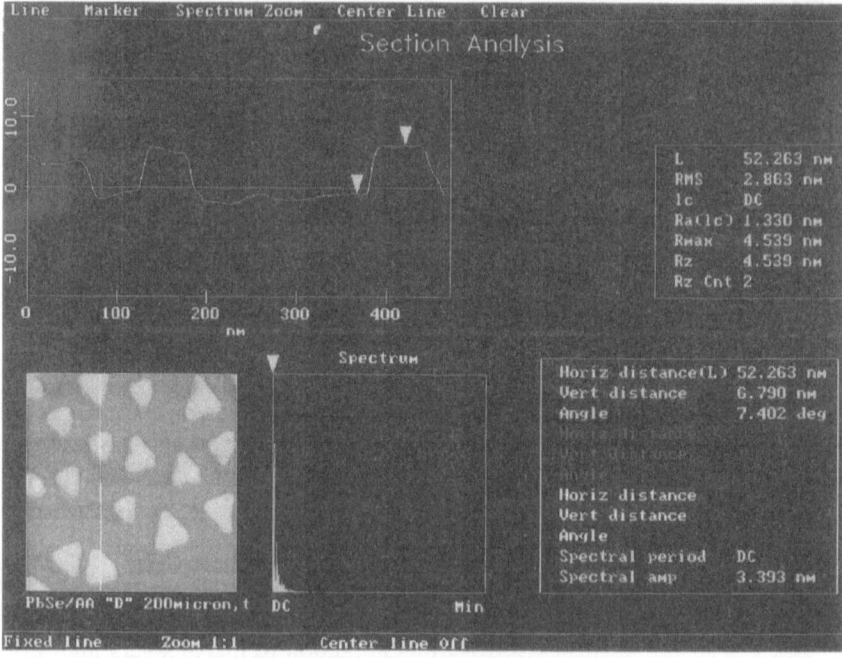

Fig. 116a. Three-dimensional side view of a PbSe particulate film imaged on a 470 nm × 470 nm section by atomic force microscopy (AFM). **b** Profile of the PbSe crystals by AFM sectioning. The *vertical line* in the image shows the position of the sectioning [648]

rate, limiting the encounter of the PbS precursors. The rate and amount of H_2S can also be controlled. These measures have ensured the formation of a critically sized nucleus and the subsequent ion-by-ion heteroepitaxial growth of the PbS crystals.

Significantly, equilateral triangular PbS crystals have been grown under compressed monolayers in the same orientation (see Fig. 114).

Epitaxial growth of nanocrystalline lead selenide particles under AA monolayers has also been demonstrated [648]. Exposure of an AA-monolayer-coated $Pb(NO_3)_2$ solution to H_2Se resulted in the formation of reasonably uniform crystals which constituted the particulate film. A typical three-dimensional, 470 nm by 470 nm image of PbSe crystallites is shown in Fig. 116a [648]. The thickness of these crystals was determined to be in the order of 65 Å by sectioning techniques (Fig. 116b). The equilateral triangular PbSe crystals were not entirely uniform; along their corners they were somewhat thicker than in their middles

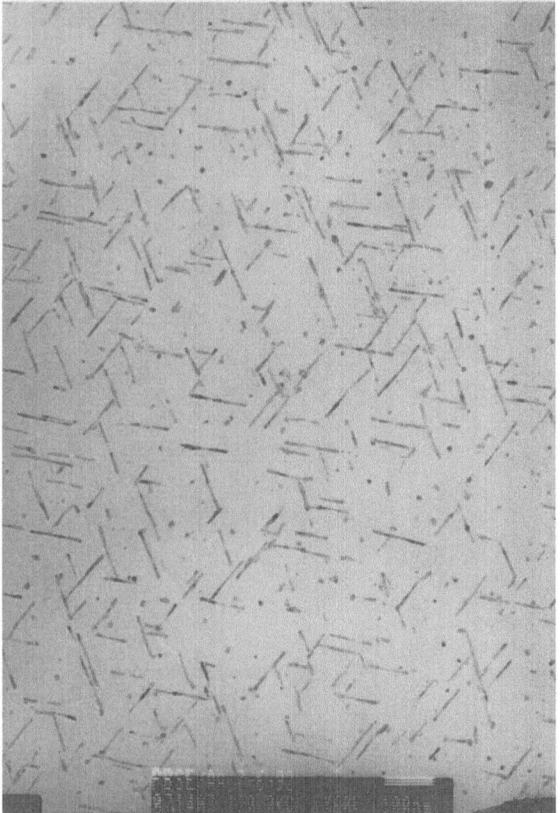

Fig. 117. TEM of a PbSe particulate film. The film was formed by the 25 min infusion of H_2Se (50 µl) over an AA monolayer, kept at 40 mN/m, which was floating on an aqueous 5.0×10^{-4} M $Pb(NO_3)_2$ solution in the Lauda trough. The PbSe particulate film was transferred to an amorphous-carbon-coated 200-mesh copper grid. The *bar* = 200 nm [648]

(Fig. 116a). Interestingly, almost all crystals were aligned in their edge directions. Electron diffraction of a 2-µm-diameter area (Fig. 116) showed diffraction spots corresponding to the $\{220\}$, $\{422\}$, and $\{440\}$ planes, indicating that all crystals line up and have their (1 1 1) planes perpendicular to the electron beam.

In some cases, particularly at high monolayer surface pressures, formation of rod-like PbSe particles was observed (Fig. 117) [648]. The thickness of the rods was typically 10 nm or less and the length was of the order of 100 nm. In addition to the rods, there were also many dot-like particles in the 10-nm size range (Fig. 117). Generation of rod-like semiconductor particles under monolayers provides a potential approach to nanofabricated quantum wires.

Semiconductor particle formation between Langmuir-Blodgett (LB) films is an alternative and promising approach to band-gap engineering [662–666]. The ease of formation of self-assembled (SA) films [183–185, 203–246] promoted the incorporation of semiconductors into these media [663, 710].

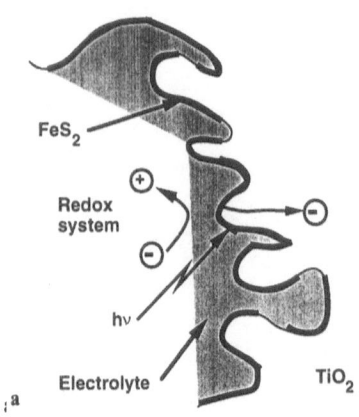

a

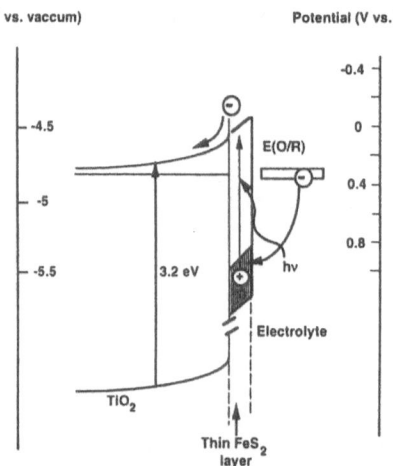

b

Fig. 118a. Schematic of the electron injection process at a highly porous TiO$_2$ electrode surface (only a patchy and fractured pyrite layer can be expected on the highly porous TiO$_2$ layer). **b** Surface TiO$_2$/FeS$_2$ energy diagram at pH = 1, indicating the injection of electrons from the valence band of FeS$_2$ into the conduction band of TiO$_2$, and the reaction of holes with the redox system [714]

Polyions, polymers, and polymeric membranes also provide suitable compartments for semiconductor particles and particulate films [668–682]. With improved chemical and physical methodologies, polymer chemists will soon be able to construct supramolecular, dimensionally controlled assemblies which will rival the sophistication of LB and SA films.

Particularly promising is the development of nanoporous ceramic semiconductor membranes [692–695]. They not only possess all of the advantages of ceramic materials, but they may also be efficient light harvesters having large surface areas which could provide sites for sensitizers [108, 711–714]. Indeed, FeS_2 particles, deposited into (and onto) a porous TiO_2 electrode, sensitized photoelectron conversion well (Fig. 118) [714].

5 Conductors and Superconductors

Metallic conductance arises in materials with partially filled bands; the entire population of charge carriers in these bands can move to higher energies under the application of an electric field. Conductivities of metals lie in the range of 10^4 to 10^6 $S\,cm^{-1}$ (where S stands for Siemens – the derived SI unit of conductance, related to the derived SI unit of the electric resistance, the ohm, Ω, by $S = 1/\Omega$); in contrast, the conductivity of an insulator can be as low as 10^{-14} $S\,cm^{-1}$. Conductivity in metals increases with decreasing temperature and reaches a plateau value at sufficiently low temperature. Some metals, at a critical temperature (slightly above 0 K) undergo a second order transition to a superconducting state. Superconductors have infinite d.c. conductivity (i.e. zero resistivity) but display normal resistance above a certain frequency. They are perfectly diamagnetic and totally exclude the magnetic flux. Superconductivity has been described in terms of highly coordinated motion of electron pairs (Cooper pairs) by Bardeen, Cooper and Schrieffer in their BSC theory [715, 716].

The discovery of synthetic electrical conductors and high-temperature superconductors constitutes a major breakthrough [457, 717]. Polysulfur nitride, $(SN)_x$, for example, has a room temperature conductivity in the order of 10^3 $S\,cm^{-1}$ and becomes superconducting below 0.3 K [718]. The conductivity of polyacetylene (about 10^{-5} $S\,cm^{-1}$) has been increased up to 10^4 $S\,cm^{-1}$ by converting it to an ionic complex through exposure to an oxidizing or a reducing agent. The process, in the language of solid-state physics, corresponds to a p-type (using an oxidizing agent) or an n-type (using a reducing agent) doping. Conducting polypyrrole, polythiophene, and a host of other polymers have been developed by analogous doping techniques [718]. Polarons and bipolarons (i.e. cation and dication radicals associated with lattice distortions) have been postulated to be responsible for polymeric conductivity [719]. Perceived similarities between bipolarons and Cooper pairs have prompted the intense search for polymeric superconductors.

Bulk crystalline radical ion salts and electron donor-electron acceptor charge transfer complexes have been shown to have room temperature d.c. conductivities up to 500 S cm^{-1} [457, 720, 721]. Tetrathiafulvalene (TTF), tetraselenofulvalene (TST), and bis-ethyldithiotetrathiafulvalene (BEDT-TTF) have been the most commonly used electron donors, while tetracyano p-quinodimethane (TCNQ) and nickel 4,5-dimercapto-1,3-dithiol-2-thione {Ni(dmit)$_2$} have been the most commonly utilized electron acceptors (see Table 8). Metallic behavior in charge transfer complexes is believed to originate in the facile electron movements in the partially filled bands and in the interaction of the electrons with the vibrations of the atomic lattice (phonons). Lowering the temperature causes fewer lattice vibrations and increases the intermolecular orbital overlap and, hence, the conductivity. The good correlation obtained between the position of the maximum of the charge transfer absorption band (proportional to

Table 8. Charge-transfer complex-forming donors and acceptors

Donors

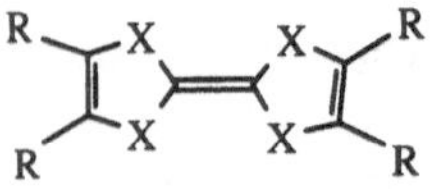

X	R		
S	H	TTF	tetrathiafulvalene
Se	H	TSF	tetraselenofulvalene
S	Me	TMTTF	tetramethyltetrathiafulvalene
Se	Me	TMTSF	tetramethyltetraselenofulvalene

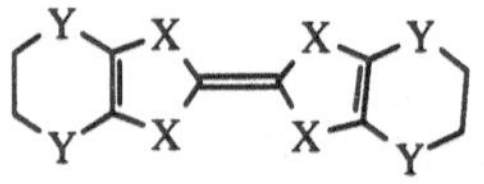

X	Y		
S	S	BEDT-TTF	bisethylenedithiotetrafulvalene

Acceptors

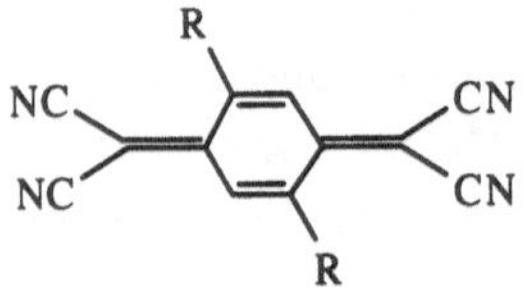

R = H	TCNQ	tetracyano p-quinodimethane
R = Me	2,5-DMTCNQ	2,5-dimethyl-tetracyano p-quinodimethane

Table 8. Contd.

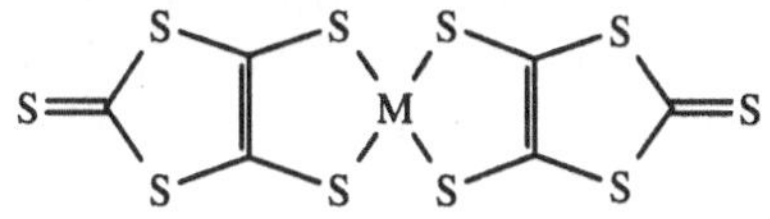

M = Au	Au(dmit)$_2$
M = Ni	Ni(dmit)$_2$
	H$_2$dmit = 4,5-dimercapto-1,3-dithiol-2-thione

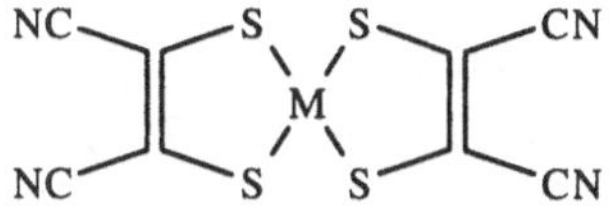

M = Au	Au(mnt)$_2$
M = Ni	Ni(mnt)$_2$
	H$_2$(mnt) = maleonitriledithiolate or *cis*-2,3-dimercapto-2-butenedinitrile

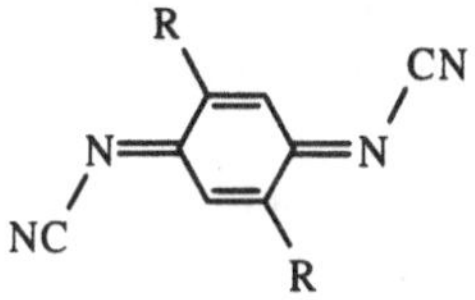

R = H	DCNQ I	*N,N'*-dicyanoquinone diimine
R = Me	2,5 DCNQ I	2,5-dimethyl-*N,N'*-dicyanoquinone diimine

the optical gap) and the d.c. conductivity for a number of charge transfer complexes [721] supports the postulated electronic structure of the complex.

Emphasis will be placed, in this section, on the development of conducting and superconducting Langmuir–Blodgett (LB) and self-assembled (SA) films [68, 720–722]. There are several advantages of LB and SA films. In the first place, they can be formed from a large variety of functionalized surfactants at room temperature and atmospheric pressure. Secondly, the properties can be tailored via control of the molecular architecture. Finally, there are a large variety of physical and chemical methodologies which may be utilized for the characterization of LB and SA films at the molecular level.

Conductivities can be measured across (i.e. perpendicular or normal to the plane of the aligned surfactants) or in the plane (i.e. parallel to the plane of the aligned surfactants) of LB and SA films. Surfactants possess large apolar regions, have high dielectric constants, and can, thus, be considered as insulators. Placing well-characterized, and presumably defect-free, LB films between two electrodes allows, therefore, the investigation of electron tunneling and photoconductivities as a function of the film thickness and molecular organization [106, 137, 723, 724]. Examination of the conductivities of long chain carboxylic acids, spanning a distance of 24–30 Å between the electrodes in LB films, has revealed a much greater increase of resistance with increasing distance than that expected for Ohmic behavior [725]. Conductivity, measured at such small electrode separation distances, has been attributed, therefore, to quantum mechanical tunneling. More practically, being thin dielectrics, LB and SA films provide an entry into the fabrication of advanced devices based on metal-insulator-semiconductor (MIS), metal-insulator-metal (MIM), superconductor-insulator-superconductor (Josephson) junctions.

Construction of LB films having lateral d.c. conductivities is a burgeoning activity. Results of published work are summarized in Table 9 [726–772]. The first formation of a conducting LB film was reported by French workers in 1985 [726]. Non-conducting LB films were formed from N-docosylpyridinium TCNQ. Subsequent exposure to iodine vapor resulted, however, in the lateral conductivities in the order of 0.1 S cm^{-1} [726]. The initially formed LB film was shown to consist of (TCNQ$^-$.)$_2$ dimers whose molecular planes were almost parallel to the substrate. Iodination resulted in the development of a brown-purple color, the partial oxidization of the radical anions to TCNQ$°$ and, most importantly, a dramatic rearrangement of the LB film. In iodine-doped films, the TCNQ molecules have been shown to assume a position almost perpendicular to the substrate [721, 773].

Conducting LB films have also been prepared form a variety of charge transfer salts by a number of other workers (see Table 9). Salts have been formed from both short- and long-chain electron donors and acceptors or from a combination of long- and short-chain species either within one monolayer or between two monolayers. A variety of doping methods have also been utilized [722, 744]. Emphasis have been placed on the characterization of a given preparation [774, 775] and on the optimization of the lateral d.c. conductivity. Small changes in both the molecular structure and composition of the surfactants was found to affect profoundly the orientation and properties of the LB film formed. For example, appreciable structural differences have been found between iodine-doped N-docosylpyridinium TCNQ and octadecylpyridinium TCNQ LB films. Absorption and infrared spectroscopic and low-angle X-ray diffraction measurements of the latter LB film have been interpreted in terms of interdigitated, inclined multilayers (Fig. 119) [728]. In contrast, TCNQ has been proposed to orient perpendicularly to the substrate in iodine-doped, N-docosylpyridinium LB films (see above) [721].

Advantage has been taken of organometallic complexes with long-chain alkyl-ammonium or pyridinium salts to increase the dimensionality and, hence, the lateral conductance [746, 747]. Indeed, the highest reported room temperature conductivity to date (25 S cm^{-1}) was measured in bromine-doped LB films prepared from tridecylmethyl-ammonium Au-(dmit)$_2$ [741, 742].

Table 9. Conducting Langmuir–Blodgett films

Preparation	Characterization	Results	Reference
Langmuir–Blodgett (LB) films prepared from N-docosylpyridinium and TCNQ, transferred to substrates, and exposed to iodine vapor	Visible and infrared spectra and low-angle X-ray diffraction	Lateral d.c. conductivity was about 10^{-1} S cm^{-1}	726, 729
Monolayers of octadecylpyridinium$^+$-(TCNQ-1,45)$^-$ formed and compressed on a glyceryl subphase, shifted to a water subphase, and transferred to solid substrates to form LB films	Visible and infrared spectra	Lateral d.c. conductivity was 10^{-4} S cm^{-1}	730
LB films prepared from trimethylocta-decylphosphonium TCNQ, transferred to substrates, and exposed to iodine vapor	FTIR	Lateral d.c. conductivity was estimated (from energy of the maximum of the charge transfer absorption band) to be about 20–50 S cm^{-1}	731
LB films prepared from TCNQ octadecyl-dimethylsulfonium, octadecyl-methyl-ethyl-sulfonium, and octadecyl-trimethyl-phosphonium salts, transferred to substrates, and exposed to iodine vapor	Absorption, infrared spectra, and ESR spectra	Lateral d.c. conductivity was estimated (from energy of the maximum of the charge transfer absorption band) to be about 20–50 S cm^{-1}	731, 732, 733
LB films prepared from ethylene dithio-dioctadecylthio-tetrathiafulvalene TCNQ, transferred to substrates, and exposed to iodine vapor	Absorption, infrared spectra, and ESR spectra	Lateral d.c. conductivity was determined to be 5×10^{-2} S cm^{-1}	734, 735
LB films prepared from mixtures of ethylnedithiodiocta-decylthio tetrathiafulvalene with ω-tricosenoic acid (or with stearic acid), transferred to substrates, and exposed to iodine vapor	Absorption and infrared spectra and low-angle X-ray scattering	Lateral d.c. conductivity was determined to be 0.2–1 S cm^{-1}	736, 737, 738
LB films prepared from N-docosylpyridinium (TCNQ)$_2$ and transferred to substrates	Absorption and infrared spectra	Lateral d.c. conductivity was determined to be 10^{-2} S cm^{-1}, even without iodine doping	739, 740
Monolayers prepared from TMTTF octadecyl TCNQ and TTF-octadecyl TCNQ on glycerin subphase	Surface-pressure/surface-area measurements and in situ conductivity	Lateral d.c. conductivity was of the order of 1 S cm^{-1}	741

Table 9. Contd.

Preparation	Characterization	Results	Reference
LB films prepared from a mixture of TMTTF and octadecyl-TCNQ acceptor and transferred to solid substrates	Absorption and infrared spectroscopy	Lateral d.c. conductivity = 0.1 S cm^{-1}; conductivity normal to the film = 10^{-14} S cm^{-1}	742
LB films prepared from an azabenzene-containing surfactant donor-TCNQ acceptor and transferred to solid substrates	Absorption spectra, polarized microscopy, and conductivity measurements	Sharp discontinuity in temperature-dependent lateral d.c. conductivity was observed	743, 744, 745
LB films prepared from tridecylmethyl-ammonium Au-(dmit)$_2$ and H$_2$dmit = 4,5-dimercapto-1,3-dithiol-2-thione, transferred to hydrophobized glass substrates, and oxidized (by Br$_2$ or electrochemically)	Absorption spectra and temperature-dependent conductivity measurements	Lateral d.c. conductivity increased with increasing temperature; at room temperature, lateral d.c. conductivity = 25 S·cm^{-1}	746, 747
LB films prepared from polyimide	Absorption spectra and conductivity	Polyimide film pyrolyzed (to graphite-like polymer) at 100°C in vacuum had lateral d.c. conductivity up to 600 S cm^{-1}	748, 749, 750
LB films prepared from a 3:2 charge-transfer complex of TMTTF and TCNQ and transferred to cadmium-arachidate (five layers) precoated sheets of poly-ethylene terephthalate	ESR	Preferential in plane orientation of the charge transfer complex and rearrangement of substrate on aging were established	751
LB films prepared from N-octadecyl pyridinium TCNQ and transferred to glass or calcium fluoride substrates	Absorption and infrared spectra	Lateral d.c. conductivity at room temperature = (2.0 ± 0.2) 10^{-2} S cm^{-1}, activation energy = 0.13 ± 0.005 eV (in the 100–300 K range)	752
LB films prepared from 1-tetrathiafulvalenyl-1-o1 TCNQ and transferred to glass or calcium fluoride substrates	Absorption and infrared spectra and cyclic voltammetry	Lateral d.c. conductivity at room temperature = 10^{-3} S cm^{-1}; activation energy = 0.27 eV (in the 80–320 K range)	753
LB films prepared from alternating layers of long-chain TCNQ and long-chain TTF and transferred to glass substrates	Absorption spectra and conductivity measurements	Lateral d.c. conductivity at room temperature = (5 ± 1) 10^{-3} S cm^{-1}; activation energy = 0.26 ± 0.01 eV (in the 100–300 K range)	754

LB films prepared from O-hexadecylthio-carboxy-tetrafulvalene, transferred to substrates, and exposed to iodine	Absorption, infrared, and photo-electron spectroscopy and low-angle X-ray diffraction	Lateral d.c. conductivity after iodine doping = 1.0 ± 0.2 S cm^{-1}	755, 756
LB films prepared from (N-octadecylpyridinium)$_2$-Ni(dmit)$_2${H$_2$dmit = 4,5-dimercapto-1,3-dithiol-2-thione}, transferred to glass substrates, and exposed to iodine	Absorption and infrared spectroscopy and electrical measurements	Room temperature lateral d.c. conductivity prior and subsequent to iodine doping = 6×10^{-6} S cm^{-1} and 0.2–0.8 S cm^{-1}; films remained stable for at least one month	757, 758
Monolayers of electrochemically prepared poly(3-dodecylthiophene) transferred successively to substrates, by horizontal lifting to give LB films, and exposed to iodine	Absorption spectroscopy, quartz crystal microbalance and electrical measurements and femtosecond degenerate four-wave mixing studies of third-order optical non-linearity	Lateral d.c. conductivity of iodine-doped 1-, 20-, and 40-layer films was determined to be 0.014, 0.20, and 0.51 S cm^{-1}, respectively	759
LB films prepared from poly(p-phenylene vinylene), transferred to substrates, and doped with SO$_3$	Absorption and infrared spectroscopy and electrical measurements	Lateral and through d.c. conductivity of doped, 300-layer films were determined to be 0.5 S cm^{-1} and 4×10^{-6} S cm^{-1}	760
LB films prepared from 2-trifluoromethyl-5,6;11,12-bisepidithiotetraacetate, transferred to substrates, and exposed to iodine	Absorption spectra and conductivity measurements	Lateral d.c. conductivity of iodine-doped film was 1.2×10^{-4} S cm^{-1}	761
LB films prepared based on a copper(II) dibenzotetra-aza [14] annulene derivative, transferred to substrates, and exposed to iodine	Absorption and infrared spectra and X-ray diffraction	Lateral d.c. conductivity of iodine-doped film was about 1×10^{-4} S cm^{-1}	762
LB films prepared from a mixture of 2-n-octyloxy-5,6,11,12-tetrathiotetracene and methyl arachidate (mol ratio = 2.3:1), transferred to substrates, and doped by iodine	Absorption spectra, conductivity, and X-ray diffraction measurements	Method of doping (exposure of LB film to iodine vapor, immersion of LB films in aqueous KI$_3$ solution, or anodic oxidation of LB films deposited on indium tin oxide glass) determined properties of the films. Lateral d.c. conductivity varied between 10^{-4} and 10^{-2} S cm^{-1}	763
LB films prepared from iron(III) stearate, transferred to glass substrates, dried, exposed to hydrochloric acid in a desiccator, and subsequently exposed to pyrrole vapor at 0.3–0.4 Torr	Absorption and infrared spectra, X-ray diffraction, and conductivity measurements	Lateral d.c. conductivities fell within the range of 0.1–2 S cm^{-1} although values as high as 10 S cm^{-1} have been obtained	764

Table 9. Contd.

Preparation	Characterization	Results	Reference
LB films prepared from hexadecyl-TCNQ TMTTF and (heptadecyl-dimethyltetrathia-fulvalene)$_2$ TCNQ and from their mixtures	Electron microscopy and electron diffraction	Molecular packing determined conductivities; best lateral d.c. conductivity was 0.5 S cm^{-1}	765
LB films prepared from mixtures of donors (heptadecyldimethyltetrathiafulvalene, hexadecylbis(ethylenedithio)-tetrathiafulvalene, and hexadecylethylene-dithiopropylenedithiotetra-thiafulvalene) and acceptors (hexadecyl-TCNQ and heptadecyloxycarbonyl-TCNQ)	Electron microscopy, electron diffraction, and conductivity measurements	Molecular packing and conductivities were examined	766
LB films prepared from metal complex salts $(R^+)_2[Ni(dmit)_2]^{2-}$, $(R^+)_2-[Ni(mnt)_2]^{2-}$, and $R^+[Ni(dmit)_2]^-$ (where $R = (CH_3)_2 (C_{12}H_{25})_2N^+$), transferred to substrates, and exposed to bromine vapor	Surface-potential and conductivity measurements	Lateral d.c. conductivities of bromide exposed films were in the 0.001 to 0.28 S cm^{-1} range	767
LB films prepared from poly(thiophene-3-acetic acid)-stearylamine and from sulfonatedpolyaniline-stearylaminepolyion complexes, transferred to substrates, and doped by acid	FTIR, X-ray diffraction, and conductivity measurements	Lateral d.c. conductivities of acid-doped films as high as 0.05 S cm^{-1} were obtained	768, 769
LB films prepared from mixtures of polyalanine and stearic acid (or perfluoro-octanoic acid), transferred to substrates, and doped with acid	FTIR and conductivity measurements	Lateral d.c. conductivities of acid-doped films as high as 1 S cm^{-1} were obtained	769
Pyrrole polymerized at the interface of non-polymerizable 3-octadecanoyl pyrrole and transferred to substrates to form LB films	FTIR and conductivity measurements	Lateral d.c. conductivities as high as 0.1 S cm^{-1} were obtained	769, 770, 771
LB films prepared from poly-thiophene-3-acetic acid stearylamine or sulfonated polyaniline stearylamine polyion complexes, transferred to substrates, and doped by SbCl$_5$	Absorption and infrared spectra, X-ray diffraction, and conductivity measurements	Lateral d.c. conductivities of doped films were moisture dependent and were as high as 2 S cm^{-1}; molecular organization consisted of randomly oriented polymers lying as extended chains parallel between the substrates and sandwiched between layers of stearylamine molecules whose chains were interdigitated	772

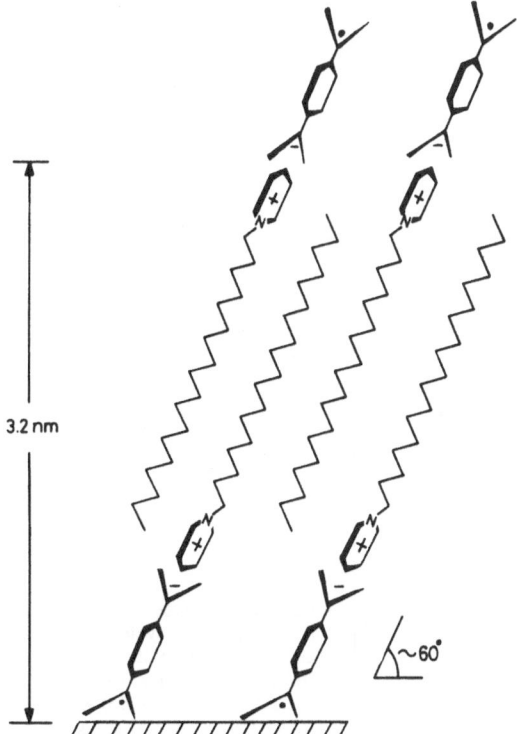

3.2 nm

~60°

Fig. 119. Schematic diagram showing the proposed average orientation in LB film form [457]

Introduction of an azabenzene unit into the surfactant backbone permitted light-induced *cis-trans* rearrangements. A change in conductivity accompanied the *cis*-to-*trans* transition. Thus, these LB films offer a basis for the construction of molecular switching devices (Fig. 120) [743–745].

Utilization of polyelectrolytes in conjunction with in situ pyrrole polymerization has provided a new approach to the formation of conducting LB films [768–772]. The method is illustrated by the interfacial polymerization of pyrrole under monolayers, formed from non-polymerizable 3-octadecanoyl pyrrole (Fig. 121) and by the sequential exposure of LB films, prepared from iron(III) stearate, to HCl gas and pyrrole vapor (Fig. 122) [772]. Spectroscopic and X-ray diffraction measurements established that the LB films consisted of interdigitated stearylamine monolayers and that the polymers were oriented between the monolayer headgroups parallel to the substrate (Fig. 123a) [772]. Heterostructured LB films were also prepared by the transfer of alternate monolayers of polythiophene and polyaniline polyion complexes to the substrate (Fig. 123b) [772]. Importantly, the lateral d.c. conductivities of moist, heterostructured LB films (up to 0.2 S cm⁻¹) were found to be consistently higher than those of the corresponding moist, homopolymeric LB films [772]. Electrical characterization of these conducting LB films was carried out by transferring them to metal-coated glass slides and then depositing a suitable electrode on their surfaces by

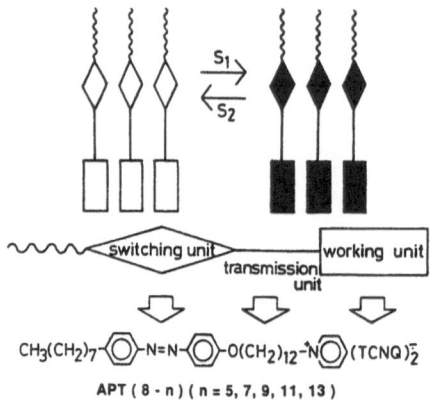

APT (8 - n) (n = 5, 7, 9, 11, 13)

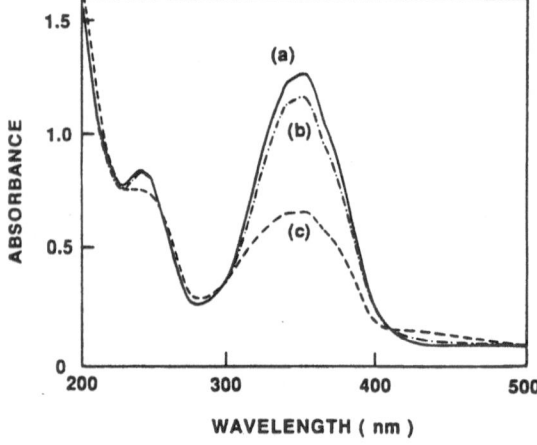

Fig. 120a. Schematic view of the photochemical switching device and the compounds used in this study. **b** Absorption spectra of the LB films of APT(8–13) irradiated with UV (365 nm) and visible (436 nm) light; before irradiation (a) and after irradiation with visible light (b) and UV light (c) [740]

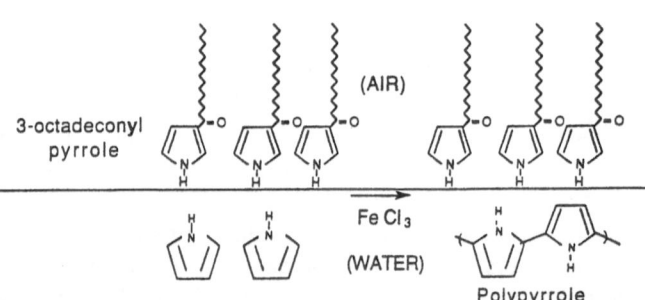

Fig. 121. Polymerization of pyrrole monomer at the air-water interface [764]

r.f. sputtering or vacuum evaporation (see Fig. 124) [771]. Current transport and capacitance measurements indicated the presence of thin, partially blocked Schottky barriers due to low barrier heights and, hence, the tunneling of charge carriers through the depletion layer. Unfortunately, experimental difficulties in

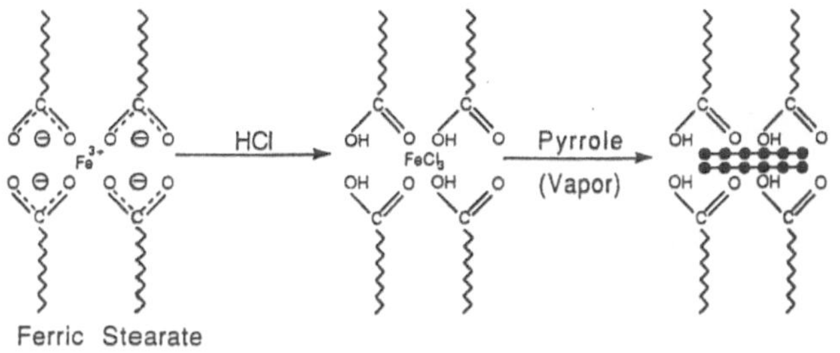

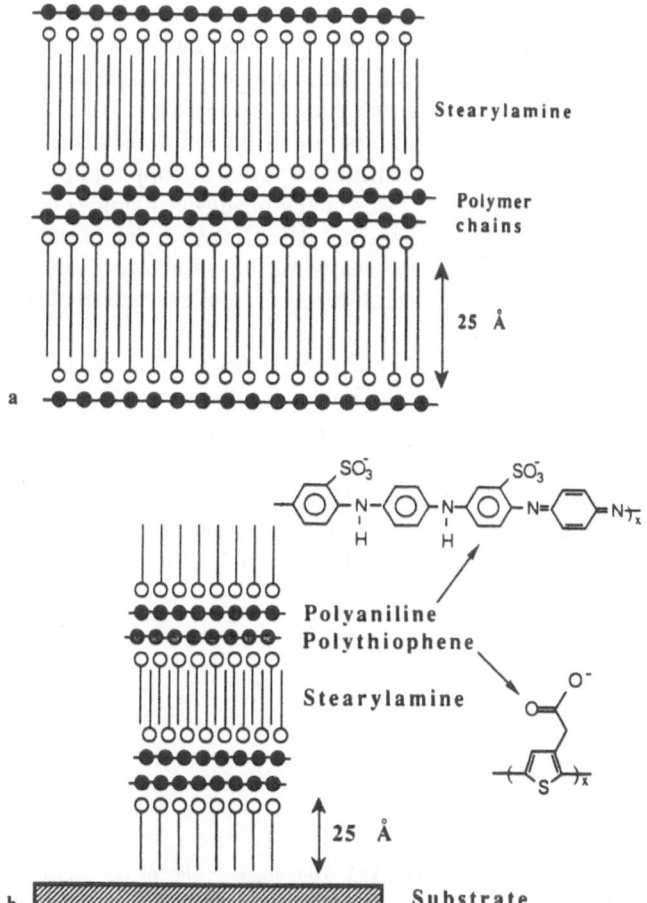

Fig. 122. Polymerization of pyrrole monomer in a preformed ferric stearate multilayer film (*black dots* represent electrically conducting polypyrrole) [764]

Fig. 123a. Simplified structure of the polyion complex LB films showing ionically bound mono-layers of conjugated polymer sandwiched between stearylamine spacer groups with interdigitated hydrocarbon tails. **b** Simplified organization of the heterostructure LB films showing alternating layers of 1:1 PTAA-StNH$_2$ and SPAn-StNH$_2$ [767]

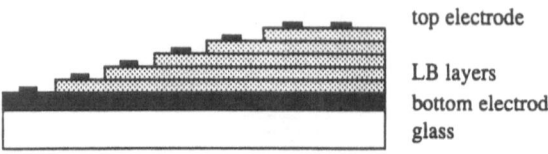

top electrode

LB layers
bottom electrode
glass

Fig. 124. A schematic sample configuration [771]

the deposition of the top electrode precluded the examination of samples which had fewer than nine monolayers [771].

Progress has also been made towards the construction of LB-film-based transistors [776] and photoelectron converting and storage devices [777]. Thus, a transistor, having finger-shaped Cr/Au drain and source electrodes, has been constructed from 13 layers of oligomeric, quinquethiophene-containing arachidic acid as the semiconductor, 30 layers of arachidic acid as the gate insulator, and 30 layers of poly (3-hexylthiophene)-containing arachidic acid as the gate (Fig. 125) [776]. Efficient photoelectron transfer and charge separation have also been recently achieved in an ultrathin heterostructured LB film formed from two layers of steroidal TCNQ (36 Å) as the acceptor, two layers of copper phthalocyanine (28 Å) as the sensitizer, two layers of steroidal p-phenylenediamine (48 Å) as the donor, and 30 layers of polyisobutylmethacrylate (330 Å) as the insulator. Deposited on ITO-sputtered glass plates and biased through a mercury contact (Fig. 126), lifetimes as long as 10 min were observed for the charge-separated species in this device [777].

Several laboratories pursue various mimetic approaches to superconductor formation. Formation of superconducting hollow and solid $YBa_2Cu_3O_x$ illustrates a published strategy [778]. Polypropylene hollow fibers were filled with a Polymer-Metal-Solution (consisting of 4.4 g of pyromellitic anhydride, 3.6 g of diaminodiphenylether, and 60 g of DMF containing 2.5 mmol of $Y(NO_3)_3 \cdot 5H_2O$, 5 mmol $Ba(NO_3)_2$, and 7.5 mmol $Cu(NO_3)_2 \cdot 2.5H_2O$ in DMF/H_2O). The solution was maintained under 1.9 bar N_2, heated to 80 °C

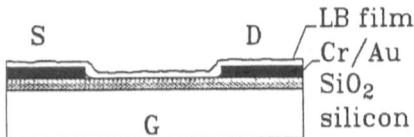

S   D   ⌐LB film
                           ⌐Cr/Au
                            SiO$_2$
      G   silicon

Fig. 125. *Top:* cross section of the transistor structure. The drain and source electrodes make ohmic contacts with the LB film. The carrier concentration and, therefore, the conductance of the LB film is controlled by the gate. *Bottom:* schematic layout of the finger shaped drain and source electrodes [776]

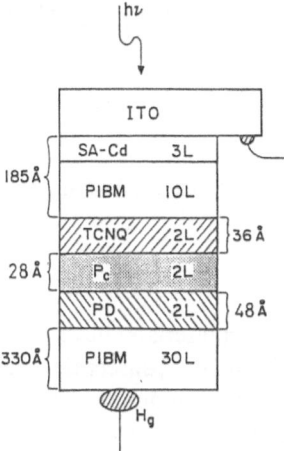

Fig. 126. Molecular structures and an A/S/D hetero LB film device structure [777]

under vacuum, heated further under a flow of N_2 (5 °C/min), and held at 160 °C for 2 h, 300 °C for 2 h, and 600 °C for 2 h. This process resulted in the formation of intermetallic compounds. Hollow fibers of $YBa_2Cu_3O_x$ were finally obtained by calcination under a flow of O_2 at 300 °C for 2 h, 600 °C for 2 h, 900 or 930 °C for 2 h (Fig. 127) [778]. Measurements of a.c. susceptibility indicated a high temperature (> 80 K) superconducting behavior for the $YBa_2Cu_3O_x$ hollow fiber [778]. Vigorous and sustained research activity in this area is fully expected.

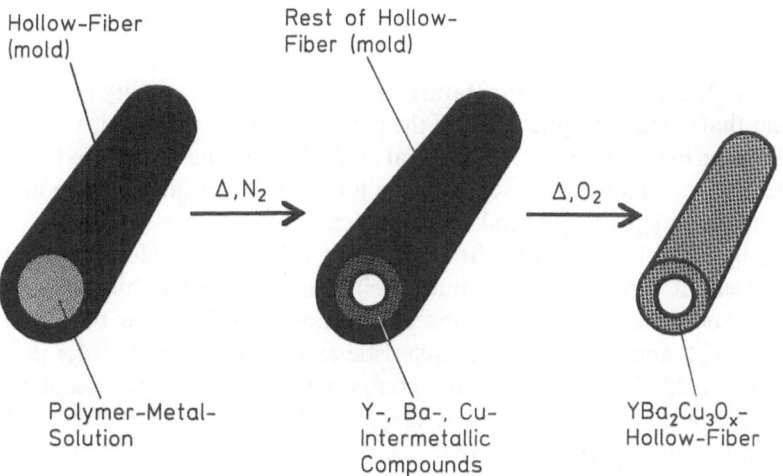

Fig. 127. Schematic representation of the processes taking place during the preparation of the ceramic hollow fibers [778]

6 Magnetism, Magnetic Particles, and Magnetic Particulate Films in Membrane-Mimetic Compartments

Magnetism is an intrinsic property of all materials. When placed in a magnetic field, H, a substance develops an amount of magnetization, M (magnetic moment per unit volume), given by:

$$M = \chi H \tag{36}$$

where χ is the magnetic susceptibility. Diamagnetic materials have their electrons in paired states and exhibit weak negative (about 10^{-6}) susceptibilities (they are repelled out of a magnetic field). Most inorganic compounds, with the exception of those containing transition metal ions, and most organic compounds, except free radicals and some charge transfer complexes, are diamagnetic. Paramagnetic materials have unpaired electrons and small positive (about 10^{-4}) susceptibilities (they are attracted into a magnetic field). Paramagnetism primarily occurs in substances whose atoms or molecules possess a permanent magnetic dipole moment. Complex transition metal salts and odd electron molecules, like NO_2 and O_2, are paramagnetic.

Although for most metals χ is independent of the temperature, for many substances it is inversely proportional to the absolute temperature, T, as expressed by Curie's law:

$$\chi = C/T \tag{37}$$

where C is the Curie constant. A more accurate expression is the Curie–Weiss law which uses a characteristic temperature θ:

$$\chi = C/(T - \theta) \tag{38}$$

It is implied in Eq. (38) that, at temperature θ, the value of susceptibility becomes infinite and that there is a finite value of the magnetic moment per unit volume (see Eq. 36), even in the absence of an external field (H = 0) at temperatures lower than θ. Below θ (a characteristic temperature for a given substance), atom and dipoles (spins) of magnetic materials are ordered.

Ferromagnetic behavior derives from the parallel alignment of electron spins, manifests itself in large positive (about 10^2) magnetic susceptibilities in a macroscopic spontaneous magnetization at zero applied field below the Curie temperature (T_C), and displays a characteristic saturation moment (M_S) in a finite applied field. Ferromagnets are composed of microscopic magnetic domains which are each spontaneously magnetized almost to saturation, but which are mutually oriented such that the net moment of the macroscopic material is zero. Application of an external field converts the microscopic ordering to a macroscopic one by alignment of the domains. With increasing temperature, the thermal energy becomes comparable to the exchange energy.

Spontaneous magnetization of ferromagnets, therefore, decreases with increasing temperature and it disappears at temperatures above T_C. Above T_C, ferromagnets become paramagnets and obey the Curie–Weiss law. Metals such as iron, cobalt, and nickel and their alloys, and γ-Fe_2O_3 are examples of ferromagnetic materials.

In antiferromagnets, the neighboring spins are aligned antiparallel, there is no net macroscopic moment in zero applied field, and the susceptibility is anisotropic below the Néel temperature, T_N. Most antiferromagnets are insulating solids (NiO and MnO, for example).

Materials in which adjacent spins are both unequal in magnitude and antiparallel in alignment are said to be ferrimagnetic. Compounds falling into this category are typically composed of two chemically different species which have two different sublattices ($FeCr_2O_4$, for example). The temperature-dependent behavior of paramagnetic, ferromagnetic, antiferromagnetic, and ferrimagnetic solids with unpaired spins (atomic dipoles) is shown in Fig. 128 [779]. Application of a magnetic field to a ferromagnetic results in the alignment of the

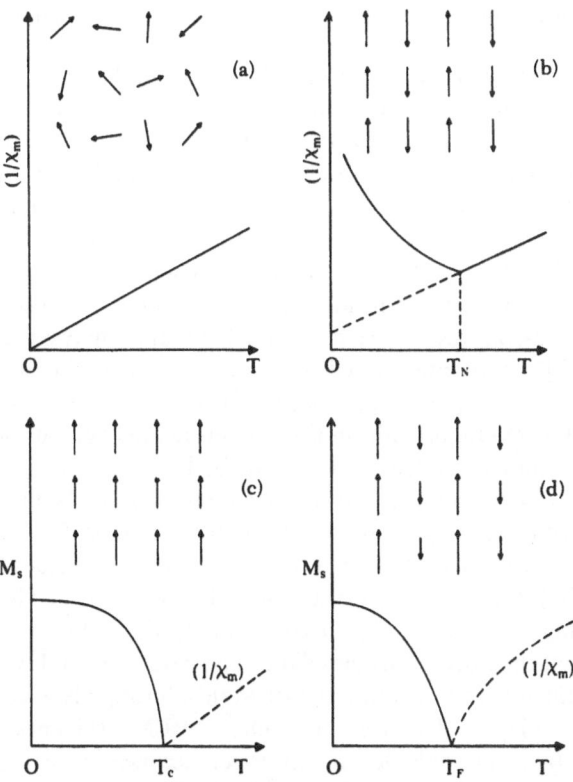

Fig. 128a–d. The low temperature ordering (if any) of neighboring dipoles and the consequent behavior of spontaneous magnetization and/or susceptibility for: **a** paramagnetism; **b** antiferromagnetism; **c** ferromagnetism; **d** ferrimagnetism [779]

ferromagnetic domains. Metamagnetism is the field-dependent transformation of an antiferromagnet to a ferromagnet. A superparamagnet is a ferromagnet whose size is less than that of a single domain.

Investigations of magnetic behavior in membrane-mimetic systems may have important theoretical and practical ramifications. The molecular organization characteristic of these systems permits, at least in principle, systematic examination of the effects of dimensionality reduction on magnetic properties [780]. The possibility also exists to construct membrane-mimetic systems from surface-active, organic, ferromagnetic molecules [781] and/or from surface-active, magnetic charge-transfer complexes [782]. Studies of modulated phase separation in monolayers and Langmuir–Blodgett films, prepared from mixtures of two surfactants, may provide insight into the behavior of magnetic domains [783]. A variety of membrane-mimetic compartments can also be used to stabilize magnetic particles and particulate films [784–798]; the confinement offered by these systems can also be exploited to influence magnetic effects on chemical processes [799–801]. In view of the obvious significance of these research areas, the paucity of published work is truly surprising. It is hoped that the ensuing survey will stimulate increased activity in the membrane-mimetic approach to the preparation, characterization, and utilization of magnetic materials.

Theory predicts the possibility of magnetic ordering in three-dimensional solids and the impossibility of such ordering in one-dimensional solids [802]. Prediction is, however, model-dependent for two-dimensional solids. Two-dimensional magnetic ordering is only compatible with the Ising model, which considers the magnetic spin interactions to be constrained in a single direction. Conversely, interaction of the magnetic spins is allowed in any direction in the Heisenberg model and in the plane in the x–y model. Magnetic ordering of manganese ions between the hydrophilic headgroups of LB films has been demonstrated [803]. Electron-spin-resonance field and lineshape analyses of manganese LB films at 2 K were interpreted in terms of predominantly antiferromagnetic behavior combined with a weak ferromagnetic contribution [804, 805]. Additional work is required, both on this and on related systems, if dimensionality reduction in mimetic systems is to be realized.

Magnetic particles have been confined to reversed micelles, surfactant vesicles, polymers, and porous vycor glasses (see Table 10). Cationic Fe_3O_4 particles have been shown to be strongly attracted onto one or both surfaces of glyceryl monooleate (GMO) bilayer lipid membranes (BLMs) [795]. The absence of capacitance changes across the GMO BLM during deposition has been interpreted to imply that the Fe_3O_4 particles did not penetrate beyond the headgroup region of the surfactants constituting the BLM (headgroup distance, $d_p = 6$–8 Å; hydrocarbon bilayer distance, $d_h = 48$ Å; BLM thickness, $d_b = 2d_p + d_h = 62 \pm 2$ Å). Incident-angle-dependent reflectance measurements led to a model for the Fe_3O_4-particle-coated GMO BLM with the following parameters: refractive index of the magnetic particles, $n_m = 1.96$; thickness of the magnetic particles on the BLM, $D_m = 55.1$ Å ($D_m = d_m + d_p$); center-to-center

Table 10. Magnetic particles and particulate films in membrane-mimetic compartments

Membrane-mimetic compartment	Incorporated particle(s)	Comments	Reference
AOT-isooctane (or cyclohexane)-H_2O and hexaoxyethylene-isooctane (or cyclohexane)-H_2O reversed micelles	Magnetite particles prepared in situ in reversed micelles from $FeCl_2/FeCl_3$ and NH_3	Coagulation of reversed-micelle-containing magnetite by acetone washing and drying led to magnetite particles which were characterized by X-ray diffraction and Mössbauer spectroscopy	784
Nematic liquid crystals prepared from 35.9% sodium dodecyl sulfate, 7.2% decanol, and 56.9% water (H_2O or D_2O)	Ferrofluid (1%) dispersed in the liquid crystal	Stable nematic II liquid crystalline ferrofluid was produced; diameter = 154 ± 9.4 Å; number of grains/cm^3 = 1.9 × 10^4, magnetization at saturation = 2.20 g/cm^3	785, 786
Single-bilayer phosphatidylcholine (14) surfactant vesicles	Magnetite prepared from equimolar Fe^{2+}/Fe^{3+} and NaOH and sonicated with the lipid	X-ray analysis established the presence of magnetite in the vesicles	787, 788
Single-bilayer dioctadecyldimethylammonium chloride (DODAC) (6) vesicles	Sonication of surfactant-stabilized magnetite with DODAC	Vesicle-incorporated magnetite influenced the outcome of benzophenone photolysis	789
Single-bilayer dihexadecyl phosphate (DHP) vesicles	Magnetite particles prepared in situ in vesicles, from Fe^{2+}/Fe^{3+} and OH^-	Particles were characterized by magnetic birefringence	790
Single-bilayer phosphatidylcholine (14) vesicles	Magnetic particles prepared in situ from Fe^{2+}/Fe^{3+} by OH^-	Particles were characterized by transmission electron microscopy, electron diffraction, and X-ray microanalysis; morphologies of intravesicular particles (spherical or disk-shaped) differed from those precipitated in bulk (acicular)	791
Vesicles prepared from phosphatidylglyceryl	Nanometer-sized, sodium-laurate-coated Fe_3O_4 particles	In the presence of vesicles, lipids adopted a bilayer configuration around the iron oxide core	792

Table 10. Contd.

Membrane-mimetic compartment	Incorporated particle(s)	Comments	Reference
Single-bilayer vesicles prepared from dimyristoylphosphatidylglyceryl, dipentadecanoylphophatidylglyceryl, and distearoylphosphatidylglyceryl	Nanometer-sized, sodium-laurate-coated Fe_3O_4 particles	Magnetic-particle-containing liposomes were used in lipid exchange studies	793
Single-bilayer DODAC (6) vesicles	Magnetite particles prepared in situ in vesicles from Fe^{2+}/Fe^{3+} and OH^-	Static and dynamic polarized and depolarized light scattering and static and time-resolved dichroic anisotropy, as well as conventional magnetization vs applied magnetic field determinations were used for characterization	794
Glyceryl monooleate (22) bilayer lipid membrane	Nanometer-sized, single-domain, Fe_3O_4 particles attached to BLMs	Electrical measurements and reflection spectroscopy were used for characterization	795
Nafion membrane	Size-quantized Fe_2O_3 particles prepared in situ	Photoelectron transfer was examined	672
Langmuir-Blodgett films prepared from arachidic acid	Nanosized, cationic Fe_2O_4 sandwiched between polar headgroups in LB films	Domain structures were recognized in magnetic particles	796
Cast multibilayers prepared from N-(ω-(trimethyl-ammonio)undecanoyl) dioctadecyl-L-glutamate bromide dispersions	Fe_3O_4 particles generated in situ	Magnetic anisotropy of particles was retained in cast multibilayer films	797
Porous vycor glass	Surface-bound, octahedrally coordinated Fe^{3+} species obtained in the photolysis of $Fe(CO)_5$ physisorbed onto the glass	Magnetically ordered materials were obtained which exhibited magnetic hyperfine fields of 370 and 425 kG	798

distance between the magnetic particles, $S_m = 57.6\,\text{Å}$ (Fig. 129) [795]. Fe_3O_4 particles were attracted so strongly to the BLM that they could not be removed by a magnet even as strong as 400 Oe. This strong attraction overcome the electrostatic repulsion between neighboring particles and permitted coverage of the BLM to the extent that a monolayer of particulate film developed. Formation of a second layer of particulate film was proposed to be precluded by particle-particle repulsions that prevailed in the absence of the attractive forces of the BLM. Magnetic domains on BLMs have been visualized by polarized videomicroscopy of the Kerr rotation (Fig. 130) [795].

Magnetic particles have also been in situ generated in cast multibilayers [797]. This method involved the sonic dispersal of a chiral long chain glutamate (20 mM) in water, the addition of aqueous 10 mM $FeCl_2$ or $FeCl_3$ (4:5 = iron chloride:surfactant mol ratio), further sonication, spreading the vesicles on fluorocarbon sheets and allowing them to stand at 25 °C in 60% relative humidity for two days, and, finally immersing the cast film into an aqueous NaOH (pH about 12) solution. The presence of magnetite (Fe_3O_4) in the cast film was established by electron diffraction. The determined saturation isothermal (0.71 emu/g) and remnant (0.11 emu/g) magnetization and the coercivity (20 Oe), as well as the observed magnetic anisotropy of the film prepared from $FeCl_2$ (Fig. 131) indicated ferromagnetic behavior [797]. In contrast, the cast bilayer prepared from $FeCl_3$ was found to be paramagnetic and isotropic [797].

Cationic magnetic Fe_3O_4 particles have been sandwiched between the polar headgroups of arachidate-ion monolayers deposited on oxidized silicon substrates [796]. Optical thicknesses of the SiO_2 layer on the substrate, the arachidate ion monolayers (A^-), and the Fe_3O_4 particles thereon have been characterized by incident-angle-dependent reflectance measurements, assuming that these components possess stratified planar structures. The determined

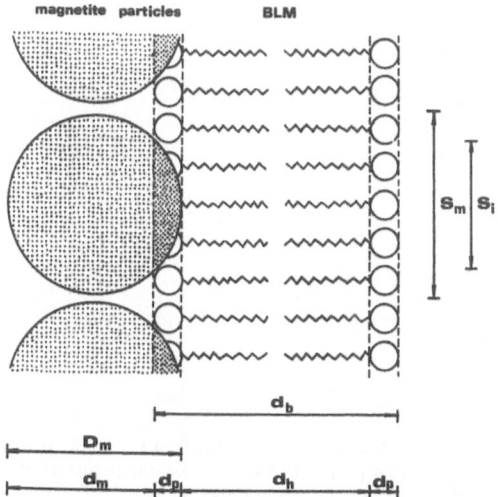

Fig. 129. Schematics of the particulate thin Fe_3O_4 on one side of the GMO BLM [794]

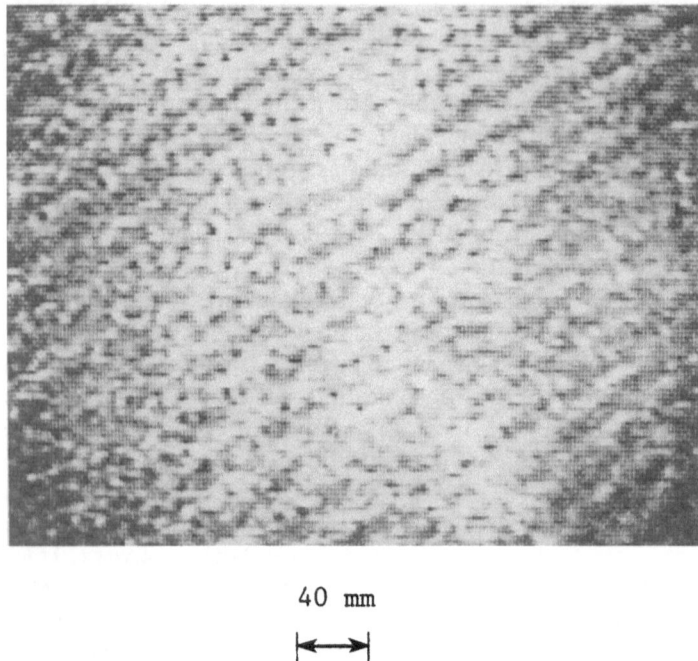

40 mm

|←——→|

Fig. 130. Maze domain pattern of the particulate Fe_3O_4 film deposited on one side of a GMO BLM [795]

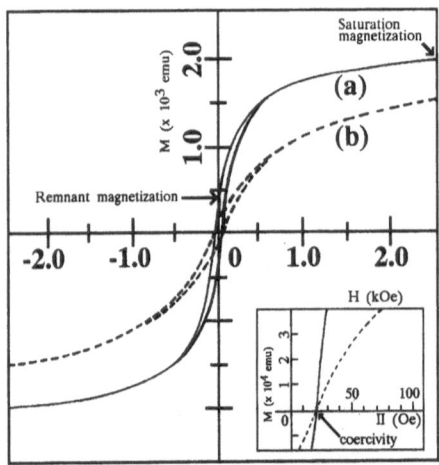

Fig. 131. VSM measurement of a Fe(II) cast film. The sample size is $3\,mm \times 9\,mm \times 0.1\,mm$. (a) Magnetic field is parallel to the film plane. (b) Magnetic field is perpendicular to the film plane [797]

values for optical thicknesses (d_j) and reflective indexes (n_j), $d_{SiO_2} = 91.5, 136.2,$ and 172.5 Å; $n_{SiO_2} = 1.46$; $d_{A^-} = 26.7$ Å; $n_{A^-} = 1.52$; $d_{Fe_3O_4} = 49.8$ Å; and $n_{Fe_3O_4} = 1.976$ (in water, system A), agreed well with values predicted by the models used. Similarly, incident-angle-dependent reflectivity measurements of seven successive units of arachidate-ion-sandwiched Fe_3O_4 particles on LB films led to a determined average thickness of 89.2 Å for a single-sandwich unit (Fig. 132) [796]. Fe_3O_4 particles were found to display intralayer and interlayer separations of 83.2 Å and 96.2 Å, respectively. These values indicated that 35% of the substrate had been covered by Fe_3O_4 particles. Transmission electron micrographs of arachidate-ion-sandwiched Fe_3O_4 particles on a cellulose-coated copper grid showed an average center-to-center separation of the particles of 91 ± 10 Å and a surface coverage of $30 \pm 15\%$. Longitudinal magneto-optical effects for a 12-sandwich unit revealed rod-shaped magnetic domain structures.

The organizational advantage of membrane structures is best illustrated by the demonstration of magnetic effects on a chemical reaction, even in the absence of an externally applied magnetic field [789]. The underlying principle of these effects is described by the chemically induced dynamic nuclear polarization model [799–801]. This theory suggests that the chemical reactivity of radical pairs will depend on the hyperfine interactions between the orbitally uncoupled electrons and the magnetic field. Photoreduction of benzophenone $[(C_6H_5)_2C=O(BP)]$, to diphenylcarbinol $[(C_6H_5)_2CHOH]$ and to minor light-absorbing products provides a good example for this phenomenon. Product formation has been explained in terms of the decay of the photolytically

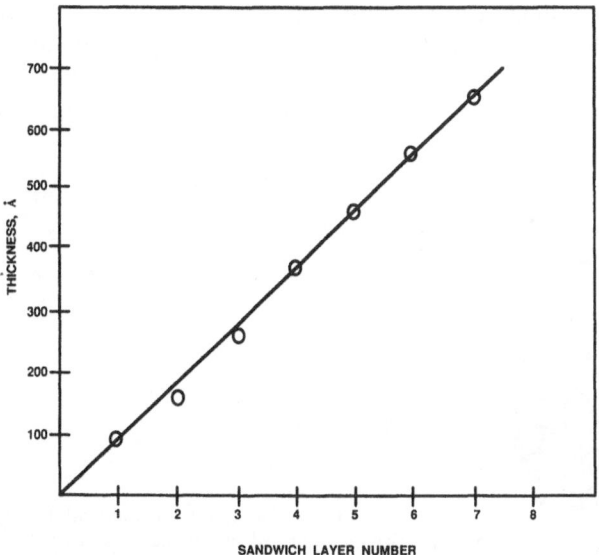

Fig. 132. Determined thickness of the LB films vs the number of sandwich units of arachidate ion-coated Fe_3O_4 particles [796]

generated benzophenone triplets ($^3BP^*$) in the presence of a hydrogen donor (RH) to a caged triplet radical pair, $^3[B\dot{P}H + \dot{R}]$. The triplet radical pair may then either undergo intersystem crossing to a caged singlet radical pair $^1[B\dot{P}H + \dot{R}]$ state or dissociate into $B\dot{P}H$ and $\dot{R}$ free radicals:

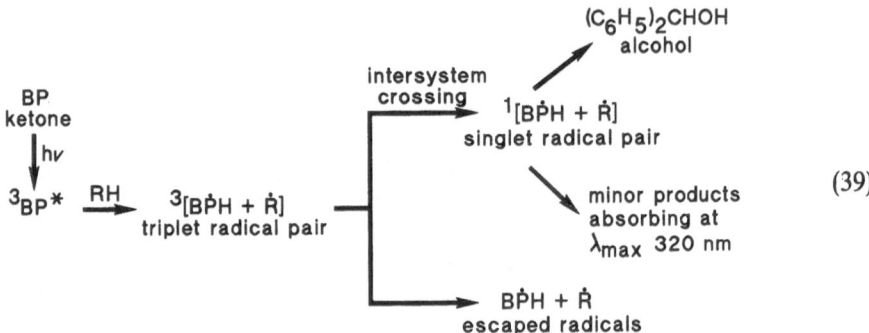

$$(39)$$

Incorporation of magnetic particles (Fe_3O_4) and benzophenone into vesicles prepared from DODAC has an effect on photolysis that is identical with that for an externally applied magnetic field of 2000 G. The percentage of triplet radical states that dissociate to form escaped radicals is increased from 56% to 87% [789]. This dramatic alteration of benzophenone photochemistry originates in the localization of benzophenone within the influence of single-domain magnetic particles; both are confined within the matrices of surfactant vesicles (Fig. 133). Magnetic particles which are in close proximity effectively split the triplet sublevels of the radical pair and reduce the rate of intersystem crossing from triplet levels. This, in turn, enhances the efficiency of radical escape as noted by the decreased production of light-adsorbing products. Thus, the observed effect of the photochemistry of benzophenone is a direct result of the magnetic moment

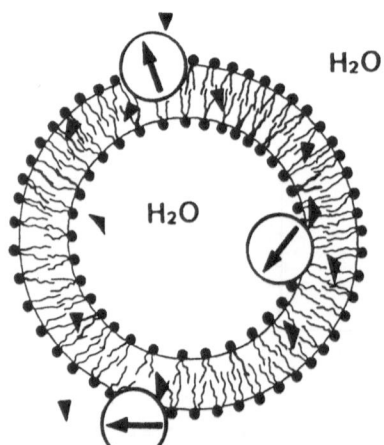

Fig. 133. Artist's conception of colloidal Fe_3O_4 particles incorporated in dioctadecyl-dimethylammonium cation vesicles containing benzophenone (▼). Positions and dimensions of Fe_3O_4 particles need not be taken literally [789]

in the vicinity of the magnetic particles. Investigations of magnetic behavior in membrane-mimetic systems are expected to test existing theories and to lead to innovative devices.

7 Advanced Ceramics

Ceramics are made by firing non-metallic inorganic solids at a high temperature. The term "ceramic", originating from the Greek *keramikos*, means potter's earth. Indeed, the birth of our civilization was heralded by the appearance of pottery. In the broader sense, traditional ceramics include such diverse materials as concrete and cement, in addition to fired clay products such as porcelain, pottery, tiles, and bricks. Advanced ceramics materials have been exploited in high-temperature light-weight engines and high-temperature nuclear reactors; as heat shields in satellites; as insulators, semiconductors, superconductors, and optical and electro-optical components; and as ultrafine fibers, artificial bones, and membranes. Their selection for these applications derives from such characteristic properties as hardness and lightness, as well as their resistance to wear, corrosion, and abrasive chemicals, even at high temperatures (1500 °C or higher). However, they are also associated with such disadvantages as brittleness and low tensile strength. These problems have been overcome by embedding ceramic particles in a secondary matrix and thereby creating, for example, ceramic-metal (cermet) and ceramic-polymer (ceramer) composites [806]. Traditional ceramics are derived from naturally occurring clay minerals, silicates and oxides. In contrast, advanced ceramics are obtained synthetically from chemicals or from highly refined, naturally occurring materials [806–811].

Conventional routes to ceramics involve precipitation from solution, drying, size reduction by milling, and fusion. The availability of well-defined monodispersed particles in desired sizes is an essential requirement for the formation of advanced ceramics. The relationship between the density of ceramic materials and the sizes and packing of their parent particles has been examined theoretically and modeled experimentally [810]. Colloid and surface chemical methodologies have been developed for the reproducible formation of ceramic particles [809–812]. These methodologies have included (i) controlled precipitation from homogeneous solutions; (ii) phase transformation; (iii) evaporative deposition and decomposition; and (iv) plasma- and laser-induced reactions.

Colloid chemists have greatly perfected the art of controlled precipitation to form uniform particles [812–814]. In particular, in so-called "forced hydrolysis", they have employed elevated temperatures, controlled the pH of the solution, and selected the most appropriate counterions. At optimal conditions, the rate of hydrolysis and, hence, the nucleation are controlled to such an extent that uniform growth and narrow particle-size distribution are achieved. Using this approach, monodispersed micron-sized aluminum, chromium, titanium, iron, and cobalt (hydr)oxide particles have been prepared [813–815]. Adjusting experimental conditions to control the release of anions from organic molecules (sulfide ion from thioacetamide or selenide ion from selenourea, for example) in the presence of metal-salt solutions also led to micron-sized particles with narrow size distributions [815]. Similarly, the controlled slow release of cations from organometallic complexes in the presence of hydroxide ions was shown to yield monodispersed micron-sized metal oxides. Colloidal copper, iron, cobalt, and nickel oxides have been prepared using this method [816–818]. Simultaneous or coprecipitation of two or more substances under highly controlled conditions can also lead to monodispersed complex colloidal particles.

Controlled phase transformation of preformed particles can also result in enhanced size control and increased monodispersity [812]. The sol-gel-phase transformation, known as the sol-gel process [809], is the best known example of

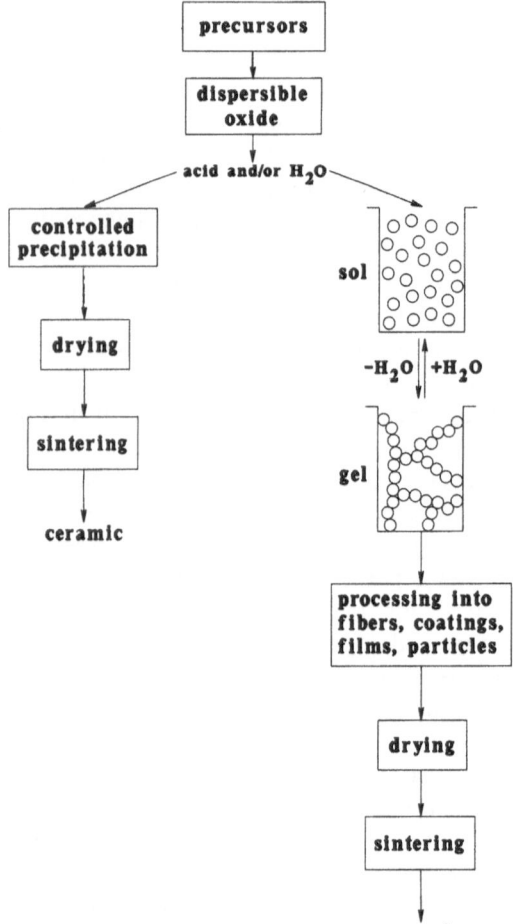

Fig. 134. Schematics of controlled precipitation and sol-gel processes to ceramic materials

this approach. The simplest sol-gel process involves (i) the conversion of a metal salt to dispersible oxide particles; (ii) the formation of a dispersed colloid (the sol) in water; (iii) the removal of water and/or anions from the sol to produce a gel in a desired morphology (i.e. the phase transformation); and (iv) drying and sintering (or calcinating) the gel to obtain the ceramic material. Simple dissolution-reprecipitation or recrystallization, under optimized conditions, have also been found to improve the monodispersities of colloidal particles [812–815]. Some of the colloid chemical routes to ceramic materials are schematically illustrated in Fig. 134.

Evaporative decomposition of solutions and spary pyrolysis have been found to be useful in the preparation of submicrometer oxide and non-oxide particles, including high temperature superconducting ceramics [819, 820]. Allowing uniform aerosol droplets (titanium ethoxide in ethanol, for example) to react with a vapor (water, for example) to produce spherical colloidal particles with controllable sizes and size distributions [821–825] is an alternative vapor phase approach. Chemical vapor deposition techniques (CVD) have also been extended to the formation of ceramic particles [825].

The membrane-mimetic approach has the potential of providing superior size, morphology, and monodispersity control for ceramic particles. The relatively meager amount of published work in this area [826–834] (see Table 11) is rather surprising. Vigorous and sustained activities, inspired by biomineralization [15–18] and modeled on the incorporation of metallic, catalytic, and semiconducting particles into membrane-mimetic compartments, are fully expected.

8 Conclusion and Prospects

It has been demonstrated in this monograph that the membrane-mimetic approach constitutes a viable route to the synthesis of advanced materials. Indeed, although still in its infancy, much has been accomplished via this approach. A wide variety of membrane-mimetic media, including monolayers, Langmuir–Blodgett and self-assembled films, reversed micelles, surfactant vesicles, polymerized vesicles, polymeric vesicles, bilayer lipid membranes, cast bilayers, and other systems, have been fruitfully employed as compartments. Nanosized metallic, catalytic, semiconducting, conducting, magnetic, and ceramic particles and particulate films have been in situ generated in these compartments using well-characterized methodologies. The application of more sophisticated techniques has facilitated closer control and has enhanced understanding of the advanced materials generated; importantly, a few nanosized electronic devices have been constructed. The rapid progress in this field has derived from research undertaken with an interdisciplinary approach.

Table 11. Ceramic materials in membrane-mimetic compartments

Membrane-mimetic compartments	Incorporated materials	Comments	Reference
Vesicles prepared from egg yolk phosphatidylcholine	An aqueous solution of 0.15 M $Cu(NO_3)_3 \cdot 3H_2O$: 0.1 M $Ba(NO_3)_2$: 0.05 M $Y(NO_3)_3 \cdot 6H_2O$: 0.0125 M $AgNO_3$ = 3:2:1:0.25 sonicated with 0.67 wt% lipid	Electron microscopy, X-ray diffraction, inductively coupled plasma emission spectra, and dynamic light-scattering measurements indicated the presence of multidomain spherical particles, containing Ag_2O, Y_2O_3, CuO, $Y(NO_3) \cdot H_2O$, $Cu(NO_3)_3 \cdot 6H_2O$, $Ba(NO_3)$, and $AgNO_3$, with mean diameters 34.8 ± 13.2 nm within the 59–99 nm diameter vesicles	826
Emulsions prepared from propylene carbonate	Alkoxides (titanium ethoxide, titanium n-propoxide, titanium n-butoxide, titanium sec-butoxide, titanium ethylhexoxide, aluminum sec-butoxide, zirconium n-propoxide, and their mixtures), typically 3 ml, sonicated with propylene carbonate, typically 17 ml, for 20–30 to form uniformly turbid emulsions. Immediate addition of water (10 mol excess) hydrolyzed the alkoxides. Precipitated oxide powders washed (in a Soxhlet extractor) with THF or 2-propanol for 21–30 h, dried, calcinated	Emulsion droplet size and composition determined particle sizes and compositions. The amorphous powders (spherical with diameters less than 1 μm) became crystalline upon calcination	827
Water-in-oil emulsions prepared from Tween, Span, Aerosol-OT, and Pluronic industrial type emulsifying agents	Yttrium nitrate dissolved in emulsion and evaporated in hot oil baths	1–2 μm unagglomerated yttrium oxide particles are obtained	828
Suspensions prepared from alginic acid, obtained from *Magrocystis pyrifera* (kelp)	High purity α-Al_2O_3 with 0.4 μm particle size	Low-viscosity, alginate-stabilized α-Al_2O_3 is prepared	829

Films cast from poly(n-butylmethacrylate) and poly(methylmethacrylate)	Methylethylketone (5 g) solutions of the polymer (10 wt%) was added to isopropanol (0.2732 g) solutions of titanium isopropoxide (0.106 g, 0.3687 mmol) and barium bis-isopropoxide (0.367 g, 0.3687 mmol); the solution was cast onto microscope slides. The polymer film was dried, removed, and hydrolyzed in boiling barium hydroxide solution	Sheets, less than 30-μm thick, of $BaTiO_3$ have been produced	830
Films cast from poly(methylmethacrylate), poly(vinylidenefluoride), poly(butylmethacrylate), and poly(vinylchloride)	Polymer and precursors ($FeCl_3$, iron(III) acetate, titanium isopropoxide, silicon tetraethoxide, and copper(II) chloride) dissolved in organic solvent (acetone, ethyl acetate) and cast onto microscopic slides; precursors in situ hydrolyzed	Miscibility of the polymer and the precursor, the rate of reaction, and precipitation kinetics determine the outcome of (hydr)oxide formation	831
Triethoxysilane-capped poly(arylene ether ketone), PEK, or poly(arylene ether sulfone), PSF	THF (5 g) mixed with 0.1 ml 10 N HCl and added slowly to 5 g of titanium isopropoxide with fast stirring. This sol was then mixed with a 10 wt% solution of triethoxysilane capped PF or PEK in THF. The final solution was cast into poly(4-methyl-pentene) Petri dishes and covered for drying and curing at 60 °C for 24 h. Films were then removed and annealed at 200 °C for 15 min	The obtained materials, ceramers (ceramic polymers), had optical dispersions between organic polymers and inorganic glasses	832, 833
Hydroxy-propyl cellulose (HPC)	Titanium tetraethoxide in ethanol is hydrolyzed in the presence of HPC	Mean particle size decreased with increasing HPC	834

Industrial exploitation of the membrane-mimetic approach to advanced materials synthesis will require much work. Superior mimetic compartments tailored for the production of a given material should be prepared by relatively the straightforward methods. Self-assembled monolayers and multilayers, formed from functionalized surfactants, appear to be promising in this respect. Sensitive size and morphology control is demanded for particles generated in the mimetic compartments. The extension of nanofabricating techniques, by scanning tunneling and atomic force microscopies, may satisfy these criteria. Increased emphasis should be placed on the preparation of composite multicomponent systems. Sandwich, size-quantized semiconductor particles should be examined in different mimetic systems. Full realization of the potential of these particles will be achieved once a fundamental understanding of size quantization and the role which mimetic compartments play in defining the chemical and physical properties of nanoparticles has been gained. Ultimately, a greater emphasis should be placed on device construction by the membrane-mimetic approach. It is sincerely hoped that the present monograph will inspire researchers to make their own contributions to this exciting and highly relevant field of study.

Acknowledgments. I am grateful to the Alexander von Humboldt Foundation for an Award, during the tenure of which this monograph was written. The gracious hospitality of Professor Arnim Henglein at the Bereich Photochemische Energieumwandlung of the Hahn-Meitner Institute (Berlin) has been especially appreciated. Thanks are also due to colleagues who have challenged and stimulated my thinking and writing. Many of them kindly sent reprints and preprints of their publications. The present version of the text has benefitted from the invaluable comments and constructive criticism of Dr. Fiona Meldrum; thanks Fee. My gratitude is also extended to Mrs. Kyunghee C. Yi who collected and checked the references which I used. An enormous amount of reformatting, editing, and proofreading is involved in a manuscript of this size. I am grateful to Mrs. Christi Ritchie for her most competent translation of my sloppy files into a respectable manuscript. Without generous support from the National Science Foundation and the U.S. Department of Energy, progress in our own laboratories would have been much slower. True credit is due, of course, to my co-workers (whose names appear in the cited publications) for their creative, skilful, and dedicated work.

9 References

1. National Research Council (1989) Materials science and engineering for the 1990s: maintaining competetiveness in the age of materials; National Academy, Washington, D.C.
2. Abbaschian R, Appleton BR, Bernstein IM, Eisenberger, PM, Langer JS, Rosenblatt, GM, Williams JC (1991) A national agenda in materials science and engineering: implementing the MS&E report; Materials Research Society, Pittsbugh, Pa
3. Seymour RB, Carraher Jr CE (1992) Polymer chemistry; Dekker, New York
4. Takagi T (1989) The concept of intelligent materials and the guidelines on R&D promotion; Science and Technology Agency, Government of Japan, Tokyo
5. Whitesides GM, Mathias JP, Seto CT (1991) Science 254: 1312
6. Ozin GA (1992) Adv Mater 4: 612
7. Lindsey JS (1991) New J Chem 15: 153
8. Henglein A (1988) Top Curr Chem 143: 115

9. Henglein A (1989) Chem Rev 89: 1861
10. Bawendi MG, Steigerwald ML, Brus LE (1990) Ann Rev Phys Chem 41: 477
11. Wang y (1991) Acc Chem Res 24: 133
12. Cahn PW (1990) Nature 348: 389
13. Gleiter H (1992) Adv Mater 4: 474
14. Gleiter H (1992) Presentation, NATO Advanced Study Institute, Portugal, June 29 to July 11, 1992
15. Westbroek P, deJong EW (1983) Biomineralization and biological metal accumulation, Kluwer, Dordrecht, Holland
16. Simkiss K, Wilbur KM (1989) Biomineralization, cell biology and mineral deposition. Academic, San Diego
17. Lowenstam HA, Weiner S (1989) On biomineralization. Oxford University Press, New York
18. Addali L, Weiner S (1992) Angew Chem Int Ed Eng 31: 153
19. Baer E, Cassidy JJ, Hiltner A (1992) In: Glasser W, Hatakayama H (eds) Viscoelasticity of biomaterials. Am Chem Symp Ser 489, Am Chem Soc, Washington, D.C., p 2
20. Baer E, Hiltner A, Keith HD (1987) Science 235: 1015
21. Wainwright SA, Boggs WD, Currey JD, Gosline JM (1982) Mechanical design in organism. Princeton U.P, Princeton N.J.
22. Vincent J (1990) Structural biocomposites. Princeton U.P. Princeton N.J.
23. Sarikaya M, Aksay IA (1992) In: Case S (ed) Structure, cellular synthesis, and assembly of biopolymers. Springer-Verlag, Amsterdam, p 1
24. Hansma PK, Drake B, Marti O, Gould SCN, Prater CB (1989) Science 243: 641
25. Bard AJ, Denault G, Lee G, Mandler D, Wipf DO (1990) Acc Chem Res 23: 357
26. Manne S, Hansma PK, Massie J, Elings VB, Gewirth AA (1991) Science 152: 183
27. Fuchs H, Schimmel T (1991) Adv Mater 3: 112
28. Kim YT, Bard AJ (1992) Langmuir 8: 1096
29. Hansma HG, Gould SAC, Hansma PK, Gaub HE, Longo ML, Zasadzinski JAN (1991) Langmuir 7: 1051
30. Chiabrera A, DiZitti E, Costa F (1989) J Phys D: Appl Phys 22: 1571
31. Prasad PN, Reinhardt BA (1990) Chem Mater 2: 600
32. Geniès EM (1991) New J Chem Mater 15: 373
33. Spangler CW, Havelka KO (1991) New J Chem 15: 125
34. Chang LL (1992) In: Esaki L (ed) Highlights in condensed matter physics and future prospects. Plenum, New York, p 83
35. Capasso F, Datta S (1990) Physics Today 43: 74
36. Carter FL (1982) Molecular electronic devices. Marcel Dekker, New York
37. Carter FL (1982) Molecular electronic devices II. Marcel Dekker, New York
38. Hameroff SR (1987) Ultimate computing: Biomolecular consciousness and nanotechnology. Elsevier, Amsterdam
39. Aviram A (1989) Molecular electronics – science and technology. United Engineering Trustees, New York
40. Hong FT (1989) Molecular electronics: biosensors ad biocomputers. Plenum, New York
41. Metzger RM, Panetta CA (1991) New J Chem 15: 209
42. Joachim C (1991) New J Chem 15: 223
43. Khalafalla SE, Reimers GW (1980) IEEE Trans Mag MAG-16: 178
44. Sze SM (1981) Physics of semicoductor devices, 2nd edn. Wiley, New York
45. Dalven R (1990) Introduction to applied solid state physics, 2nd Edn. Plenum Press, New York
46. Eckstein JN, Bozovic I, Klausmeier-Brown ME, Virshup G, Ralls KS (1992) Thin Solid Films 216: 8
47. Adamson A (1990) Physical chemistry of surfaces, 5th edn. Wiley, New York
48. Mukherjee SP (1984) Ultrastructure processing of ceramics, glasses and composites. John Wiley, New York
49. Nancolles GH (1979) Adv Colloid Interface Sci 10: 215
50. Addadi L, Berkovich-Yellin Z, Weissbuch I, vanMil J, Simon L, Lahav M, Leiserowitz L (1985) Angew Chem Int Ed Eng 24: 466
51. Bittar EE (1980) Membrance structure and function. Wiley-Interscience, New York
52. Jain MK, Wagner RC (1980) Introduction to biological membrances. Wiley-Interscience, New York
53. Robertson N (1983) The lively membranes. Cambridge University Press, Cambridge

54. Hoppe W, Lohmann W, Markl H, Ziegler H (1983) Biophysics. Springer, Berlin Heideberg New York
55. Fendler JH (1982) Membrane mimetic chemistry. Wiley, New York
56. Fendler JH (1984) Chem Eng News 62 (Jan 2): 25
57. Fendler JH (1984) Science 223: 888
58. Fendler JH, Fendler EJ (1975) Catalysis in micellar and macromolecular systems. Academic Press, New York
59. Calvin M (1978) Acc Chem Res 11: 369
60. Porter G (1978) Proc Roy Soc (London) A362: 281
61. Grätzel M (1981) Acc Chem Res 14: 376
62. Fendler JH (1980) J Phys Chem 84: 1485
63. Fendler JH (1985) J Phys Chem 89: 2730
64. Thomas JK (1980) J Phys Chem 91: 267
65. Mann S, Frankel RB, Blakemore RP (1984) Nature 310: 405
66. Torres de Araujo FF, Pires MA, Frankel RB, Bicudo CEM (1986) Biophys J 50: 375
67. Bazylinski DA, Garratt-Reed AJ, Frankel RB (1992) In: Bailey GW, Bentley J, Small JA (eds) Proceedings of the 50th Annual Meeting of the electron microscopy society of America. San Francisco Press, San Francisco, Ca., p 1022
68. Ulman A (1991) An introduction to ultrathin organic films from Langmuir–Blodgett to self-assembly, Academic Press, Boston
69. Kunitake T (1992) Angew Chem Int Ed Eng 31: 709
70. Embs F, Funhoff D, Laschewsky A, Licht U, Ohst H, Prass W, Ringsdorf H, Wegner G, Wehrmann R (1991) Adv Mater 3: 25
71. Fuhrhop J.-H, Mathieu J (1984) Angew Chem Int Ed Eng 23: 100
72. Ringsdorf H, Schlarb B, Venzmer J (1988) Angew Chem Int Ed Eng 27: 113
73. Ahlers M, Müller W, Reichert A, Ringsdorf H, Venzmer J (1990) Ang Chem Int Ed Eng 29: 1269
74. Lehn J.-M (1988) Angew Chem Int Ed Eng 27: 89
75. Ostro MJ (1987) Liposomes: From biophysics to therapeutics. Marcel Dekker, New York
76. Israelachvili JN, (1987) Acc Chem Res 20: 415
77. Israelachvili JN, McGuiggan PM (1990) J Mater Res 5: 2223
78. Klein J (1980) Nature 288: 248
79. Parker JL, Christenson HK, Ninham BW (1989) Rev Sci Intrum 60: 3135
80. Binning G, Rohrer H (1983) Surf Sci 126: 236
81. Hansma PK, Tersoff J (1987) J Appl Phys 61: R1
82. Tao NJ, DeRose JA, Lindsay SM (1993) J Phys Chem 97: 910
83. Garcia R, Sancez JJ, Soler JM, Garcia N (1987) Surf Sci 181: 69
84. Stroscio JA, Feenstra RM, Fein AP (1987) J Vac Sci Technol B 5: 923
85. Feenstra RM, Stroscio JA, Fein AP (1987) Surface Sci 181: 295
86. Fan F.-RF, Bard AJ (1990) J Phys Chem 94: 3761
87. Kuk Y, Silverman P (1989) Rev Sci Inst 60: 165
88. Prater CB, Hansma PK, Tan I.-H, Lishan DG, Hu EL (1992) J Vac Sci Technol B 10: 1211
89. Weisenhorn AL, Maivald P, Butt H.-J, Hansma PK (1992) Phys Rev B 45: 11226
90. Löche M, Möhwald H (1984) Rev Sci Instr 55: 1968
91. Moy VT, Keller DJ, McConnell HM (1988) J Phys Chem 92: 5233
92. Rice PA, McConnell HM (1989) Proc Natl Acad Sci USA 86: 6445
93. McConnell HM, Rice PA, Benvegnu DJ (1990) J Phys Chem 94: 8965
94. Henon S, Meunier J (1991) Rev Sci Instr 62: 936
95. Hönig D, Möbius D (1991) J Phys Chem 95: 4590
96. Als-Nielsen J, Möhwald H (1991) In: Ebashi S, Koch M, Rubinstein E (eds) Handbook on synchrotron radiation, Vol 4. Elsevier, BV, p 1
97. Jacquemain D, Wolf SG, Leveiller F, Deutsch M, Kjaer K, Als-Nielsen J, Lahav M, Leiserowitz L (1992) Angew Chem Int Ed Eng 31: 130
98. Möbius D, Möhwald H (1991) Adv Mater 3: 19
99. Jacquemain D, Leveiller F, Weinbach SP, Leiserowitz L, Kjaer K, Als-Nielsen J (1991) J Am Chem Soc 113: 7684
100. Tippmann-Krayer P, Möhwald H (1991) Langmuir 7: 2303
101. Israelachvili JN (1992) Intermolecular & surface forces, 2nd Edn. Academic, New York
102. Hautman J, Klein ML (1989) J Chem Phys 91: 4994
103. Bareman JP, Cardini G, Klein ML (1988) Phys Rev Lett 60: 2152

104. Gaines GL (1966) Insoluble Monolayers at liquid-gas interfaces, interscience, New York
105. Gershfeld NL (1976) Annu Rev Phys Chem 27: 349
106. Möbius D (1981) Acc Chem Res 14: 63
107. Stewart MV, Arnett EM (1982) Top Stereochem 13: 195
108. Swalen JD, Allara DL, Chandross EA, Garoff S, Israelachvili J, McCarthy TJ, Murray R, Pease RF, Rabolt JF, Wynne KJ, Yu H (1987) Langmuir 3: 932
109. Möhwald H (1990) Annu Rev Phys Chem 41: 441
110. Yi K (1993) Dissertation, Syracuse University, Syracuse, NY 13244 USA
111. Middleton SR, Iwahashi M, Pallas NR, Pethica BA (1984) Proc R Sco Lond Ser A 396: 143
112. Pallas NR, Pethica BA (1987) J Chem Soc Faraday Trans I 83: 585
113. Pallas NR, Pethica BA (1985) Langmuir 1: 509
114. Arnett EM, Chao J, Kinzig BJ, Stewart MV, Thompson O, Verbiar RJ (1982) J Am Chem Soc 104: 389
115. Arnett EM, Harvey NG, Rose PL (1989) Acc Chem Res 22: 131
116. Xu SQ, Fendler JH (1989) Macromol 22: 2962
117. Joos P (1969) Bull Soc Chim Belges 78: 207
118. Joos P, Demel RA (1969) Biochem Biophys Acta 183: 447
119. Zsako J, Tomoaia-Cotisel M, Chifu EJ (1984) Colloid Interface Sci 102: 186
120. Defay R, Prigogine I, Bellemans A, Everett DH (1966) Surface tension and adsorption. Longmans & Green, London
121. Fendler JH (1984) In: Mittal KL, Lindman B (eds) Surfactants in solution. Plenum, New York, p 1947
122. Naito K (1989) J Colloid interface sci 131: 218
123. Tredgold RH (1988) J Chim Phys 85: 1079
124. Gaines GL (1991) Langmuir 7: 834
125. Binks BP, Fletcher PDI, Phillips JS, Richardson RM (1992) Thin Solid Films 209: 280
126. Kaku M, Hsiung H, Sogah DY, Levy M, Rodriguez-Parada JM (1992) Langmuir 8: 1239
127. Wegner G (1992) Thin Solid Films 216: 105
128. Thibodeaux AF, Duran RS, Ringsdorf H, Schuster A, Skoulios A, Gramain P, Ford W (1992) In: Stroeve P, Balazs AC (eds) Macromolecular assemblies in polymeric systems, ACS symposium series 493. Washington, D.C, p 20
129. Obeng YS, Bard AJ (1991) J Am Chem Soc 113: 6279
130. Back R, Lennox B (1992) J Phys Chem 96: 8149
131. Dhathathhereyan A, Baumann U, Müller A, Möbius D (1988) Biochim Biophys Acta 944: 265
132. Hórvölgyi Z, Németh S, Fendler J (1993) Colloids and Surfaces A 71: 327
133. Blodgett KB (1935) J Am Chem Soc 57: 1007
134. Blodgett KB, Langmuir I (1937) Phys Rev 51: 964
135. Zwick MM, Kuhn H (1962) Z.Naturforsch 17a: 411
136. Kuhn H, Möbius D (1971) Angew Chem Int Ed Eng 10: 620
137. Kuhn H, Möbius D, Bücher H (1971) In: Weissberger A, Rossiter B (eds) Physical methods of chemistry, Vol I, Part 3B. John Wiley, New York, p 577
138. Möbius D (1978) Ber Bunsenges Phys Chem 82: 848
139. Möbius D (1985) Adv Chem 160: 113
140. Blinov LM (1983) Russ Chem Rev 52: 713
141. Roberts GG (1990) Langmuir–Blodgett films. Plenum Press, New York
142. Popovitz-Biro R, Hill K, Landau EM, Lahav M, Leiserowitz L, Sagiv J, Hsiung H, Meredith GR, Vanherzeele H (1988) J Am Chem Soc 110: 2672
143. Prasad PN, Williams DJ (1990) Introduction to nonlinear optical effects in molecules and polymers. John Wiley, New York
144. Hasmonay H, Vincent M, Dupeyrat M (1980) Thin Solid Films 68: 21
145. Shimomura M, Fuji K, Karg P, Frey W, Sackmann E, Meller P, Ringsdorf H (1988) Jpn J App Phys 27: L1761
146. Feigin LA, Lvov YM (1988) Macromol Chem Macromol Symp 15: 259
147. Feigin LA, Lvov YM, Troitsky VI (1989) Sov Sci Rev A Phys 11: 285
148. Grundy MJ, Musgrove RJ, Richardson RM, Roser SJ, Penfold J (1990) Langmuir 6: 519
149. Allara DL, Swalen D (1982) J Phys Chem 86: 2700
150. Cui DF, Howarth VA, Petty MC, Ancelin H, Yarwood J (1990) Thin Solid Films 192: 391
151. Takahara A, Morotomi N, Hiraoka S, Higashi N, Kunitake T, Kajiyama T (1989) Macromol 22: 617

152. Neumann AW (1964) Z Phys Chem 43: 71
153. Neumann AW, Good RJ (1979) In: Good RJ, Stromberg RR (eds) Surface and colloid science, Vol 11. Plenum Press, New York, p 31
154. Peng JB , Ketterson JB, Dutta P (1988) Langmuir 4: 1198
155. Peng JB, Abraham BM, Dutta P, Ketterson JB (1985) Thin Solid Films 134: 187
156. Buhaenko MR, Richardson RM (1988) Thin Solid Films 159: 231
157. Blake TD, Haynes JM (1969) J Colloid Interface Sci 30: 421
158. Fendler JH, Tundo P (1984) Acc Chem Res 17: 3
159. Gros L, Ringsdorf H, Schupp H (1981) Angew Chem Int Ed Eng 20: 305
160. Bader H, Dorn K, Hupfer B, Ringsdorf H (1985) Advan Polymer Sci 64: 1
161. O'Brien DF, Klingbiel RT, Specht DP, Tyminski PN (1985) Ann NY Acad Sci 446: 282
162. Regen S (1987) In: Ostro MJ (ed) Liposomes from biophysics to therapeutics. Marcel Dekker, New York, p 73
163. Rolandi R, Paradiso R, Xu SQ, Palmer C, Fendler JH (1989) J Am Chem Soc 111: 5233
164. Palmer CA (1992) Dissertation, Syracuse University, New York
165. McLean LR, Durrani AA, Whittam MA, Johnston DS, Chapman D (1983) Thin Solid Films 99: 127
166. Albrecht O, Johnston DS, Villaverde C, Chapman D (1982) Biochim Biophys Acta 687: 165
167. Day DR, Ringsdorf H (1979) Makromol Chem 180: 1059
168. Palacin S, Ruaudel-Teixier A, Barraud A (1989) J Phys Chem 93: 7195
169. Vandevyver M, Ruaudel-Teixier A, Palacin S, Bourgoin J.-P, Barraud A, Bozio R, Menghetti M, Pecile C (1990) Mol Cryst Liq Cryst 187: 327
170. Albrecht O, Cumming W, Kreuder W, Laschewsky A, Ringsdorf H (1986) Coll Polym Sci 264: 659
171. Laschewsky A (1989) Angew Chem Int Ed Eng Adv Mater 28: 1574
172. Elbert R, Laschewsky A, Ringsdorf H (1985) J Am Chem Soc 107: 4134
173. Shafrin EG, Zisman WA (1960) J Phys Chem 64: 519
174. Fowkes FM, Zisman WA (1964) Contact angle, wettability and adhesion, Adv Chem, Vol 43. American Chem Soc. Washington, D.C.
175. Whitesides GM, Laibinis PE (1990) Langmuir 6: 87
176. Evans SD, Sharma R, Ulman A (1991) Langmuir 7: 156
177. Wasserman SR, Tao Y.-T, Whitesides GM (1989) Langmuir 5: 1074
178. Bain CD, Whitesides GM (1989) Langmuir 5: 1370
179. Kim Y.-T, McCarley RL, Bard AJ (1992) J Phys Chem 96: 7416
180. Tidswell IM, Ocko BM, Pershan PS, Wasserman SR, Whitesides GM, Axe JD (1990) Phys Rev B 41: 1111
181. Ducharme D, Max JJ, Salesse C, Leblanc RM (1990) J Phys Chem 94: 1927
182. Wasserman SR, Whitesides GM, Tidswell IM, Ocko BM, Pershan PS, Axe JD (1989) J Am Chem Soc 111: 5852
183. Sagiv J (1980) J Am Chem Soc 102: 92
184. Sagiv J (1979) Isr J Chem 18: 339
185. Sagiv J (1979) Isr J Chem 18: 346
186. Tabushi I, Kurihara K, Naka K, Yamamura K, Hatekeyama H (1987) Tetrahedron Lett 28: 4299
187. Yamamura K, Hatakeyama H, Naka K, Tabushi I, Kurihara K (1988) J Chem Soc Chem Commun 79
188. Kim J.-H, Cotton TM, Uphaus RA (1988) J Phys Chem 92: 5575
189. Anderson LI, Mandenius CF, Mosbach K (1988) Tetrahedron Lett 29: 5437
190. Yamamura K, Hatekeyama H, Tabushi I (1988) Chem Lett (Jpn) 99
191. Kallury KMR, Thompson M, Tripp CP, Hair ML (1992) Langmuir 8: 947
192. Fromherz PZ (1973) Z.Naturforsch Teil C 28: 144
193. Netzer L, Sagiv J (1983) J Am Chem Soc 105: 674
194. Tillman N, Ulman A, Penner TL (1989) Langmuir 5: 101
195. Maoz R, Netzer L, Gun J, Sagiv J (1988) J Chim Phys Phys-Chim Biol 85: 1059
196. Gun J, Sagiv J (1986) J Collid Interface Sci 112: 457
197. Gun J, Iscovici R, Sagiv J (1984) J Colloid Interface Sci 101: 201
198. Maoz R, Sagiv J (1985) Thin Solid Films 132: 135
199. Tillman N, Ulman A, Schildkraut JS, Penner TL (1988) J Am Chem Soc 110: 6136
200. Cohen SR, Naaman R, Sagiv J (1986) J Phys Chem 90: 3054

201. Angst DL, Simmons GW (1991) Langmuir 7: 2236
202. Nuzzo RG, Allara DL (1983) J Am Chem Soc 105: 4481
203. Allara DL, Nuzzo RG (1985) Langmuir 1: 52
204. Nuzzo RG, Fusco FA, Allara DL (1987) J Am Chem Soc 109: 2358
205. Li TTT, Weaver M (1984) J Am Chem Soc 106: 6107
206. Li TTT, Liu HY, Weaver M (1984) J Am Chem Soc 106: 1233
207. Finklea HO, Avery S, Lynch M, Furtsch T (1987) Langmuir 3: 409
208. Porter MD, Bright TB, Allara DL, Chidsey CED (1987) J Am Chem Soc 109: 3559
209. Nuzzo RG, Zegarski BR, Dubois LH (1987) J Am Chem Soc 109: 733
210. Chidsey CED, Liu GY, Rowntree P, Scoles G (1989) J Chem Phys 91: 4421
211. Troughton EB, Bain CD, Whitesides GM, Nuzzo RG, Allara DL, Porter MD (1988) Langmuir 4: 365
212. Bain CD, Troughton EB, Tao Y-T, Evall J, Whitesides GM, Nuzzo RG (1989) J Am Chem Soc 111: 321
213. Ulman A, Tillman N (1989) Langmuir 5: 1418
214. Evans SD, Ulman A (1990) Chem Phys Lett 170: 462
215. Harris AL, Rothberg L, Dubois LH, Levinos NJ, Dhar L (1990) Phys Rev Lettt 64: 2086
216. Ulman A, Evans SD, Snyder RG, (1992) Thin solid films 210/211: 806
217. Fabianowski W, Coyle LC, Weber BA, Granata RD, Castner DG, Sadownik A, Regen SL (1989) Langmuir 5: 35
218. Laibinis PE, Whitesides GM, Allara DL, Tao Y-T, Parikh AN, Nuzzo RG (1991) J Am Chem Soc 113: 7152
219. Bertilsson L, Liedberg B (1993) Langmuir 9: 141
220. Doblhofer K, Figura J, Fuhrhop J-H (1992) Langmuir 8: 1811
221. Ulman A, Scaringe RP (1992) Langmuir 8: 894
222. Bain CD, Whitesides GM (1988) J Am Chem Soc 110: 5897
223. Bain CD, Whitesides GM (1988) J Am Chem Soc 110: 6560
224. Bain CD, Evall J, Whitesides GM (1989) J Am Chem Soc 111: 7155
225. Bain CD, Whitesides GM (1988) J Am Chem Soc 110: 3665
226. Bain CD, Whitesides GM (1988) Science 240: 62
227. Bain CD, Whitesides GM (1989) J Am Chem Soc 111: 7164
228. Bain CD, Biebuyck HA, Whitesides GM (1989) Langmuir 5: 723
229. Bain CD, Whitesides GM (1989) J Phys Chem 93: 1670
230. Strong L, Whitesides GM (1988) Langmuir 4: 546
231. Ulman A (1989) J Mat Ed 11: 205
232. Stewart KR, Whitesides GM, Godfried HP, Silvera IF (1986) Rev Sci Instrum 57: 1381
233. Widrig CA, Chung C, Porter MD (1991) J Electroanal Chem 310: 335
234. Widrig CA, Alves CA, Porter MD (1991) J Am Chem Soc 113: 2805
235. Evans SD, Urankar E, Ulman A, Ferris N (1991) J Am Chem Soc 113: 4121
236. Sabatani E, Rubinstein I, Maoz R, Sagiv J (1987) J Electroanal Chem 219: 365
237. Sabatani E, Rubinstein I (1987) J Phys Chem 91: 6663
238. Rubinstein I, Steinberg S, Tor Y, Shanzer A, Sagiv J (1988) Nature 332: 426
239. Steinberg S, Rubinstein I (1992) Langmuir 8: 1183
240. Bareman JP, Cardini G, Klein ML (1989) Mat Res Soc Symp Proc 141: 411
241. Walczak MM, Chung CK, Stole SM, Widrig CA, Porter MD (1991) J Am Chem Soc 113: 2370
242. Coyle LC, Danilov YN, Juliano RL, Regen SL (1989) Chem Mater 1: 606
243. Lee H, Kepley LJ, Hong H-G, Mallouk TE (1988) J Am Chem Soc 110: 618
244. Creager SE, Collard DM, Fox MA (1990) Langmuir 6: 1617
245. Gomez M, Li J, Kaifer AE (1991) Langmuir 7: 1797
246. Sheen CW, Shi J-X, Martensson J, Parikh AN, Allara DL (1992) J Am Chem Soc 114: 1514
247. Laibinis PE, Hickman JJ, Wrighton MS, Whitesides GM (1989) Science 245: 845
248. Kumar A, Biebuyck HA, Abbott NL, Whitesides GM (1992) J Am Chem Soc 114: 9188
249. Prime KL, Whitesides GM (1991) Science 252: 1164
250. Wennerstrom H, Lindman B (1979) Phys Rev 52: 1
251. Brown GH, Wolken JJ (1979) Liquid crystals and biological structures. Academic Press, New York
252. Noël C, Navard P (1991) Prog Polym Sci 16: 55
253. Casey WH (1991) J Colloid Interface Sci 146: 582
254. Wendorff JH (1991) Angew Chem Int Ed Eng 30: 405

255. Giroud-Godquin AM, Maitlis PM (1991) Angew Chem Int Ed Eng 30: 375
256. Fendler JH (1976) Acc Chem Res 9: 153
257. Eicke HF (1980) Top Curr Chem 87: 85
258. Luisi PL, Straub E (1984) Reversed Micelles: Biological and technological relevance of amphiphilic structures in apolar media. Plenum, New York
259. Kertes AS, Gutmann H (1975) Surf Colloid Sci 8: 193
260. Luisi PL, Magid LJ (1986) CRC Crit Rev Biochem 20: 409
261. Biomolecules in Organic Solvents (1991) CRC Press, Boca Raton, Florida
262. Walde P, Giuliani AM, Boicelli CA, Luisi PL (1980) Chem Phys Lipids 53: 265
263. El Seoud OA, Chinelatto AM (1983) J Colloid Interface Sci 95: 163
264. El Seoud OA (1989) Adv Coll Interf Sci 30: 1
265. Smith RE, Luisi PL (1980) Helv Chim Acta 63: 2302
266. O'Connor CJ (1987) In: Eiche H, Parfitt GD (eds) Interfacial phenomena in apolar solvents. Marcel Dekker, New York, p 188
267. Armstrong DW, Seguin R, McNeal CJ, Macfarlane RD, Fendler JH (1978) J Am Chem Soc 100: 4605
268. Martinek K, Levashov AV, Klyachko NL, Berezin IV (1977) Doklady Akad Nauk SSSR 236: 920
269. Barbaric S, Luisi PL (1981) J Am Chem Soc 103: 4239
270. Martinek K, Levashov AV, Klyachko NL, Pantin VI, Berezin IV (1981) Biochim Biophys Acta 657: 277
271. Martinek K, Levashov AV, Khmelnitsky YL, Berezin IV (1982) Science 218: 889
272. Khmelnitski YL, Dien FK, Semenov AN, Martinek K (1984) Tetrahedron 40: 4425
273. Luthi P, Luisi PL (1984) J Am Chem Soc 106: 7285
274. Klyachko NL, Levashov AV, Pshezhetsky AAV, Bogdanova NG, Berezin IV, Martinek K (1986) Eur J Biochem 381: 1
275. Martinek K, Berezin IV, Khmelnitski YL, Klyachoko NL, Levasov AV (1987) Biocatalysis 1: 9
276. Escamilla E, Ayala G, Tuena de Gómez-Puyou M, Gómez-Puyou A, Millan L, Darszon A (1989) Arch Biochem Biophys 272: 332
277. Karpe P, Ruckenstein E (1991) J Colloid Interface Sci 141: 534
278. Ruckenstein E, Karpe P (1991) J Phys Chem 95: 4869
279. Balny C, Douzou P (1979) Biochimie 61: 445
280. Douzou P, Keh E, Balny C (1979) Proc Natl Acad Sci USA 76: 681
281. Langevin D (1988) Acc Chem Res 21: 255
282. Atkinson PJ, Grimson MJ, Heenan RK, Howe AM, Robinson BH (1989) J Chem Soc Chem Commun 1807
283. Luisi PL, Scartazzini R, Haering G, Schurtenberger P (1990) Colloid Polym Sci 268: 356
284. Quellet C, Eicke HF, Xu G, Hauger Y (1990) Macromol 23: 3347
285. Schurtenberger P, Scartazzini R, Magid LJ, Leser ME, Luisi PL (1990) J Phys Chem 94: 3695
286. Kilfiker R, Eicke HF, Steeb C, Hofmeier U (1991) J Phys Chem 95: 1478
287. Schurtenberger P, Magid L, King SM, Lindner P (1991) J Phys Chem 95: 4173
288. Mascolo G, Giutini M, Luisi PL, Lang J (1990) J Colloid Interface Sci 140: 401
289. Kunieda H, Nakamura K, Evans DF (1991) J Am Chem Soc 113: 1051
290. Kunieda H, Nakamura K, Infante MR, Solans C (1991) Adv Mater 4: 291
291. Bachmann PA, Walde P, Luisi PL, Lang J (1990) J Am Chem Soc 112: 8200
292. Bachmann PA, Walde P, Luisi PL, Lang J (1991) J Am Chem Soc 113: 8204
293. Fendler JH (1990) Acc Chem Res 13: 7
294. Lasic DD, Kidric J, Zagorc S (1987) Biochim Biophys Acta 896: 117
295. Kaler EW, Murthy AK, Rodriguez BE, Zasadzinski JAN (1989) Science 245: 1371
296. Hauser H, Mantsch HH, Casal HL (1990) Biochem 29: 2321
297. Lasic DD (1982) Biochim Biophys Acta 692: 501
298. Lasic DD (1987) J Theoret Biol 124: 35
299. Lasic DD (1988) Biochem J 256: 1
300. Kunitake T, Okahata Y (1977) J Am Chem Soc 99: 3860
301. Kunitake T, Okahata Y, Tamaki K, Kumamura F, Takayanagi M (1977) Chem Lett (Jpn) 387
302. Okuyama K, Soboi Y, Iijima N, Hirabayashi K, Kunitake T, Kajiyama T (1988) Bull Chem Soc Jpn 61: 1485
303. Okuyama K, Iijima N, Hirabayashi K, Kunitake T, Kusunoki M (1988) Bull Chem Soc Jpn 61: 2337

304. Kunitake T, Okahata Y, Shimomura M, Yasunami SI, Takarabe K (1981) J Am Chem Soc 103: 5401
305. Kunitake T, Kimizuka N, Higashi N, Nakashima N (1984) J Am Chem Soc 106: 1978
306. Kimizuka N, Ohira H, Tanaka M, Kunitake T (1990) Chem Lett (Jpn) 29
307. Führhop J-H, David H-H, Mathieu J, Liman V, Winter H-J, Boekema E (1986) J Am Chem Soc 108: 1785
308. Kunitake T, Higashi N (1985) J Am Chem Soc 107: 692
309. Caffrey M, Moynihan D, Hogan J (1991) Chem Phys Lipids 57: 275
310. Lerebours B, Watzke HJ, Fendler JH (1990) J Phys Chem 94: 1632
311. Riquelme G, Lopez E, Garcia-Segure LM, Ferragut JA, Gonalez-Ros JM (1990) Biochem 29: 11215
312. Faucon JF, Mitov MD, Méléard P, Bivas I, Bothorel P (1989) J Physique 50: 2389
313. Méléard P, Mitov MD, Faucon JF, Bothorel P (1990) Europhys Lett 11: 355
314. Evans E, Needham D (1987) J Phys Chem 91: 4219
315. Evans EA, Skalak R (1980) Mechanics and Thermodynamics of Biomembranes, CRC Press, Boca Raton, Fl
316. Tundo P, Kurihara K, Kippenberger DJ, Politi M, Fendler JH (1982) Angew Chem Int Ed Eng 21: 81
317. Rafaeloff R, Tricot Y-M, Nome F, Tundo P, Fendler JH (1985) J Phys Chem 89: 1236
318. Regen SL, Samuel NKP, Khurana JM (1985) J Am Chem Soc 107: 5804
319. Sadownik A, Stefely J, Regen SL (1986) J Am Chem Soc 108: 7789
320. Chang EL, Gaber BP, Sheridan JP (1982) Biophys J 39: 197
321. Wong M. Thompson TE (1982) Biochemistry 21: 4133
322. Kurihara K, Fendler JH (1983) J Chem Soc Commun: 1188
323. Reed W, Guterman L, Tundo P, Fendler JH (1984) J Am Chem Soc 106: 1897
324. Nome F, Reed W, Politi M, Tundo P, Fendler JH (1984) J Am Chem Soc 106: 8086
325. Regen SL, Singh A, Oehme G, Singh M (1982) J Am Chem Soc 104: 791
326. Kunitake T, Okahata Y, Yasunami S (1981) Chem Lett (Jpn) 1397
327. Mutz M, Bensimon D (1991) Phys Rev A43: 4525
328. Izawa H, Arakawa M, Konto T (1986) Biochim Biophys Acta 855: 2439
329. Hartmann W, Galla H-J (1978) Biochim Biophys Acta 509: 474
330. Carrier D, Dufourq J, Faucon J-F, Pézolet M (1985) Biochim Biophys Acta 820: 131
331. Walter A, Steer CJ, Blumenthal R (1986) Biochim Biophys Acta 861: 319
332. Gad AE, Elyashiv G, Rosenberg N (1986) Biochim Biophys Acta 860: 314
333. Borden KA, Eum KM, Langley KH, Tirell DA (1987) Macromol 20: 454
334. Takada M, Yuzuriha T, Katayama K, Iwamoto K, Sunamoto J (1984) Biochim Biophys Acta 802: 237
335. Aliev KV, Ringsdorf H, Schlarb B, Leister K.-H (1984) Makromol Chem Rapid Commun 5: 345
336. Regen SL, Shin J-S, Yamaguchi K (1984) J Am Chem Soc 106: 2446
337. Regen SL, Shin J-S, Hainfeld JF, Wall JS (1984) J Am Chem Soc 106: 5756
338. Fukuda H, Diem T, Stefely J, Kezdy FJ, Regen SL (1986) J Am Chem Soc 108: 2321
339. Brady JE, Evans DF, Kachar B, Ninham BW (1984) J Am Chem Soc 106: 4279
340. Ringsdorf H, Schlarb B (1986) Polymer Prep 27: 127
341: Ringsdorf H, Schlarb B, Tyminsky PN, O'Brien DF (1988) Macromol 21: 671
342. Kunitake T (1983) In: Ise N, Tabushi I (eds) An introduction to specialty polymers. Cambridge University Press, London, p 174
343. Okahata Y (1986) Acc Chem Res 19: 59
344. Okahata Y (1987) In: Ottenbrite RM, Utracki LA, Inoue S (eds) Current topics in polymer science. Hanser Publishers, New York, p 65
345. Okahata Y, Ariga K, Seki T (1988) J Am Chem Soc 110: 2495
346. Laschewsky A, Ringsdorf H, Schmidt G, Schneider J (1987) J Am Chem Soc 109: 788
347. Frey W. Schneider J, Ringsdorf H, Sackmann E (1987) Macromol 20: 1312
348. Kunitake T, Nakashima N, Takarabe K, Nagai M, Tsuge A, Yanagi H (1981) J Am Chem Soc 103: 5945
349. Elbert R, Folda T, Ringsdorf H (1984) J Am Chem Soc 106: 7962
350. Jayasuriys N, Bosak S, Regen SL (1990) J Am Chem Soc 112: 5851
351. Bangham AD, Horne RW (1964) J Mol Biol 8: 660
352. Bangham AD, Hill MW, MIller NGA (1974) In: Korn ED (ed) Methods in membrane biology. Plenum, New York, p 1

353. Papahadjopoulos D, Vail WJ, Jacobson K, Poste G (1975) Biochim Biophys Acta 394: 483
354. Fonteijn TAA, Hoekstra D, Engberts JBFN (1992) Langmuir 8: 2437
355. Yager P, Schoen PE (1984) Mol Cryst Liq Cryst 106: 371
356. Schnur JM, Price R, Schoen P, Jager P, Calvert J, Singh A (1987) Thin Solid Films 152: 181
357. Singh A. Schoen PE, Schnur JM (1988) J Chem Soc Chem Commun 1222
358. Yager P, Schoen P, Davies CA, Price R, Singh A (1985) Biophys J 48: 899
359. Yager P, Price RR, Schnur JM, Schoen PE, Singh A, Rhodes D (1988) Chem Phys Lipids 46: 171
360. Yager P, Schoen PE, Gregor Jr JH, Price RR, Singh A (1985) Biophys J 48: 899
361. Georger JH, Singh A, Price RR, Schnur JM, Yager P, Schoen PE (1987) J Am Chem Soc 109:
     6169
362. Markowitz M, Singh A (1991) Langmuir 7: 16
363. Rudolph AS, Ratna BR, Kahn B (1991) Nature 352: 52
364. Ou-Yang Z-c (1990) Phys Rev A41: 4517
365. Ou-Yang Z-c, Helfrick W (1989) Phys Rev A39: 5280
366. Lu M-H, Lando JB, Mann Jr A, Petschek RG, Rosenblatt C (1991) Langmuir 7: 1988
367. Ihara H, Takafuji M, Hirayama C, O'Brien DF (1992) Langmuir 8: 1548
368. Nakashima N, Asakuma S, Kunitake T (1984) Chem Lett (Jpn) 1709
369. Yamada K, Ihara H, Ide T, Fukumoto T, Hirayama C (1984) Chem Lett (Jpn) 1713
370. Frankel DA, O'Brien DF (1991) J Am Chem Soc 113: 7436
371. Fuhrop J-H Blumtritt P, Lehmann C, Luger P (1991) J Am Chem Soc 113: 7437
372. Fuhrhop J-H, Krull M (1991) In: Schneider H-J, Dürr H (eds) Frontiers in supramolecular and
     organic photochemistry. VCH Weinhein, Germany, p 223
373. Fuhrop J-H, Schnieder P, Rosenberg J, Boekema E (1987) J Am Chem Soc 109: 3387
374. Fuhrhop J-H, Boettcher C (1990) J Am Chem Soc 112: 1768
375. Boettcher C, Boekema EJ, Fuhorhop J-H (1990) J Microscopy 160: 173
376. Fuhrhop J-H, Demoulin C, Rosenberg J, Boettcher C (1991) J Am Chem Soc 112: 2827
377. Fuhrhop J-H, Sevenson S, Boettcher C, Rössler E, Vieth H-M (1990) J Am Chem Soc 112: 4307
378. Fuhrhop J-H Demoulin C, Boettcher C, Köning J, Siggel U (1992) J Am Chem Soc 114: 4159
379. Pfannemüller B, Welte W (1985) Chem Phys Lipids 37: 227
380. Zabel V, Müller-Fahrnow A, Hilgenfeld R, Saenger W, Pfannemüller B, Enkelmann V, Welte W
     (1986) Chem Phys Lipids 39: 313
381. Nakashima N, Asakuma S, Kunitake T (1985) J Am Chem 107: 509
382. Fuhrhop J-H, Schnieder P, Boekema E, Helfrich W (1988) J Am Chem Soc 110: 2861
383. Tomalia DA, Naylor AM, Goddard III WA (1990) Angew Chem Int Ed Eng 29: 138
384. Newkome GR, Baker GR, Arai S, Saunders MJ, Russo PS, Theriot KJ, Moorefield CN, Rogers
     LE, Miller JE, Lieux TR, Murray ME, Philips B, Pascal L (1990) J Am Chem Soc 112: 8458
385. Tien HT (1974) Bilayer lipid membranes (BLM) Theory and practice. Marcel Dekker, New
     York
386. Krysinski P, Tien HT (1986) Prog Surf Sci 23: 317
387. White SH (1986) In: Miller C (ed) Ion channel reconstitution. Plenum, New York, p 3
388. Fettiplace R, Gordon LGM, Hladky SB, Requena J, Zingsheim HP, Haydon DA (1975) In:
     Korn ED (ed) Methods in membrane biology. Plenum, New York, p 1
389. Fettiplace R, Haydon DA (1980) Phys Rev 60: 510
390. White SH (1970) Biophys J 10: 1127
391. Montal M, Mueller P (1972) Proc Natl Acad Sci USA 69: 3561
392. White SH, Petersen DC, Simon S, Yafuso M (1976) Biophys J 16: 481
393. Schindler H (1979) Biochim Biophys Acta 555: 316
394. Schindler H (1980) FEBS Letters 122: 77
395. Schuerholz T, Schindler H (1983) FEBS Letters 152: 187
396. Heyn S-P, Egger M, Gaub HE (1990) J Phys Chem 94: 5073
397. Schindler H, Quast U (1980) Proc Natl Acad Sci USA 77: 3052
398. Schindler H, Rosenbusch JP (1978) Proc Natl Acad Sci USA 75: 3751
399. Salesse C, Ducharme D, Leblanc RM (1987) Biophys J 52: 351
400. Miller C, (1986) Ion channel reconstitution. Plenum, New York
401. Hille B (1984) Ionic channels of exitable membranes. Sinaver Associates, Sunderland, Ma
402. Sackmann B, Neher E (1983) Singh channel recording. Plenum, New York
403. Rolandi R, Flom SR, Dillon I, Fendler JH (1987) Prog Colloid Polymer Sci 73: 134
404. Benz R, Prass W, Ringsdorf H (1982) Angew Chem Int Ed Eng 21: 368
405. Benz R, Elbert R, Prass W, Ringsdorf H (1986) Eur Biophys J 14: 83

406. Higashi N, Kunitake T, Möllerfeld J, Ringsdorf H (1988) Chem Lett (Jpn) 13
407. Kato S, Kunitake T (1991) Chem Lett (Jpn) 261
408. Borle F, Sänger M, Sigrist H (1991) Biochim Biophys Acta 1066: 144
409. Zhao XK, Fendler JH (1986) J Phys Chem 90: 3886
410. Zhao XK, Fendler JH (1988) J Phys Chem 92: 3350
411. Zhao XK, Picard G, Fendler JH (1988) J Phys Chem 92: 7161
412. Yogev D, Todorov AT, Fendler JH (1991) J Phys Chem 95: 3892
413. Picard G, Denicourt N, Fendler JH (1991) J Phys Chem 95: 3705
414. Picard G, Schneider-Henriquez JE, Fendler JH (1991) J Phys Chem 94: 510
415. Petrov AG (1975) In: Physical and chemical basis of biological information transfer. Plenum New York, p 111
416. Derzhanski A, Petrov AG, Todorov AT, Hristova K (1990) Liq Cryst 7: 439
417. Ochs AL, Burton RM (1974) Biophys J 14: 473
418. Petrov AG, Bivas I (1984) Prog Surf Sci 16: 389
419. Petrov AG, Sokolov VS (1986) Fur Biophys J 13: 139
420. Petrov AG, Derzhanski A (1976) J Phys Suppl 37: 155
421. Derzhanski A, Petrov AG, Pavloff YV (1981) J Phys Lett 42: 119
422. Todorov AT, Petrov AG, Brandt MO, Fendler JH (1991) Langmuir 7: 3127
423. Petrov AG, Seleznev A, Derzhanski A (1979) Acta Phys Pol A 55: 385
424. Derzhanski A (1989) Phys Lett A 139: 170
425. Nakashima N, Ando R, Kunitake T (1983) Chem Lett (Jpn) 1577
426. Kunitake T, Shimomura M, Kajiyama T, Harada A, Okuyama K, Takayanagi M (1984) Thin Solid Films 121: L89
427. Shimomura M, Ando R, Kunitake T (1983) Ber Bunsenges Phys Chem 87: 1134
428. Higashi N, Kunitake T (1984) Polymer J (Tokyo) 16: 583
429. Shimomura M, Kunitake T (1984) Polymer J (Tokyo) 16: 187
430. Kunitake T, Tsuge A, Nakashima N (1984) Chem Lett (Jpn) 1783
431. Nakashima N, Kunitake M, Kunitake T, Tone S, Kajiyama T (1985) Macromol 18: 1515
432. Regen AL, Kirszensztejn P, Singh A (1983) Macromol 16: 335
433. Albrecht O, Laschewsky A (1984) Macromol 17: 1292
434. Regen SL, Foltynowicz Z, Yamaguchi K (1984) Macromol 17: 1293
435. Higashi N, Kajiyama T, Kunitake T, Prass W, Ringsdorf H, Takahara A (1987) Macromol 20: 29
436. Takahara A, Higashi N, Kunitake T, Kajiyama T (1988) Macromol 21: 2443
437. Fukuta K, Hami Y, Shimizu R, Kunitake T (1992) Thin Solid Films 210/211: 828
438. Asakuma S, Kunitake T (1989) Chem Lett (Jpn) 2059
439. Okahata Y, Ebato H (1989) Anal Chem 61: 2185
440. Sakata K, Kunitake T (1990) J Chem Soc Chem, Commun 504
441. Ishikawa Y, Kunitake T (1986) J Am Chem Soc 108: 8300
442. Hamachi I, Noda S, Kunitake T (1990) J Am Chem Soc 112: 6744
443. Kunitake T (1991) Polymer J (Tokyo) 23: 613
444. Asakuma S, Okada H, Kunitake T (1991) J Am Chem Soc 113: 1749
445. Ishikawa Y, Kunitake T (1991) J Am Chem Soc 113: 621
446. Sakata K, Kunitake T (1992) Thin Solid Films 210/211: 26
447. Okada H, Sakata K, Kunitake T (1990) Chem Mater 2: 89
448. Okahata Y, Lim H-J, Nakamura G, Hachiya S (1983) J Am Chem Soc 105: 4855
449. Liu J-M, Yang SC (1991) J Chem Soc Chem Commun 1529
450. DeRossi D, Kajiwara K, Osada Y, Yamauchi A (1991) Polymer gels: fundamentals and biomedical applications. Plenum, New York
451. Winter HH (1991) MRS Bulletin, August: 44
452. Hoffmann AS (1991) MRS Bulletin, August: 42
453. Osada Y, Ikuzaki H, Hori H (1992) Nature 355: 242
454. Flory PJ (1953) Principles of polymer chemistry. Cornell University Press, Ithaca, New York
455. Fischer EW. Shulz RC, Sillescu H (1991) Chemistry and physics of macromolecules. VCH Weinheim, Germany
456. Martin CR, Van Dyke LS (1992) In: Murray RW (ed) Molecular design of electrode surfaces. John Wiley, New York, p 403
457. Bryce MR (1991) Chem Soc Rev 20: 355
458. Frommer JE (1986) Acc Chem Res 19: 2

459. Kelker H, Hatz R (1980) Handbook of liquid crystals. Verlag Chemie, Weinheim Germany
460. Kepler RG, Anderson RA (1992) Advan Phys 41: 1
461. Williamson DJ (1984) Angew Chem Int Ed Eng 23: 690
462. Caro J, Finger J, Kornatowski J, Richter-Mendau J, Werner L, Zibrowius B (1992) Adv Mater 4: 273
463. Breck DW (1984) Zeolite Molecular Sieves, Wiley-Interscience, New York
464. Szostak R (1989) Molecular sieves, principles and identification. Van Nostrand Reinhold, New York
465. Mallouk TE, Lee H (1990) Symp Mol Arch 67: 829
466. Enzel P, Bein T (1989) J Chem Soc Chem Commun 1326
467. Enzel P, Bein T (1989) Angew Chem Int Ed Eng 28: 1692
468. Enzel P, Bein T (1992) Chem Mater 4: 819
469. Davis ME, Saldarriage C, Montes C, Graces J, Crowder C (1988) nature 331: 698
470. Annen MJ, Young D, Davis ME, Cavin OB, Habbard CR (1991) J Phys Chem 95: 1380
471. Estermann M, McCusker LB, Baerlocher C, Merrouche A, Kessler H (1990) Nature 352: 320
472. Huo Q, Xu R, Li S, Ma Z, Thomas JM, Jones RH, Chippindale AM (1992) J Chem Soc Chem Commun 875
473. Beck JS, Vartuli JC, Roth WJ, Leonowicz ME, Kresge CT, Schmitt KD, Chu CT-W, Olson DH, Sheppard EW, McCullen SB, Higgins JB, Schlenker JL (1992) J Am Chem Soc 114: 10834
474. Dutta PK, Robins D (1991) Langmuir 7: 1048
475. Curl RF, Smalley RE (1991) Sci Am Ocotober: 54
476. Kroto HW, Allaf AW, Balm SP (1991) Chem Rev 91: 1213
477. Iijima S (1991) Nature 354: 56
478. Bard AJ, Mallouk T (1992) In: Techniques of chemistry, Series XXII. Wiley, New York, p 270
479. Theng BKG (1974) The chemistry of clay-organic reactions. Adam Hilger, London
480. Van Olphen H (1977) An introduction to clay colloid chemistry, 2nd Edn. Wiley Interscience, New York
481. Newmann ACD (1987) Chemistry of clays and clay minerals. Mineralogical society, London
482. Dékány I, Szántó F, Weiss A, Lagaly G (1985) Ber Bunsenges Phys Chem 89: 62
483. Ogawa M, Handa T, Kuroda K, Kato C, Tani T (1992) J Phys Chem 96: 8116
484. Kotov NA, Putyera K, Fendler JH, Thombácz E, Dékány I (1993) Colloids & Surfaces, in press
485. Rusling JF (1991) Acc Chem Res 24: 75
486. Urabe K, Kouno N, Sakurai H, Izumi Y (1991) Adv Mater 3: 558
487. Turro NJ, Cheng CC, Mahler W (1984) J Am Chem Soc 106: 5022
488. Honda K, Chiba K, Tsuchida E, Frank AJ (1989) J Mater Sci 24: 4004
489. Smotkin ES, Brown Jr RM, Rabenberg LK, Salomon K, Bard AJ, Campion A, Fox MA, Mallouk TE, Webber SE, White JM (1990) J Phys Chem 94: 7543
490. Ballarin B, Brumlik CJ, Lawson DR, Liang W, Van Dyke LS, Martin CR (1992) Anal Chem 64: 2647
491. Foss Jr CA, Tierney MJ, Martin CR (1992) J Phys Chem 96: 9001
492. Baumeister W, Vogell W (1980) Electron microscopy at molecular dimensions. Springer, Berlin Heidelberg New York
493. Clark NA, Douglas K, Rothschild KJ (1986) Appl Phys Lett 48: 676
494. Anderson JR (1975) Structure of metallic catalysis. Academic London
495. Somorjai GA (1981) Chemistry in two dimension: surfaces. Cornell University, Ithaca, NY
496. Somorjai GA (1985) Science 227: 902
497. Bönnemann H, Brijoux W, Brinkman R, Dinjus E, Jousen T, Korall B (1991) Angew Chem Int Ed Eng 30: 1312
498. Boudart M (1969) Adv Catal 20: 153
499. Farin D, Avnir D (1988) J Am Chem Soc 110: 2039
500. Fendler JH (1987) Chem Rev 87: 877
501. Träger F, zu Putlitz G (1986) Metal clusters. Springer, Berlin Heidelberg New York
502. Rademan K (1989) Ber Bunsenges Phys Chem 93: 653
503. Schmid G (1990) Endevavour 14: 172
504. Charlé K.-P, Frank F, Schulze W (1984) Ber Bunsenges Phys Chem 88: 350
505. Stevens AD, Symons MCR (1989) J Chem Soc Faraday Trans I 85: 1439
506. Henglein A, Mulvaney P, Linnert T (1991) Faraday Discuss 92: 31
507. Mulvaney P (1992) In: Mackay RA Texter J (eds) Electrochemistry in colloids and dispersions. VCH Publishers, New York, p 345
508. Henglein A (1993) Israel J Chem 33: 77

509. Weiser HB (1933) Inorganic colloid chemistry, vol 1, The colloidal elements. Wiley, New York
510. Kerker M (1969) The scattering of light and other electromagnetic radiation. Academic Press, New York
511. Mostafavi M, Marignier JL, Amblard J, Belloni J (1989) Z Phys D 12: 31
512. Mulvaney P, Henglein A (1990) J Phys Chem 94: 4182
513. Linnert T, Mulvaney P, Henglein A, Weller H (1990) J Am Chem Soc 112: 4657
514. Mostafavi M, Keghouche N, Delcourt M-O (1990) Chem Phys Lett 169: 81
515. Mostafavi M, Keghouche N, Delcourt M.-O, Belloni J (1990) Chem Phys Lett 167: 193
516. Linnert T, Mulvaney P, Henglein A (1991) Ber Bunsenges Phys Chem 95: 838
517. Doremus RH (1965) J Chem Phys 42: 414
518. Genzel L, Martin TP, Kreibig U (1975) Z Physik B21: 339
519. Cohen RW, Cody GD, Coutts MD, Abeles B (1973) Phys Rev B8: 3689
520. Wertheim GK (1989) Z Phys D 12: 319
521. Charlé K-P, Schulze W, Winter B (1989) Z Phys D 12: 471
522. Creighton JA, Eadon DG (1991) J Chem Soc Farady Trans 87: 3881
523. Lee MH, Dobson PJ, Cantor B (1992) Thin Solid Films 219: 199
524. Doyle WT (1958) Phys Rev 111: 1067
525. Tausch-Treml R, Henglein A, Lilie J (1978) Ber Bunsenges Phys Chem 82: 1335
526. Henglein A, Tausch-Treml R (1981) J Colloid Interface Sci 80: 84
527. Mossert S, Henglein A, Janata E (1989) J Phys Chem 93: 6791
528. Ershov BG, Janata E, Michaelis M, Henglein A (1991) J Phys Chem 95: 8996
529. Henglein A, Janata E, Fojtik A (1992) J Phys Chem 96: 4734
530. Henglein A, Linnert T, Mulvaney P (1990) Ber Bunsenges Phys Chem 94: 1449
531. Mulvaney P, Linnert T, Henglein A (1991) J Phys Chem 95: 7843
532. Henglein A, Mulvaney P, Linnert T, Holzwarth A (1992) J Phys Chem 96: 2411
533. Mulvaney P, Giersig M, Henglein A (1993) J Phys Chem 97: 6334
534. Mulvaney P, Giersig M, Henglein A (1993) J Phys Chem 97: 7061
535. Bohren CF, Huffman DR, (1983) Absorption and scattering of light by small pratcles. Wiley Interscience, New York
536. Johnston PB, Christy RW (1974) Phys Rev B9: 5056
537. Kreibig UZ (1970) Z Physik 234: 307
538. Henglein A, Holzwarth A, Mulvaney P (1992) J Phys Chem 96: 8700
539. Boutonnet M, Kizling J, Mintsa-Eya V, Choplin A, Touroude R, Marie G, Stenius P (1987) J Catal 103: 95
540. Claerbout A, B. Nagy J (1991) Stud Surf Catal 63: 705
541. Itoh H, Miura MO, Okamoto R, Kikuchi E (1991) Bull Chem Soc Jpn 64: 333
542. Barnickel P, Wokaun A, Sager W, Eicke H-F (1992) J Colloid Interface Sci 148: 80
543. B. Nagy J, Derouane EG, Lufimpadio N, Ravet I, Verfaillie JP (1981) In: Millal KL (ed) Surfactants in solution, Vol 10. Plenum Press, New York, p 1
544. B Nagy J, Claerbout A (1991) In: Mittal KL, Shah DO (eds) Surfactants in solution, Vol 11. Plenum Press, New York, p 363
545. B Nagy J (1989) Colloids and Surfaces 35: 201
546. Rosier D, Dallons JL, Jannes G, Puttemans JP (1987) Acta Chim Hung 124: 57
547. Jannes G, Puttemans JP, Vanderwegen P (1989) Catalysis Today 5: 265
548. B Nagy J, Bodart-Ravet I, Derouane EG (1989) Faraday Discuss Chem Soc 87: 189
549. Hirai H (1979) J Macromol Sci Chem A13: 633
550. Subramanian S, Nedeljkovic JM, Patel RC (1992) J Colloid Interface Sci 150: 81
551. Olsen AW, Kafafi ZH (1991) J Am Chem Soc 113: 7758
552. Deschamps A, Lagier J-P, Fievet F, Aeiyach S, Lacaze P-C (1992) J Mater Chem 2: 1213
553. Khatouri J, Mostafavi M, Amblard J, Belloni J (1992) Chem Phys Lett 191: 351
554. Toshima N, Harada M, Yamazaki Y, Asakura K (1992) J Phys Chem 96: 9927
555. Xu B, Kevan L (1991) J Phys Chem 95: 1147
556. Michalik J, Wasowicz T, van der Pol A, Reijerse EJ, de Boer E (1992) J Chem Soc Chem Commun 29
557. Heinrich JL, Curtis CL, Credo GM, Kavanagh KL, Sailor MJ (1992) Science 255: 66
558. Terasaki O, Yamazaki K, Thomas JM, Ohsuna T, Watanabe D, Sanders JV, Barry JC (1987) Nature 330: 58
559. Parise JB, MacDougall JE, Herron N, Farlee R, Sleight AW, Wang Y, Bein T, Moller K, Moroney LM (1988) Inorg Chem 27: 221
560. Malla PB, Ravindranathan P, Komarneni S, Roy R (1991) Nature 351: 555

561. Breitscheidel B, Zieder J, Schubert U (1991) Chem Mater 3: 559
562. Kurihara K, Kizling J, Stenius P, Fendler JH (1983) J Am Chem Soc 105: 2574
563. Kurihara K, Fendler JH (1983) J Am Chem Soc 105: 6152
564. Kurihara K, Fendler JH, Ravet I, B Nagy J (1983) J Mol Catal 34: 325
565. Horváth O, Fendler JH (1992) J Phys Chem 96: 9591
566. Zhao XK, Fendler JH (1990) J Phys Chem 94: 3384
567. Kotov NA, Zaniquelli MED, Fendler JH (1993) Presentation at the Electrochemical Society Meeting, Hawaii, May 16–21
568. Zhao XK, Fendler JH (1990) unpublished results
569. Raether H (1988) Surface plasmons on smooth and rough surfaces and gratings. Springer, Berlin Heidelberg New York
570. Kotov N, Meldrum F, Fendler JH (1993) unpublished results
571. Kamat PV, Dimitrijević NM (1990) Solar Energy 44: 83
572. Steigerwald ML, Brus LE (1990) Acc Chem Res 23: 183
573. Wang Y, Herron N (1991) J Phys Chem 95: 525
574. Kamat PV (1991) In: Grätzel M, Kalyanasundaram K (eds) Kinetics and catalysis in microheterogeneous systems, Surfactant Science Series vol 38. Marcel Dekker, New York, p 375
575. Weller H (1993) Angew Chem Int Ed Eng 32: 41
576. Weller H (1993) Adv Mater 5: 88
577. Weller H, Fojtik A, Henglein A (1985) Chem Phys Lett 117: 485
578. O'Neil M, Marohn J, McLendon G (1990) J Phys Chem 94: 4356
579. Eychmüller A, Hasselbarth A, Katsikas L, Weller H (1991) Ber Bunsengesell Phys Chem 1: 79
580. Shen YR (1989) Annu Rev Phys Chem 40: 327
581. Brus LE (1983) J Chem Phys 79: 5566
582. Brus LE (1984) J Chem Phys 80: 4403
583. Schmidt HM, Weller H (1986) Chem Phys Lett 129: 615
584. Kayanuma Y (1986) Solid State Commun 59: 405
585. Wang Y, Suna A, Mahler W, Kasowski R (1987) J Chem Phys 87: 7315
586. Wang Y, Herron N (1990) Phys Rev B: Condens Matter 42: 7253
587. Lippens PE, Lanoo M (1989) Phys Rev B: Condens Matter 39: 935
588. Nosaka Y (1991) J Phys Chem 95: 5054
589. Rama Krishna MV, Friesner RA (1991) J Chem Phys 95: 8309
590. Einevoll GT (1992) Phys Rev B: Condens Matter 45: 3410
591. Eychmüller A, Katsikas L, Weller H (1990) Langmuir 6: 1605
592. Herron N, Wang Y, Eckert H (1990) J Am Chem Soc 112: 1322
593. Wang Y, Herron N (1991) Research Chem Intermed 15: 17
594. Herron N, Suna A, Wang Y (1992) J Chem Soc Dalton Trans 2329
595. Spanhel L, Haase M, Weller H, Henglein A (1987) J Am Chem Soc 109: 5649
596. Katsikas L, Eychmüller A, Giersig M, Weller H (1990) Chem Phys Lett 172: 201
597. Eychmüller A, Hässelbarth A, Katsikas L, Weller H (1991) J Lumines 48&49: 475
598. Wang Y, Mahler W (1987) Opt Commun 61: 233
599. Hilinski E, Lucas P, Wang Y (1988) J Chem Phys 89: 3435
600. Wang Y, Suna A, McHugh J, Hilinski E, Lucas P, Johnson RD (1990) J Chem Phys 92: 6927
601. Baral S, Fojtik A, Weller H, Henglein A (1988) J Am Chem Soc 108: 375
602. Henglein A, Kumar A, Janata E, Weller H (1986) Chem Phys Lett 132: 133
603. Haase M, Weller H, Henglein A (1988) J Phys Chem 92: 4706
604. Wang Y (1991) J Phys Chem 95: 1119
605. Kortan AR, Hull R, Opila RL, Bawendi MG, Steigerwald ML, Caroll PJ, Brus LE (1990) J Am Chem Soc 112: 1327
606. Hässelbarth A, Eychmüller A, Eichberger R, Giersig M, Weller H (1993) J Phys Chem (in press)
607. Hässelbarth A, Eychmüller A, Weller H (1993) Chem Phys Lett (in press)
608. Gopidas KR, Bohorquez M, Kamat PV (1990) J Phys Chem 94: 6435
609. Kamat PV (1991) J Am Chem Soc 113: 9705
610. Kietzman R, Willig F, Weller H, Vogel R, Nath DN, Eichberger R, Liska P, Lehnert J (1991) Mol Cryst Liq Cryst 194: 169
611. Meyer M, Wallberg C, Kurihara K, Fendler JH (1984) J Chem Soc Chem Commun 90
612. Lianos P, Thomas JK (1986) Chem Phys Lett 125: 299
613. Petit C, Pileni MP (1988) J Phys Chem 92: 2282

614. Petit C, Lixon P, Pileni MP (1990) J Phys Chem 94: 1598
615. Motte L, Petit C, Boulanger L, Lixon P, Pileni MP (1992) Langmuir 8: 1049
616. Tricot Y-M, Rafaeloff R, Emeren Å, Fendler JH (1985) ACS Symp Ser 278: 99
617. Modes S, Lianos P (1989) J Phys Chem 93: 5854
618. Towey TF, Khan-Lodhi A, Robinson BH (1990) J Chem Soc Faraday Trans 86: 3757
619. Dannhauser T, O'Neil M, Johansson K, Whitten D, McLendon G (1986) J Phys Chem 90: 6074
620. Barzykin AV, Fox MA (1993) private communication
621. Steigerwald ML, Alivisatos AP, Gibson JM, Harris TD, Kortan R, Muller AJ, Thayer AM, Duncan TM, Douglass DC, Brus LE (1988) J Am Chem Soc 110: 3046
622. Alivisatos AP, Harris TD, Carroll PJ, Steigerwald ML, Brus LE (1989) J Chem Phys 90: 3463
623. Sachleben JR, Wooten EW, Emsley L, Pines A, Colvin VL, Alivisatos AP (1992) Chem Phys Lett 198: 431
624. Tricot Y-M, Fendler JH (1986) J Phys Chem 90: 3369
625. Watzke HJ, Fendler JH (1987) J Phys Chem 91: 854
626. Tricot Y-M, Fendler JH (1984) J Am Chem Soc 106: 2475
627. Tricot Y-M, Fendler JH (1984) J Am Chem Soc 106: 7359
628. Tricot Y-M, Emeren Å, Fendler JH (1985) J Phys Chem 89: 4721
629. Tricot Y-M, Fendler JH (1986) In: Pelizetti E, Serpone N (eds), Homogeneous and heterogeneous photocatalysis. D. Reidel, Dordrecht, Holland, p 241
630. Rafaeloff R, Tricot Y-M, Nome F, Fendler JH (1985) J Phys Chem 89: 533
631. Youn H-C, Tricot Y-M, Fendler JH (1987) J Phys Chem 91: 581
632. Youn H-C, Baral S, Fendler JH (1988) J Phys Chem 92: 6320
633. Chang A-C, Fendler JH (1989) J Phys Chem 93: 2538
634. Chang A-C, Pfeiffer WF, Guillaume B, Baral S, Fendler JH (1990) J Phys Chem 94: 4284
635. Tricot Y-M, Manassen J (1988) J Phys Chem 92: 5239
636. Tricot Y-M, Porat Z, Manassen J (1991) J Phys Chem 95: 3242
637. Khramov MI, Parmon VN (1993) J Photochem Photobiol, in press
638. Zhao XK, Yuan Y, Fendler JH (1990) J Chem Soc Chem Commun 1248
639. Zhao XK, Xu S, Fendler JH (1991) Langmuir 7: 520
640. Zhao XK, Fendler JH (1991) Chem Mater 3: 168
641. Zhao XK, Fendler JH (1991) J Phys Chem 95: 3716
642. Yuan Y, Cabasso I, Fendler JH (1990) Chem Mater 2: 226
643. Yi KC, Fendler JH (1990) Langmuir 6: 1519
644. Zhao XK, McCormick LD, Fendler JH (1991) Chem Mater 3: 922
645. Zhao XK, McCormick LD, Fendler JH (1991) Langmuir 7: 1255
646. Zhao XK, McCormick LD, Fendler JH (1992) Adv Mater 4: 93
647. Zhao XK, Yang J, McCormick LD, Fendler JH (1992) J Phys Chem 96: 9933
648. Yang J, Fendler JH, Jao T.-C, Laurion T (1993) J Electron Microscopy Tech (in press)
649. Baral S, Zhao XK, Rolandi R, Fendler JH (1987) J Phys Chem 91: 2701
650. Zhao XK, Baral S, Rolandi R, Fendler JH (1988) J Am Chem Soc 110: 1012
651. Baral S, Fendler JH (1989) J Am Chem Soc 111: 1604
652. Zhao XK, Baral S, Fendler JH (1990) J Phys Chem 94: 2043
653. Kutnik J, Tien HT (1987) Photochem Photobiol 46: 413
654. Rolandi R, Ricci D (1990) Prog Colloid Polym Sci 81: 222
655. Rolandi R, Ricci D, Brandt O (1992) J Phys Chem 96: 6783
656. Kimizuka N, Miyoshi T, Ichinose I, Kunitake T (1991) Chem Lett (Jpn) 2039
657. Tien HT, Chen J-W (1990) Int J Hydrogen Ener 15: 563
658. Smotkin ES, Lee C, Bard AJ, Campion A, Fox MA, Mallouk TE, Webber SE, White JM (1988) Chem Phys Lett 152: 265
659. Xu S, Zhao XK, Fendler JH (1990) Adv Mater 2: 183
660. Leloup J, Ruaudel-Teixier A, Barraud A (1992) Thin Solid Films 210/211: 407
661. Grieser F, Furlong DN, Scoberg D, Ichinose I, Kimizuka N, Kunitake K (1992) J Chem Soc Faraday Trans 88: 2207
662. Du Z, Zhang Z, Zhao W, Zhu Z, Zhang J, Jin Z, Li T (1992) Thin Solid Films 210/211: 404
663. Peng X, Wei Q, Jiang Y, Chai X, Li T, Shen J (1992) Thin Solid Films 210/211: 401
664. Peng X, Zhang Y, Yang J, Zou B, Xiao L, Li T (1992) J Phys Chem 96: 3412
665. Zhu R, Min G, Wei Y, Schmitt HJ (1992) J Phys Chem 96: 8210
666. Peng X, Guan S. Chai X, Jiang Y, Li T (1992) J Phys Chem 96: 3170
667. Colvin VL, Goldstein AN, Alivisatos AP (1992) J Am Chem Soc 114: 5221

668. Makhmadmurodov A, Gruzdkov YA, Savino EN, Parmon VN (1986) Kinetika i Kataliz 27: 133
669. Finlayson MF, Park KH, Kakuta N, Bard AJ, Campion A, Fox MA, Webber SE, White JM (1988) J Lumines 39: 205
670. Smotkin ES, Brown Jr RM, Rabenberg LK, Salomon K, Bard AJ, Campion A, Fox MA, Mallouk TE, Webber SE, White JM (1990) J Phys Chem 94: 7543
671. Tennakone K, Ileperuma OA, Bandara JMS, Thaminimulla CTK, Ketipearachchi US (1991) J Chem Soc Chem Commun 579
672. Miyoshi H, Tanaka K, Uchida H, Yoneyama H, Mori H, Sakata T (1990) J Electroanal Chem 295: 71
673. Miyoshi H, Yamachika M, Yoneyama H, Mori H (1990) J Chem Soc Faraday Trans 86: 815
674. Miyoshi H, Nippa S, Uchida H, Mori H, Yoneyama H (1990) Bull Chem Soc Jpn 63: 3380
675. Gopidas KR, Kamat P (1990) Mater Lett 9: 372
676. Yanagida S, Enokida T, Shindo A, Shiragami T, Ogata T, Fukumi T, Sakaguchi H, Sakata T (1990) Chem Lett (Jpn) 1773
677. Wang Y, Herron N, Mahler W, Suna A (1989) J Opt Soc Am B6: 808
678. Wang Y, Herron N (1992) Chem Phys Lett 200: 71
679. Dalas E, Sakkopoulos S, Kallitsis J, Vitoratos E, Koutsoukos PG (1990) Langmuir 6: 1356
680. Yamamoto T, Taniguchi A, Dev S, Kubota E, Osakada K, Kubota K (1991) Colloid Polym Sci 269: 969
681. Yuan Y, Cabasso I, Fendler JH (1990) Macromol 23: 3198
682. Yuan Y, Fendler JH, Cabasso I (1992) Chem Mater 4: 312
683. Willner I, Eichen Y (1987) J Am Chem Soc 109: 6862
684. Goren Z, Willner I, Nelson AJ, Frank AJ (1990) J Phys Chem 94: 3784
685. Rajh T, Vucemilovic MI, Dimitrijevic NM, Micic OI, Nozik AJ (1988) Chem Phys Lett 143: 305
686. Chepic DI, Efros AL, Ekimov AI, Ivanov MG, Kharchenko VA, Kudriavtsev IA, Yazeva TV (1990) J Lumines 47: 113
687. O'Regan B, Moser J, Anderson M, Grätzel M (1990) J Phys Chem 94: 8720
688. Champagnon B, Andrianasolo B, Duval E (1991) J Chem Phys 94: 5237
689. Minti H, Eyal M, Reisfeld R, Berkovic G (1991) Chem Phys Lett 183: 277
690. Sabate J, Anderson MA, Kikkawa H, Edwards M, Hill Jr CG (1991) J Catal 127: 167
691. Rosenberg I, Brock JR, Heller A (1992) J Phys Chem 96: 3423
692. Sakohara S, Tickanen LD, Anderson MA (1992) J Phys Chem 96: 11086
693. Xu Q, Anderson MA (1989) Mat Res Soc Symp Proc 132: 41
694. Spanhel L, Anderson MA (1990) J Am Chem Soc 112: 2278
695. Spanhel L, Anderson MA (1991) J Am Chem Soc 113: 2826
696. Liu X, Thomas JK (1989) Langmuir 5: 58
697. MacDougall JE, Eckert H, Stucky GD, Herron N, Wang Y, Moller K, Bein T, Cox D (1989) J Am Chem Soc 111: 8006
698. Wang Y, Herron N (1987) J Phys Chem 91: 257
699. Wang Y, Herron N (1988) J Phys Chem 92: 4988
700. Uchida H, Ogata T, Yoneyama H (1990) Chem Phys Lett 173: 103
701. Stramel RD, Nakamura T, Thomas JK (1988) J Chem Soc Faraday Trans I 84: 1287
702. Liu X, Thomas JK (1989) J Colloid Interface Sci 129: 476
703. Cao G, Rabenberg LK, Nunn CM, Mallouk TE (1991) Chem Mater 3: 149
704. Miyoshi H, Mori H, Yoneyama H (1991) Langmuir 7: 503
705. Kotov NA, Fendler JH (1993) unpublished results
706. Dameron CT, Reese RN, Mehra RK, Kortan AR, Carroll PJ, Steigerwald ML, Brus LE, Winge DR (1989) Nature 338: 5996
707. Dameron CT, Winge DR (1990) Inorg Chem 29: 1343
708. Meldrum FC, Wade VJ, Nimmo DL, Heywood BR, Mann S (1991) Nature 349: 684
709. Bigham SR, Coffer JL (1992) J Phys Chem 96: 10581
710. Sagiv J (1993) unpublished results, private communication
711. Vlachopoulos N, Liska P, Augustynski J, Grätzel M (1988) J Am Chem Soc 110: 1216
712. O'Regan B, Grätzel M (1991) Nature 353: 737
713. Spanhel L, Weller H, Henglein A (1987) J Am Chem Soc 109: 6632
714. Ennaoui A, Fiechter S, Tributsch H, Giersig M, Vogel R, Weller H (1992) J Electrochem Soc 139: 2514
715. Bardeen J, Cooper LN, Schrieffer JR (1957) Phys Rev 108: 1175

716. Lynton EA (1971) Superconductivity, Third Ed., Halsted, New York
717. Adrian FJ, Cowan DO (1992) Chem Eng News (Dec 21): 24
718. Skotheim TJ (1986) Handbook of conducting polymers. Marcel Dekker, New York
719. Bredas JL, Street GB (1985) Acc Chem Res 18: 309
720. Fuchs H, Ohst H, Prass W (1991) Adv Mat 3: 10
721. Richard J, Delhaes P, Vandevyver M (1991) New J Chem 15: 137
722. Vandevyver M (1992) Thin Solid Films 210/211: 240
723. Polymeropoulos EE (1978) Solid State Comm 28: 883
724. Polymeropoulos EE, Möbius D, Kuhn H (1978) J Chem Phys 68(8): 3918
725. Kuhn H (1979) J Photochem 10: 111
726. Ruaudel-Teixier A, Barraud A, Vandevyver M, Belbeoch B, Roulliay M (1985) J Chim Phys 82: 711
727. Richard J, Vandevyver M, Lesieur P, Ruaudel-Teixier A, Barraud A, Bozio R, Pecile C (1987) J Chem Phys 86(4): 2428
728. Dhindsa AS, Davies GH, Bryce MR, Yarwood J, Lloyd JP, Petty MC, Lvov YM (1989) J Mol Elect 5: 135
729. Vandevyver M, Barraud A, Lesieur P, Richard J, Ruaudel-Teixier A (1986) J Chim Phys 83: 599
730. Richard J, Barraud A, Vandevyver M, Ruaudel-Teixier A (1988) Thin Solid Films 159: 207
731. Barraud A, Lesieur P, Richard J, Ruaudel-Teixier A, Vandevyver M, Lequan M, Lequan RM (1988) Thin Solid Films 160: 81
732. Vandevyer M, Richard J, Barraud A, Ruaudel-Teixier A, Lequan M, Lequan RM (1987) J Chem Phys 87(11): 6754
733. Barraud A, Lequan M, Lequan RM, Lesieur P, Richard J, Ruaudel-Teixier A, Vandevyver M (1987) J Chem Soc Chem Commun 797
734. Richard J, Vandevyver M, Barraud A, Morand JP, Lapouyade R, Delhaes P, Jacquinot JF, Roulliay M (1988) J Chem Soc Chem Commun 754
735. Morand JP, Lapouyade R, Delhaes P, Vandevyver M, Richard J, Barraud A (1988) Synthetic Metals 27: B569
736. Vandevyver M, Roulliay M, Bourgoin JP, Barraud A, Gionis V, Kakoussis VC, Mousdis GA, Morand JP, Noel O (1991) J Phys Chem 95: 246
737. Vandevyver M, Roulliay M, Bourgoin JP, Barraud A, Morand JP, Noel O (1991) J Colloid Interface Sci 141: 459
738. Richard J, Vandevyver M, Barraud A, Morand JP, Delhaes P (1989) J Colloid Interface Sci 129: 254
739. Nakamura T, Matsumoto M, Takel F, Tanaka M, Sekiguchi T, Manda E, Kawabata Y (1986) Chem Lett (Jpn) 709
740. Matsumoto M, Nakamura T, Takei F, Tanaka M, Sekiguchi T, Mizuno M, Manda E, Kawabata Y (1987) Synthetic Metals 19: 675
741. Nakamura T, Takei F, Tanaka M, Matsumoto M, Sekiguchi T, Manda E, Kawabata Y, Saito G (1986) Chem Lett (Jpn) 323
742. Kawabata Y, Nakamura T, Matsumoto M, Tanaka M, Sekiguchi T, Komizu H, Manda E, Saito G (1987) Synthetic Metals 19: 663
743. Tachibana H, Komizu H, Nakamura T, Matsumoto M, Tanaka M, Manda E, Kawabata Y, Kato T (1989) Chem Lett (Jpn) 841
744. Tachibana H, Azumi R, Nakamura T, Matsumoto M, Kawabata Y (1992) Chem Lett (Jpn) 173
745. Tachibana H, Nishio Y, Nakamura T, Matsumoto M, Manda E, Niino H, Yabe A, Kawabata Y (1992) Thin Solid Films 210/211: 293
746. Nakamura T, Kojima K, Matsumoto M, Tachibana H, Tanaka M, Manda E, Kawabata Y (1989) Chem Lett (Jpn) 367
747. Miura YF, Takenaga M, Kasai A, Nakamura T, Nishio Y, Matsumoto M, Kawabata Y (1992) Thin Solid Films 210/211: 306
748. Akatsuka T, Tanaka H, Toyama J, Nakamura T, Kawabata Y (1990) Chem Lett (Jpn) 975
749. Akatsuka T, Tanaka H, Toyama J, Nakamura T, Matsumoto M, Kawabata Y (1991) Chem Lett (Jpn) 1351
750. Akatsuka T, Tanaka H, Toyama J, Nakamura T, Matsumoto M, Kawabata Y (1992) Thin Solid Films 210/211: 458
751. Ikegami K, Kuroda S.-i, Tabe Y, Saito K, Saito M, Sugi M, Nakamura T, Tachibana H, Matsumoto M, Kawabata Y (1992) Thin Solid Films 210/211: 303
752. Dhindsa AS, Bryce MR, Lloyd JP, Petty MC (1987) Synthetic Metals 22: 185
753. Dhindsa AS, Pearson C, Bryce MR, Petty MC (1989) J Phys D: Appl Phys 22: 1586

754. Pearson C, Dhindsa AS, Bryce MR, Petty MC (1989) Synthetic Metals 31: 275
755. Dhindsa AS, Badyal JP, Bryce MR, Petty MC, Moore AJ, Lvov YM (1990) J Chem Soc Chem Commun 970
756. Dhindsa AS, Bryce MR, Ancelin H, Petty MC, Yarwood J (1990) Langmuir 6: 1680
757. Dhindsa AS, Badyal JP, Pearson C, Bryce MR, Petty MC (1991) J Chem Soc Chem Commun 322
758. Pearson C, Dhindsa AS, Petty MC, Bryce MR (1992) Thin Solid Films 210/211: 257
759. Logsdon PB, Pfleger J, Prasad PN (1988) Synthetic Metals 26: 369
760. Nishikata Y, Kakimoto M.-a, Imai Y (1988) J Chem Soc Chem Commun 1040
761. Wegmann A, Tieke B, Mayer CW, Hilti B (1989) J Chem Soc Chem Commun 716
762. Wegmann A, Hunziker M, Tieke B (1989) J Chem Soc Chem Commun 1179
763. Tieke B, Wegmann A, Fischer W, Hilti B, Mayer CW, Pfeiffer J (1989) Thin Solid Films 179: 233
764. Rosner RB, Rubner MF (1991) J Chem Soc Chem Commun 1449
765. Troitsky VI, Berzina TS, Sotnikov PS, Ujinova TV, Neiland OY (1990) Thin Solid Films 187: 337
766. Ayrapetiants SV, Berzina TS, Shikin SA, Troitsky VI (1992) Thin Solid Films 210/211: 261
767. Taylor DM, Gupta SK, Underhill AE, Wainwright CE (1992) Thin Solid Films 210/211: 287
768. Cheung JH, Punkka E, Rikukawa M, Rosner RB, Royappa AT, Rubner MF (1991) ACS Polym Preprints, Polym Mat Sci Eng 64: 263
769. Cheung JH, Punkka E, Rikukawa M, Rosner RB, Royappa AT, Rubner MF (1992) Thin Solid Films 210/211: 246
770. Hong K, Rosner RB, Rubner MF (1990) Chem Mater 2: 82
771. Punkka E, Rubner MF (1992) Thin Solid Films 213: 117
772. Royappa AT, Rubner MF (1992) Langmuir 8: 3168
773. Vandevyver M, Bourgoin J-P, Perez X, Veber M, Jallabert C, Strezelecka H (1992) J Phys D: Appl Phys 25: 284
774. Bourgoin J-P, Ruaudel-Teixier A, Vandevyver M, Roulliay M, Barraud A, Lequan M, Lequan R-M (1992) Thin Solid Films 210/211: 250
775. Dourthe C, Izumi M, Garrigou-Lagrange C, Buffeteau T, Desbat B, Delhaes P (1992) J Phys Chem 96: 2812
776. Paloheimo J, Stubb H, Yli-Lahti P, Dyreklev P, Inganäs O (1992) Thin Solid Films 210/211: 283
777. Naito K, Miura A, Azuma M (1992) Thin Solid Films 210/211: 268
778. Chien JCW, Gong BM, Yang Y, Cabrera I, Effing J, Ringsdorf H (1990) Adv Mater 2: 305
779. Blakemore JS (1985) Solid state physics, 2nd edn. Cambridge University Press, New York
780. Siegmann HC (1992) J Phys Condens Matter 4: 8395
781. Buchachenko AL (1990) Russ Chem Rev 59: 307
782. Miller JS, Epstein AJ, Reiff WM (1988) Chem Rev 88: 201
783. Seul M (1990) Mat Res Soc Symp Proc 177: 399
784. Gobe M, Kigiro K-N, Kandori K, Kitahara A (1983) J Colloid Interface Sci 93: 293
785. Liebert L, Martinet A (1979) J Phys Lett 40: L-363
786. Liebert L, Martinet A (1980) IEEE Trans Magn MAG-16: 266
787. Mann S, Skarnulis AJ, Williams RJP (1979) J Chem Soc Chem Commun 1067
788. Mann S, Hannington JP (1988) J Colloid Interface Sci 122: 326
789. Herve P, Nome F, Fendler JH (1984) J Am Chem Soc 106: 8291
790. Watzke H, Szezurek A, Fendler JH (1986) unpublished results
791. Mann S, Hannington JP, Williams RJP (1986) Nature (London) 324: 565
792. de Cuyper M, Joniau M (1991) Langmuir 7: 647
793. de Cuyper M, Joniau M (1990) Biochim Biophys Acta 1027: 172
794. Reed W, Fendler JH (1986) J Appl Phys 59: 2914
795. Zhao XK, Herve PJ, Fendler JH (1989) J Phys Chem 93: 908
796. Zhao XK, Xu S, Fendler JH (1990) J Phys Chem 94: 2573
797. Kunitake T (1992) Thin Solid Films 210/211: 48
798. Mendoza EA, Wolkow E, Sunil D, Wong P, Sokolov J, Rafailovich MH, den Boer M, Gafney HD (1991) Langmuir 7: 3046
799. Turro N, Kraeutler B (1980) Acc Chem Res 13: 369
800. Boxer SG, Chidsey ED, Roelofs MG (1983) Annu Rev Phys Chem 34: 389
801. Gould IR, Turro NJ, Zimmt MB (1984) In: Advances in physical organic chemistry. Academic Press, London, p 1

802. deJongh LJ, Miedema AR (1974) Adv Phys 23: 1
803. Pomerantz M (1980) In: Dash JG, Ruvalds J (eds) Phase transitions in surface films. Plenum Publishing Corporation, New York, p 317
804. Pomerantz M (1987) In: Mittal KL (ed) Surface and colloid science in computer technology. Plenum Publishing Corporation, New York, p 361
805. Haseda T, Yamakawa H, Ishizuka M, Okuda Y, Kubota T, Hata M, Amaya K (1977) Solid State Comm 24: 599
806. Ulrich DR (1990) C&E 68: 28
807. Segal D (1991) Chemical synthesis of advanced ceramic materials. Cambridge University Press, Cambridge
808. Brinker CJ, Clark DE, Ulrich DR (1986) Better ceramics through chemistry. Mat Res Soc Symp Proc vol 73. Materials Research Society, Pittsburgh
809. Brinker CJ, Scherer GW (1990) Sol-gel science, The physics and chemistry of sol-gel processing. Academic Press, Boston
810. Aksay IA (1991) In: Vincenzini P (ed) Ceramics today – tomorrow's ceramics, Materials science monographs, 66A. Elsevier Science Publishers, Amsterdam, p 49
811. Rhine WE, Bowen HK (1991) In: Vincenzini P (ed) Ceramics today – tomorrow's ceramics, Materials science monographs, 66B. Elsevier Science Publishers, Amsterdam, p 749
812. Matijević E (1987) In: Vincenzini P (ed) High tech ceramics, materials science monographs, 38A. Elsevier Science Publishers, Amsterdam, p 441
813. Matijević E (1981) Acc Chem Res 14: 22
814. Matijević E (1985) Annu Rev Mater Sci 15: 483
815. Matijević E (1986) Langmuir 2: 12
816. McFadyen P, Matijević E (1973) J Colloid Interface Sci 44: 95
817. Sapieszko RS, Matijević E (1980) J Colloid Interface Sci 74: 405
818. Sapieszko RS, Matijević E (1980) Corrosion 36: 522
819. Kodas TT (1989) Angew Chem Adv Mater 101: 814
820. Sproson DW, Messing GL (1987) In: Messing GL, Mazdiyashi KS, McCauley JW, Haber RA (eds) Advanced ceramics, vol 21, Ceramic powder science. American Ceramic Society, Westerville, Ohio, p 99
821. Visca M, Matijević E (1979) J Colloid Interface Sci 68: 308
822. Ingebrethsen BJ, Matijević E (1980) J Aerosol Sci 11: 271
823. Ingebrethsen BJ, Matijević E Partch RE (1983) J Colloid Interface Sci 95: 228
824. Flagan RC (1988) In: Messing G, Fuller Jr E, Hausner H (eds) Ceramic transactions, vol 1A, American Ceramic Society, Westerville, Ohio, p 229
825. Wachtman Jr JB, Haber RA (1986) Chem Eng Prog 82: 39
826. Liu H, Graff GL, Hyde M, Sarikaya M, Aksay IA (1991) Proc Mater Res Soc Vol 218, Alper M, Calvert P, Frankel R, Rieke P, Tirrell D (eds) Materials Research Society, Pittsburgh, Pa, p 115
827. Hardy AB, Rhine WE, Bowen HK (1993) J Am Ceram Soc 76(1): 97
828. Akinc M, Richardson K (1986) Mat Res Soc Symp Proc 73: 99
829. Pellerin N, Graff GL, Treadwell DR, Staley JT, Aksay IA (1993) Biomimetics (in press)
830. Calvert P, Broad A (1990) Mat Res Soc Symp 174: 61
831. Calvert PD, Broad RA (1989) In: Culbertson BM (ed) Contemporary topics in polymer science. Multiphase macromolecular systems, Vol 6. Plenum Press, New York, p 95
832. Wang B, Wilkes GL, Hedrick JC, Liptak SC, McGrath JE (1991) Macromol 24: 3449
833. Huang H, Orler B, Wilkes GL (1987) Macromol 20: 1322
834. Jean JH, Ring TA (1986) Mat Res Soc Symp Proc 73: 85

10 Glossary of Terms

Adaptive Materials
Man-made materials which can, to different degrees, self assemble, self diagnose, self repair, and recognize and discriminate physical and/or chemical stimuli, and, at the extreme, which have the capability of learning and self replicating. The term is often considered to be synonymous with "intelligent materials".

Advanced Materials
Man-made materials having superior mechanical, thermal, electrical, optical, and other desirable
   properties.
Aerosols
Small Particles (10^{-6} to 10^{-8} m range) dispersed in air.
Antiferromagnets
Antiferromagnets have neighboring magnetic dipoles aligned antiparallel; there is no net macro-
   scopic moment in zero applied field and the susceptibility is anisotropic below the Néel
   temperature, T_N (See Fig. 128).
Aqueous Micelles
Aggregates generated in the spontaneous and dynamic association of *ca.* 50–100 surfactant molecules
   above a characteristic surfactant concentration, labeled the critical micelle concentration (CMC).
Band-gap
The energy difference between the valence and conduction bands in semiconductors. It is related to
   the absorption edge of the semiconductor.
Bilayer Lipid Membranes (BLMs)
Bimolecularly thick films formed at a pinhole (1–2 mm) separating two aqueous solutions. The BLM
   is formed by painting the surfactant (or lipid), dissolved in a hydrocarbon solvent, across
   the pinhole and allowing it to thin to two layers of closely packed molecules which are apposed
   tail-to-tail.
Biological Membranes
Organized bi- and multilayers, composed of lipids, proteins, and other components, which constitute
   the internal and external cell walls. Biological membranes allow selective transport of materials
   into and out of cells and, thus , mediate numerous biochemical processes.
Biomaterials
Materials derived from biological substances.
Biomineralization
In vivo formation of inorganic crystals and/or amorphous particles in biological systems.
Cast multibilayers
Ultrathin films formed by the controlled evaporation of SUVs (small unilamellar vesicles) and
   MLVs (multilamellar vesicles) on substrates.
Colloids
Suspensions in which the dispersed phase is so small that the gravitational force can be neglected.
   Particles with linear dimensions in the 10^{-6} to 10^{-8} m range are considered to be colloidal.
Diamagnetic Materials
Materials which contain electrons in paired states and exhibit weak negative (about 10^{-6})
   susceptibilities (they are pushed out of a magnetic field).
Epitaxy
Lattice matching between two crystalline phases.
Ferrimagnets
Ferrimagnetism occurs in materials possessing antiferromagnetically aligned magnetic dipoles of
   unequal spins (see Fig. 128).
Ferromagnets
Ferromagnets have their spins aligned parallel and exhibit large positive (about 10^2) magnetic
   susceptibilities. They possess a macroscopic spontaneous magnetization at zero applied field
   below the Curie temperature, T_C, and have a characteristic saturation moment, M_s, in a finite
   applied field (see Fig. 128).
Flexoelectricity (Curvature-induced electricity)
Generation of microvolt-range transmembrane (trans-BLM) potential differences through the
   periodic bending of BLMs.
Gel
A continuous solid phase of a substance enclosing a continuous liquid phase of the same or of a
   different substance. Gels can be formed from sols.
Genetically Engineered Materials
Materials obtained from biological substances via genetic engineering.
Ghost Vesicles
Cast bilayers formed in the removal of surfactants from polymer-coated vesicles.

Hierarchically Organized Materials

Components (or subunits) of a complex material arranged for optimal overall performance and constructed from and organized into discrete subunits (ranging from the atomic to the molecular scale).

Hydrosols

Small particles (10^{-6} to 10^{-8} m range) dispersed in water.

Intelligent Materials

Man-made materials which can, to different degrees, self assemble, self diagnose, self repair, and recognize and discriminate physical and/or chemical stimuli, and, at the extreme, which have the capability of learning and self-replicating. The term is often considered to be synonymous with "adaptive materials".

Langmuir Blodgett (LB) Film

A monolayer or several monolayers transferred to solid substrates. Repeated withdrawal and dipping of a hydrophilic substrate through the monolayer leads to the buildup of a substrate-head-tail-tail-head-head Y-type multilayer LB film. Alternatively, consecutive dippings (i.e. no alternating withdrawal through the monolayer) of a hydrophobic substrate results in a substrate-tail-head-tail-head, or X-type, multilayer deposition. Consecutive withdrawals (i.e. no alternating dipping through the monolayer) of a hydrophilic substrate produces a substrate-head-tail-head-tail Z-type multilayer LB film (See Fig. 10).

Langmuir Film (or Monolayer)

A monomolecular layer of surfactants (or lipids) floating on an aqueous solution (the subphase), usually in a Langmuir film balance (the trough).

Liposomes

Vesicles prepared from naturally occurring lipids. The term "vesicles" is sometimes used synonymously with the term "liposomes".

Liquid Crystals

The liquid-crystalline state is an intermediate between the solid and the liquid states. Liquid crystals are condensed states with spontaneous anisotropy.

Magnetic Particles

Separate nanometer- to micron-sized colloidal magnetic particles, often dispersed in solutions; they are usually stabilized by polyions or polymers.

Magnetic Particulate Films

Physically interconnected colloidal magnetic particles supported by a monolayer, a BLM, or a solid substrate.

Membrane Mimicking

Construction of artificial systems which imitate the essential functions of biological membranes: molecular compartmentalization, organization, and discrimination. Advanced materials are in situ generated in, or incorporated into, membrane-mimetic systems. A molecular-level understanding of the membrane-mimetic hosts and the advanced-material guests is a given.

Metallic Particles

Separate nanometer- to micron-sized colloidal metals, often dispersed in solutions; they are usually stabilized by polyions or polymers.

Metallic Particulate Films

Physically interconnected colloidal metal particles supported by a monolayer, a BLM, or a solid substrate.

Microemulsions

Clear, stable dispersions of particles whose sizes are larger than those of micelles. Conceptually, they can be derived from miscelles by increasing the surfactant concentration above the CMC (or above the region where reversed miscelles predominate) and/or by adding a third component (alcohol, for example) which leads to the formation of larger aggregates: oil-in-water (o/w) or water-in-oil (w/o) microemulsions.

Monolayer (or Langmuir film)

A monomolecular layer of surfactants (or lipids) floating on an aqueous solution (the subphase), usually in a Langmuir film balance (the trough).

Multilamellar Vesicles (MLVs)

Closed, onion-like, 1000- to 8000-Å-diameter multibilayer aggregates, formed upon the swelling of thin surfactant (or lipid) films in water.

Nanoparticle Fabrication by Physical Methods
Particles are generated by the size reduction of bulk materials by engineering-type (physical) manipulation.
Nanoparticle Formation by Chemical Methods
Particles are generated from their molecular components by chemical synthesis.
Nanosized Materials
Materials which have dimensions in the 1–100 nm range.
Nanosized Metallic Particles
Nanosized particles (5–10 nm in diameter for silver) possessing metallic properties, i.e. conducting electricity and having sharp plasmon absorption bands.
Nanostructured Materials
Materials which have dimensions in the 1–100 nm range.
Nanostructures
Materials which have dimensions in the 1–100 nm range.
Non-metallic(Silver) Clusters
Small aggregates of silver ($D < 2$ nm) having sharp absorption bands in the 300–350 nm region and non-metallic properties.
Organoclay Complexes(Pillared clay minerals)
Clays (aluminosilicates) in which the sodium cations have been partially or fully exchanged by surfactants.
Organosols
Small particles (10^{-6} to 10^{-8} m range) dispersed in an organic solvent.
Paramagnetic Materials
Materials which contain unpaired electrons and small positive (about 10^{-4}) susceptibilities (they are pulled into a magnetic field) (see Fig. 128).
Phase Transition
Generally refers to the change from a crystalline to a liquid-like state in membranes and membrane-mimetic systems. The temperature (or the range of temperatures) at which the crystalline phase is converted to the liquid phase is referred to as the phase transition temperature.
Polymeric Monolayers
Monolayers prepared either from pre-polymerized surfactants or from polymers.
Polymeric Vesicles
Vesicles prepared from pre-polymerized surfactants.
Polymerized Vesicles
SUVs or MLVs formed from monomeric surfactants containing polymerizable moieties either in their headgroups or in their bilayers which are subsequently polymerized. SUVs prepared from surfactants which contain polymerizable headgroups can be polymerized both at the inner and at the outer surfaces or selectively at one of these surfaces.
Quantum Crystallites
Nanosized particles which exhibit three-dimensional size quantization, i.e. particles in which the movement of the exciton is restricted in all three directions.
Quantum Dots
Nanosized particles which exhibit three-dimensional size quantization, i.e. particles in which the movement of the exciton is restricted in all three directions.
Quantum Wells
Nanosized particles which exhibit one-dimensional size quantization, i.e. particles in which the exciton is free to move in only two dimensions.
Quantum Well Wires
Nanosized particles which exhibit two-dimensional size quantization, i.e. particles in which the exciton has only one-dimensional mobility.
Quasi-metallic(Silver) Particles
Silver particles in the 100Å $> D >$ 20 Å size domain having their absorption bands blue shifted from the plasmon band position in the bulk metal.
Reversed Micelles
Surfactant aggregates in non-polar solvents. Their formation requires water; thus, they can be considered to be surfactant-entrapped water pools in non-polar solvents.
Self-assembled(SA) Films
Monomolecularly thick films are spontaneously formed upon the immersion of a solid substrate (most commonly gold) into a surfactant solution (most commonly thiols).

Self-assembled(SA) Multilayers
Multilayers formed upon the covalent or non-covalent attachment of subsequent layer(s) of surfactants to the SA film.
Self-organized Materials
Materials capable of spontaneously associating to larger units.
Semiconductor Particles
Separate nanometer- to micron-sized colloidal semiconductors dispersed in solutions; they are usually stabilized by polyions or polymers.
Semiconductor Particulate Films
Physically interconnected colloidal semiconductor particles supported by a monolayer, a BLM, or a solid substrate.
Semiconductors
Bulk crystalline or amorphous solid-state materials whose conductivity is intermediate between metals and insulators and whose resistance decreases with increasing temperature. The valance band of an undoped semiconductor is completely filled, whereas its conduction band is empty. The energy difference between the valence and conduction bands (the band-gap) defines a semiconductor (see Fig. 95).
Size-Quantized Materials
Materials whose dimensions (typically in the 20–80 Å range for semiconductor particles) are comparable to the length of the Broglie electron, the wavelength of the Broglie electron, the wavelength of phonons, and the mean free paths of excitons. Size quantization can occur in one, two, or three dimensions and manifests in altered optical, electronic, and chemical properties.
Small Unilamellar Vesicles (SUVs)
Closed, spherical, single-bilayer, 300- to 600-Å-diameter surfactant and/or phospholipid aggregates dispersed in aqueous solutions. Ultrasonic dispersal of multilamellar vesicles (MLVs) or employing procedures such as French Press filtration result in SUV formation.
Smart Materials
Man-made materials which can, to different degrees, self assemble, self diagnose, self repair, and recognize and discriminate physical and/or chemical stimuli, and, at the extreme, which have the capability of learning and self-replicating. The term is sometimes used synonymously with "adapted materials" and "intelligent materials".
Sols
Small particles (10^{-6} to 10^{-8} m range) dispersed in a liquid (hydrosols in water, organosols in an organic solvent) or gaseous (aerosol) medium.
Superconductors
Superconductors have infinite d.c. conductivity (i.e. zero resistivity) but display normal resistance above a certain frequency. Superconductors are perfectly diamagnetic and totally exclude the magnetic flux.
Supramolecular Materials
Synthetic or naturally occurring structures generated by the association of smaller units; e.g. ferritin, whose quaternary structure is hollow sphere, is the product of the association of 24 subunits.
Surfactants
Molecules possessing distinct hydrophobic (water-repelling) and hydrophilic (water-attracting) regions. The terms "surface-active agents", "detergents", and "amphiphiles" are often used synonymously with the term "surfactants".
Two-dimensional Excitons
Nanosized particles which experience one-dimensional size quantization, i.e. particles in which the excitons are free to move only in two dimensions (quantum wells, for example).
Zero-Dimensional Excitons
Nanosized particles which experience three-dimensional size quantization, i.e. particles in which the movement of the exciton is restricted in all three directions.

Appendix

An attempt has been made, in the main body of the monograph (Chapters 3–7), to cite primary publications exhaustively and to provide ample references to secondary and tertiary sources. No convenient time exists, of course, for terminating the review of any scientific activity, let alone one which is focused upon a vigorously emerging new discipline. Indeed, research in all areas related to the "Membrane-Mimetic Approach to Advanced Materials" has continued to be pursued at an exponential rate since the submission of the present manuscript to the publisher (February, 1993). The purpose of this appendix is to highlight the most significant publications which appeared during the first nine months of 1993. They will be listed according to the sections which were used in the monograph. Advantage will be taken of this opportunity to include a few important earlier references which had been overlooked. Once again, the timely delivery of the page-proofs and the appendix to the publisher limits the additions. It is my hope, however, to produce a revised and updated version of this monograph within the next couple of years. In the meanwhile, I am looking forward to seeing the burgeoning of advanced materials research and to being inspired by the highly innovative publications which will undoubtedly appear in the open literature. I shall welcome preprints and reprints from scientists who are active in the area, as well as their constructive criticism.

Chapter 2.1

Surface pressure vs surface area isotherms have been investigated for stable Langmuir-films formed from soluble (*Bombyx mori*) silk [1].

Chapter 2.2

FTIR investigations of LB films have been reviewed [2].

Chapter 2.3

Self assembly of non-centrosymmetric hybrid superlattices containing metal ions and unidirectionally aligned dipolar organic chromophores as intercalated guests [3] and zirconium octadecylphosphonate LB films [4] have been described. Patterns of self-assembled monolayers have been imaged by scanning electron microscopy [5, 6].

Chapter 2.6

Formatiion of rods, tubules, and ribbons from micelles and vesicles and their characterization in both the fluid and solid states have been surveyed [7].

Chapter 2.7

Converse flexoelectric effects (i.e. voltage-generated curving) have been demonstrated in uranyl-acetate-stabilized phosphatidylserine BLMs by real-time stroboscopic interferometric measurements; the obtained satisfactory agreement between the converse and the direct (i.e. curvature-generated voltage) flexoelectric coefficients have been in accord with the Maxwell relationship [8].

Chapter 2.9

Reports of building up two-dimensional polymers have been published by several research groups [9–12]. Additional reports [13–15] and a review [16] have appeared on stimuli-responsive polymer gels and their application to chemomechanical systems. The preparation and application of new monosized polymer particles have been reviewed [17].

Chapter 2.10

The physical-chemical properties of a synthetic gallophosphate molecular sieve, the 30-Å supercage "cloverite", have been assessed [18]. Instead of attempting to list the burgeoning number of fullerene publications, attention is drawn to the formation and characterization of fullerene-like nanocrystals of tungsten disulfide [19, 20]. Preparation, characterization and utilization of carbon nanotubes have been the subject of a number of reports from several laboratories [21–27].

Chapter 2.11

The different membrane-mimetic compartments have been utilized in a large variety of applications. Possible commercial applications of monolayer and multilayer LB films have been surveyed [28]. They, along with other mimetic systems, have been employed in the construction of sensors and biosensors [29–35], electronic, bioelectronic, and molecular electronic devices [36, 37], and pyroelectric materials [38, 39], as well as having served as templates in two-dimensional protein crystallization [40], as supporting matrices for catalysts and chromatographic materials [41–43], as immobilizing agents for lipids and enzymes on metallic surfaces [44, 45], and as components of solar energy converting devices [46, 47].

Chapter 3.1

Absorption spectra of Ag^0, and Ag_2^+ have been redetermined by pulse radiolysis [48] and silver particle growth in solution has been summarized [49, 50].

Chapter 3.2

Preparation of non-aqueous dispersions of colloidal silver by phase transfer has been described [51] and advantage has been taken to form monodisperse, 7.0-nm-diameter silver particles by simultaneously reducing Ag^+ and partially oxidizing Ag_n particles (radiolytic "push-pull" reduction method) [52]. The surface chemistry of nanosized silver particles has continued to receive attention [53, 54].

Chapter 4.1

A colloid chemical approach to CdS/HgS/CdS spherical quantum wells was described [79]. Size-dependent third-order non-linear susceptibilities of CdS clusters were investigated [80]. Reviews appeared on size-quantized nanocrystalline semiconductor films [81] and on the quantum size effects and electronic properties of semiconductor microcrystallites [82].

Chapter 4.2

Preparation and chemical and photophysical properties of size-quantized HgS/CdS particles were reported [83, 84]. Single-crystal germanium quantum wires [85], CdTe [86], and HgTe [86] were prepared in solution. Size and monodispersity control of ultrasmall CdS particles were achieved by thiol capping [87] and by size selective precipitation [88]. Preparation, TEM, and X-ray scattering analysis of a glassy network of CdSe nanocrystallites connected by molecular bridges were reported [89].

Table 12. Metallic and catalytic particles and particulate films in membrane-mimetic compartments

Membrane-mimetic compartment	Incorporated particles	Comments	Reference
AOT-isooctane-H_2O reversed micelles	150 Å Cu particles	Mixing Cu-AOT $\{[Cu(AOT)_2] = 10^{-3}$ M in [AOT] $= 0.25$ M$\}$ with NaBH$_4$-AOT $\{[NaBH_4] = 2 \times 10^{-3}$ M in [AOT] $= 0.25$ M$\}$ in isooctane led to Cu-particle formation (monitored by absorption spectra and TEM). Particle diameters did not change greatly with the water content in the reversed micelles; but at low $[H_2O]$ to [AOT] ratios isolated clusters formed, while at higher ratios particles became more aggregated	55
Neutral diazafluorenone Schiff-base functionalized monolayers floating on aqueous silver nitrate solution	Ag	Dendritic silver particulate films generated electrochemically at the monolayer surface	56
Poly(N-vinyl-2-pyrrolidone), PVP	Bimetallic Ag-Pd particles	Photo reduction of aqueous $(0.1-3.0)10^{-3}$ M Ag$_2$ [Pd(C$_2$O$_4$)$_2$] in the presence of 5.0×10^{-3} M PVP resulted in bimetallic (not uniform) Ag-Pd particles	57
Poly(N-vinyl-2-pyrrolidone), PVP	Bimetallic Pd-Pt particles	Refluxing mixed solutions of palladium(II) chloride and hexachloroplatinic(IV) acid in ethanol-water (1:1, v/v) in the presence of PVP resulted in the formation of well-dispersed, stable, polymer-protected Pd-Pt particles	58
Porous alumina	Au/composites	Nanosized gold and oriented gold particles were prepared by an electrochemical method	59
Microdomains of block copolymer/ homopolymer blends	25-Å-diameter Pd clusters	Metal-ion precursors, introduced into cast thin films of polymer microdomains, are reduced by high pressure hydrogen	60, 61
Nafion	7.4-nm-diameter Ag particles	Silver ions are spontaneously reduced in basic air-saturated solutions of propanol in the presence of powdered Nafion	62

Material	Product	Method	Ref.
Poly(tetrafluoroethylene), PTFE	Highly ordered, perfectly aligned copper particles	Rubbing a silicon wafer with a PTFE rod across one direction resulted in the deposition of highly ordered and aligned thin strings of PTFE which nucleated the deposition of Cu From a butanolic $Cu(NO_3)_2$ solution	63
Poly(vinylpyrrolidine), PVP	Nanoscale bimetallic Pd-Cu	Refluxing mixtures of palladium acetate and copper acetate in 2-ethoxyethanol in the presence of PVP for two hours led to Pd-Cu	64
Carbon nanotubes	Cu	Copper-doped carbon nanotubes were prepared by the carbon-arc method in the presence of copper	65
Rhapidosomes (proteinaceous tubular microstructures)	Ni	Electroless deposition led to the metallization of the nanotubules	66
Carbon nanotubes	Pb	Annealing of the tubules in the presence of liquid lead resulted in the opening of the capped tubules and in the subsequent filling of the tubules with molten materials through capillary action	67
Fullerenes	Fullerene-Rh nanocomposites	Soultions of mixed fullerene extract (a commercial mixture of C_{60} & C_{70}) and $[1,5-COD)RhCl]_2$ in toluene were atomized and passed through a hot-wall reactor to give fullerene/Rh nanocomposites	68
Polyethylene thin films and C_{60}-polyethylene thin films	Au	Finely divided 2.0- to 5.0-nanometer Au particles were prepared in an ionized-cluster-beam time-of-flight mass spectrometer system	69
Zeolites	Ag	Ag_n^+ (n = 5–13) was generated by the photolysis of Ag ions entrapped in the zeolite supercages	70
Montmorillonite (clays)	Pd	Reduction of Pd(II) complexes intercalated in montmorillonite afforded highly dispersed Pd crystallites	71
Clays	Ni, Cu	Reduction of metal oxides, intercalated between the clay layers (pillared clays), led to metal-intercalated clay nanocomposites	72

Table 12. Contd.

Membrane-mimetic compartment	Incorporated particles	Comments	Reference
Tetraethoxysilane (TEOS)	Au	Dipping glass substrates into a mixture of TEOS:H_2O:HCl:$HAuCl_4$ at a constant rate of 3 cm/min, subsequent to 10 minutes of heating at 500°C (sol-gel process), led to Au particles whose sizes (80 Å to 340 Å) depended on the composition of the mixture	73
Titania and silica glass thin films	Au, Pt	Photoreduction of $HAuCl_4$ and K_2PtCl_4 in ethanol-water in the presence of poly(N-vinyl-2-pyrrolidone) or poly(methyl vinyl ether led to metal particles (sizes depended on solvent composition; the smallest, 2.8 nm in diameter, was obtained in 100% alcohol) which were mixed with Ti(i-OC_3H_7)$_4$ and acetylacetone under N_2. Subsequent to 30 minutes of stirring, exposure to moisture produced TiO_2-embedded metal particles	74
Ormosil (silanol terminated polydimethyl siloxane)	Ag, Au, Cu, Pt	Metal-cluster-modified ormosil was prepared by the sol-gel process	75
Xerogels	Ag, Au	Two steps: (1) gamma radiation initiation of metal-particle growth in a solution containing metal ions and multifunctional silanes; (2) xerogels with metal clusters grafted on an oxide network prepared via hydrolysis and condensation	76
SiC, titania, alumina and polymer particles	Cu, Ni	Submicrometer ceramic powders were coated by electroless deposition	77
Single-compartment egg yolk phosphatidylcholine vesicles	Au	Nanosized Au particles precipitated in the vesicles	78

Table 13. Semiconductor particles and particulate films in membrane-mimetic compartments

Membrane-mimetic compartment	Incorporated semiconductor(s)	Comments	Reference
LB films formed from cadmium icosanoate and nonacosa-10,12-diynoate	CdS	Mass changes accompanying the exposure of cadmium icosanoate and nonacosa-10,12-diynoate to H_2S were followed by an electrochemical quartz crystal microbalance	93
LB films formed from fluoroalkylacrylates and their copolymers	Ag/GaAs	Ag/LB film/GaAs structures were formed and their photogalvanic properties were examined	94
Mono- and multiparticulate CdS layers	CdS	Monoparticulate layers of dodecylbenzenesulfonic-acid-stabilized CdS nanoclusters were constructed, sequentially transferred to solid substrates by the LB technique, and characterized	95
Cadmium lauryl sulfate micelles	CdS	Photoelectron transfer from nanosized CdS to viologens in aqueous micelles was investigated	96
Na-Cd-AOT isooctane H_2O reversed micelles	CdS	Mixed reversed micelles favored the formation of monodisperse CdS particles	97
Sodium hexametaphosphate AOT isooctane H_2O reversed micelles	CdS	Photoluminescence quenching of thiophenol-capped, sizequantized CdS (by KI and methylviologen) is examined	98
AOT heptane H_2O reversed micelles	CdS	CdS cluster growth by picosecond spectroscopy is examined	99
Polyethyleneimine functionalized by N-(2-carboxyethyl)-N'-methyl-4,4'-bipyridinium, PEI-MVP^{2+}	TiO_2	Effective photoelectron transfer was observed	100
Polyvinylcarbazole, PVK	CdS	Photoconductivity of PVK doped by CdS nanoclusters was investigated	101, 102
Polyaniline films	TiO_2	Light images were formed in TiO_2-containing polyaniline films	103

Table 13. Contd.

Membrane-mimetic compartment	Incorporated semiconductor(s)	Comments	Reference
Methyltetracyvlododecene and bTan(ZnPh$_2$) [where bTan = 2,3-trans-bis(ter-butylimido)methyl-nornborn-5-ene] block copolymers	ZnS	Clusters of 30-Å ZnS were generated in the organometallic domains and were characterized	104
Calixarene	CdS	Size-quantized (36-Å-diameter) CdS was generated in calixarenes	105, 106
Porous glass	CdS, CdS$_x$Se$_{1-x}$	Electro-optical effects (Stark effects) of CdS nanocrystallites were investigated	107, 108, 109
Porous silica glass (formed by sol-gel method)	Zn$_{1-x}$Mn$_x$S magnetic semiconductor clusters	Three-dimensionally confined Zn$_{1-x}$Mn$_x$S magnetic semiconductor clusters were prepared	110
Zeolites	TiO$_2$	Preparation, characterization, and photoreactivity of titanium (IV) oxide encapsulated inside sodium or ammonium exchanged zeolite, mordenite, or potassium zeolites were described	111
Polyvinylidenefluoride (70%)-polymethylmethacrylate (30%) blend	TiO$_2$	Controlled hydrolysis of metal alkoxides in a polymeric matrix led to TiO$_2$ nanoparticles	112
Polysilsesquioxane xerogels	CdS	Size-quantized CdS was grown in the xerogels	113
Sol-gel processed TiO$_2$, doped by organic compounds having non-linear optical properties	TiO$_2$	Acidic TiO$_2$ gels, doped by 4-(dimethylamino)-4′-nitrosostilbene and 4-(2-(4-hydroxyphenyl)ethenyl)-N-methyl-pyridinium iodide, were spread on a spinning ITO glass plate. Another ITO plate was placed on the field and an electric field was applied prior to drying in a vacuum. Second-order optical activity was observed	114
Xerogels	Cds, ZnS, and PbS	Nanosized semiconductor particles prepared in Xerogels	115

Sol-gel processing	Cds and CdS-PbS sandwich semiconductors	Semiconductor clusters were prepared via multifunctional inorganic-organic sol-gel processing. The reaction of hexamethyldisilylthiane with metal alkoxide (in THF + alcohol) produced silane functionalized metal sulfide clusters and inorganic/organic network formers. Hydrolysis and condensation produced viscous liquids from which thin films or monoliths were prepared. Crosslinking at $T < 100\,°C$ resulted in materials of variable spectral response, thickness, and optical density	116
Ormosil	CdS	An organically modified silicate (ormosil) containing 28 wt % polydimethylsiloxane and 72 wt % silica was used as a matrix to fabricate CdS-doped glassy nanocomposites by the sol-gel method	117
Au-sputtered metal membranes	CdSe, CdTe	Arrays of semiconductor microcylinders were electrodeposited in the 2000-Å-diameter pores of the Au-treated metal membrane	118

Table 14. Conducting Langmuir-Blodgett films

Preparation	Characterization	Results	Reference
Potassium-doped LB film prepared from 50 layers of fullerene (C_{60}) and deposited on a poly(ethylene terephtalate) substrate	Microwave absorption or low-field signal	Superconductivity was observed at or below 8.1 K	121
LB film prepared from stearic acid and containing TBTTF-Zn(dmit)$_2$ charge transfer complex (and oxidized by iodine exposure)	Absorption, FTIR spectroscopy, lowangle X-ray diffraction, conductivity measurements	Maximum conductivities perpendicular and parallel to the LB film were determined to be 10^{-3} S cm^{-1} and 4.3×10^{-1} S cm^{-1}	122
LB films prepared from iron(III) stearate, transferred to solid substrates, dried, exposed to hydrochloric acid in a desicator, and subsequently exposed to pyrrole vapor	Absorption spectroscopy, scanning electron microscopy, and conductivity measurements	Lateral conductivity as high as 1.25 S cm^{-1} was measured	123

Table 15. Magnetic particles and particulate films in membrane-mimetic compartments

Membrane-mimetic compartment	Incorporated particles	Comments	Reference
LB films prepared from stearate ions	20-and 70-Å α-Fe$_2$O$_3$ particles were incorporated between the headgroups of LB films	Molecular orientation and structure was investigated by FTIR absorption and linear dichroic measurements	124
Tubules prepared from phospholipids	Ni-coated tubules	Saturation magnetization was determined	125

Table 16. Ceramic materials in membrane-mimetic compartments

Membrane-mimetic compartments	Incorporated materials	Comments	Reference
Unilamellar vesicles prepared from mixtures of cetyl trimethylammonium tosylate and sodium dodecyl benzene sulfonate	δ-Al_2O_3 prepared in vesicles	Morphologies and sizes of δ-Al_2O_3 formed in vesicles were investigated by TEM and X-ray diffraction measurements; use of vesicles as nanoreactors for the precipitation of ceramic particles was examined	129
Polyacrylic acid	Hybrid organic/inorganic network polymers were formed via the reaction of polyacrylic acid with tin(IV), titanium(IV), and silicon (IV) alkoxides and subsequent hydrolysis to form mesoporous materials. Treatment by nitric acid removed the polyacrylate template and produced microporous inorganic hydrous metal oxides	Surface areas characterized by BET measurements	130

Chapter 4.3

Surface effects on the properties of cadmium selenide quantum dots were investigated [90].

Chapter 4.4

Platinized and sensitized (by ruthenium polypyridyl complexes) layered alkali-metal titanates, niobates, and titaniobates were used as photocatalysts for H_2 and I_3^- production [91]. The use of reversed micelles as microreactors was reviewed in a feature article [92].

Chapter 5

Intercalation of polyaniline between MoS_2 layers produced nanoscale molecular composites with unusual charge transport properties [119]. Recent advances in the preparation, characterization, and utilization of conducting polymers intercalated into layered solids were surveyed [120].

Chapter 7

Reviews appeared on the synthesis of nanocomposite organoceramics [126], on hybrid nanocomposite glass-organopolymers [127], and on the sol-gel processing of transition-metal alkoxides for electronics [128].

References

1. Muller WS, Samuelson LA, Fossey SA, Kaplan DL (1993) Langmuir 9: 1857
2. Takenaka T, Umemura J (1991) In: Durig JR (ed) Vibrational spectra and structure. Elsevier, Amsterdam, vol 19
3. Maoz R, Yam R, Berkovic G, Sagiv J (1993) In: Ulman A (ed) Organic thin films and surfaces. Academic, Cambridge, vol 1
4. Byrd H, Pike JK, Talham DR (1993) Chem Mater 5: 709
5. Lópex GP, Biebuyck HA, Whitesides GM (1993) Langmuir 9: 1513
6. Wollman WE, Frisbie CD, Wrighton MS (1993) Langmuir 9: 1517
7. Fuhrhop J-H, Helfrich W (1993) Chem Rev 93: 1565
8. Todorov AT, Petrov AG, Fendler JH (1994) J Phys Chem, in press
9. Stupp SI, Son S, Lin HC, Li LS (1993) Science 259: 59
10. Lefevre D, Porteu F, Balog P, Roulliay M, Zalczer G, Palacin S (1993) Langmuir 9: 150
11. Porteu FI, Palacin S, Ruaudel-Teixier A, Barraud A (1992) Thin Solid Films 210/211: 769
12. Porteu F, Palacin S, Ruaudel-Teixier A, Barraud A (1991) Makromol Chem Macromol Symp 46: 37
13. Okuzaki H, Osada Y (1993) Journal of Intelligent Material Systems and Structures 4: 50
14. Smela E, Inganäs O, Pei Q, Lundström I (1993) Adv Mater 5: 630
15. Osada Y, Ross-Murphy SB (1993) Scientific American 82

16. Osada Y (1993) Prog Polym Sci 18: 187
17. Ugelstad J, Berge A, Ellingsen T, Schmid R, Nilsen T-N, Mork PC, Stenstad P, Hornes E, Olsvik O (1992) Prog Polym Sci 17: 87
18. Bedard RL, Bowes CL, Coombs N, Holmes AJ, Jiang T, Kirkby SJ, Macdonald PM, Malek AM, Ozin GA, Petrov S, Plavac N, Ramik RA, Steele MR,Young D (1993) J Am Chem Soc 115: 2300
19. Tenne R, Margulis L, Genut M, Hodes G (1992) Nature 360: 444
20. Tenne R, Margulis, L, Hodes G (1993) Adv Mater 5: 386
21. Iijima S, Ichihashi T (1993) Nature 363: 602
22. Li ZG, Fagan PJ, Liang L (1993) Chem Phys Lett 207: 148
23. Wallenberger FT, Nordine PC (1993) Science 260: 66
24. Tsang SC, Harris PJF, Green MLH (1993) Nature 362: 520
25. Ajayan PM, Ebbesen TW, Ichihashi T, Iijima S, Tanigaki K, Hiura H (1993) Nature 362: 522
26. Ruoff RS, Tersoff J, Lorents DC, Subramoney S, Chan B (1993) Nature 364: 514
27. Bethune DS, Kiang CH, de Vries MS, Gorman G, Savoy R, Vazquez J, Beyers R (1993) Nature 363: 605
28. Petty MC (1992) Thin Solid Films 210/211: 417
29. Steizle M, Weissmuller G, Sackmann E (1993) J Phys Chem 97: 2774
30. Pearson C, Gibson JE, Moore AJ, Bryce MR, Petty MC (1993) Electronics Lett 29: 1377
31. Zhu DG, Petty MC, Harris M (1990) Sensor and Actuators B 2: 265
32. McCoy CH, Wrighton MS (1993) Chem Mater 5: 914
33. Sauer Th, Caseri W, Wegner G, Vogel A, Hoffmann B (1990) J Phys D: Appl Phys 23: 79
34. Zhu D-G, Cui D-F, Petty MC (1993) Sensors and Actuators B 12: 111
35. Tien HT (1990) Adv Mater 2: 316
36. Azumi R, Matsumoto M, Kawabata Y, Kuroda S, Sugi M, King LG, Crossley MJ (1992) J Am Chem Soc 114: 10662
37. Peterson IR (1992) In: Göpel W, Ziegler Ch (eds) Nanostructures based on molecular materials. VCH Verlag, Weinheim, Germany
38. Petty M, Tsibouklis J, Davis F, Hodge P, Petty MC, Feast WJ (1992) J Phys D: Appl Phys 25: 1032
39. Jones CA, Petty MC, Roberts GG (1988) IEEE Transactions on Ultrasonics, Ferroelectrics, and Frequency Control 35: 736
40. Yoshimura H, Matsumoto M, Endo S, Nagayama K (1990) Ultramicroscopy 32: 265
41. Murrell LL (1977) In: Burton JJ, Garten RL (eds) Advances in materials in catalysis. Academic, New York, chap 8
42. Gokak DT, Ram RN (1989) J Mol Catal 49: 285
43. Hernan P, del Pino C, Ruiz-Hitzky E (1992) Chem Mater 4: 49
44. Kallury KMR, Lee WE, Thompson M (1992) Anal Chem 64: 1062
45. Brennan JD, Brown RS, Foster D, Kallury KMR, Krull UJ (1991) Anal Chim Acta 255: 73
46. O'Regan B, Grätzel M (1991) Nature 353: 737
47. Fujihira M (1992) In: Göpel W, Ziegler Ch (eds) Nanostructures based on molecular materials. VCH Verlag, Weinheim, Germany
48. Ershov BG, Janata E, Henglein A, Fojtik A (1993) J Phys Chem 97: 4589
49. Henglein A (1993) J Phys Chem 97: 5457
50. Ershov BG, Janata E, Henglein A (1993) J Phys Chem 97: 339
51. Hirai H, Aizawa H, Shiozaki H (1992) Chem Lett 1527
52. Gutiérrez M, Henglein A (1993) J Phys Chem 97: 11368
53. Vukovic VV, Nedeljkovic JM (1993) Langmuir 9: 980
54. Linnert T, Mulvaney P, Henglein A (1993) J Phys Chem 97: 679
55. Lisiecki I, Boulanger L, Lixon P, Pileni MP (1992) Progr Coloid Polym Sci 89: 103
56. Tai Z, Zhang G, Qian X, Xiao S, Lu Z, Wei Y (1993) Langmuir 9: 1601
57. Torigoe K, Esumi K (1993) Langmuir 9: 1664
58. Toshima N, Yonezawa T, Kushihashi K (1993) J Chem Soc Faraday Trans 89(4): 2537
59. Foss Jr CA, Hornyak GL, Stockert JA, Martin CR (1993) Mat Res Soc Symp Proc 286: 431
60. Chan YNgC, Schrock RR (1993) Chem Mater 5: 566
61. Chan YNgC, Schrock RR, Cohen RE (1992) J Am Chem Soc 114: 7295
62. Huang Z-Y, Mills G, Hajek B (1993) J Phys Chem 97: 11542
63. Kuipers EW, Doornkamp C, Wieldraaijer W, van den Berg RE (1993) Chem Mater 5: 1367
64. Bradley JS, Hill EW, Klein C, Chaudret B, Duteil A (1993) Chem Mater 5: 254

65. Hwang J-H, Hsu W-K, Mou C-Y (1993) Adv Mater 5: 643
66. Pazirandeh M, Baral S, Campbell JR (1992) Biomimetics 1: 41
67. Ajayan PM, Iijima S (1993) Nature 361: 333
68. Gurav AS, Duan Z, Wang L, Hampden-Smith MJ, Kodas TT (1993) Chem Mater 5: 214
69. Xue ZQ, Liu WM, Liu YW, Gao HJ, Zhao XY, Wu QD, Pang SJ, Zhu C, Ma Z, Shen J (1993) Thin Solid Films 228: 304
70. Ozin GA, Hugues F, Mattar SM, McIntosh DF (1983) J Phys Chem 87: 3445
71. Crocker M, Buglass JG, Herold RHM (1993) Chem Mater 5: 105
72. Malla PB, Komarneni S (1993) Mat Res Soc Symp Proc 286: 323
73. Kozuka H, Sakka S (1993) Chem Mater 5: 222
74. Ohtaki M, Ohshima Y, Eguchi K, Arai H (1992) Chem Lett 2201
75. Li C-Y, Tseng JY, Lechner C, Mackenzie JD (1992) Mat Res Soc Symp Proc 272: 133
76. Gacoin T, Chaput F, Boilot JP, Jaskierowicz G (1993) Chem Mater 5: 1150
77. Garg AK, De Jonghe LC (1993) J Mat Sci 28: 3427
78. Meldrum FC, Heywood BR, Mann S (1993) J Colloid Interf Sci 66: 161
79. Eychmüller A, Mews A, Weller H (1993) Chem Phys Lett 208: 59
80. Wang Y, Herron N (1992) Internatl J Nonlinear Opt Phys 1: 683
81. Hodes G (1993) Isr J Chem 33: 95
82. Yoffe AD (1993) Adv. Phys 42: 173
83. Eychmüller A, Hässelbarth A, Weller H (1992) J Luminesc 53: 113
84. Hässelbarth A, Eychmüller A, Eichberger R, Giersig M, Mews A, Weller H (1993) J Phys Chem 92: 5333
85. Heath JR, LeGoues FK (1993) Chem Phys Lett 208: 263
86. Müllenborn M, Jarvis Jr RF, Yacobi BG, Kaner RB, Coleman CC, Haegel NM (1993) Appl Phys A 56: 317
87. Nosaka Y, Ohta N, Fukuyama T, Fujii N (1993) J Colloid Interf Sci 155: 23
88. Chemseddine A, Weller H (1993) Ber Bunsenges Phys Chem 97: 636
89. Majetich SA, Carter AC, McCullough RD (1993) Mat Res Soc Symp Proc 286: 87
90. Carter AC, Majetich SA (1993) Mat Res Soc Symp Proc 286: 81
91. Kim YI, Atherton SJ, Brigham ES, Mallouk TE (1993) J Phys Chem 97: 11802
92. Pileni MP (1993) J Phys Chem 97: 6961
93. Furlong DN, Urquhart R, Grieser F, Tanaka K, Okahata Y (1993) J Chem Soc Faraday Trans 89(12): 2031
94. Znamensky DA, Yusupov RG, Mislavsky BV (1992) Thin Solid Films 219: 215
95. Tian Y, Wu C, Fendler JH (1994) J Phys Chem, in press
96. Jain TK, Billoudet F, Motte L, Lisiecki I, Pileni MP (1992) Progr Colloid Polym Sci 89: 106
97. Motte L, Lebrun A, Pileni MP (1992) Progr Colloid Polym Sci 89: 99
98. Chandler RR, Coffer JL (1993) J Phys Chem 97: 9767
99. Barzykin AV, Fox MA (1993) Isr J Chem 33: 21
100. Willner I, Eichen Y, Frank AJ, Fox MA (1993) J Phys Chem 97: 7264
101. Wang Y, Herron N (1992) Chem Phys Lett 200: 71
102. Wang Y, Herron N, Harmer M, Suna A (1992) Mat Res Soc Symp Proc 272: 205
103. Kuwabata S, Takahashi N, Hirao S, Yoneyama H (1993) Chem Mater 5: 437
104. Sankaran V, Yue J, Cohen RE, Schrock RR, Silbey RJ (1993) Chem Mater 5: 1133
105. Chandler RR, Coffer JL, Gutsche CD, Alam I, Yang H, Pinizzotto RF (1992) Mat Res Soc Symp Proc 272: 265
106. Coffer JL, Chandler RR, Gutsche CD, Alam I, Pinizzotto RF, Yang H (1993) J Phys Chem 97: 696
107. Ekimov AI, Efros AlL, Shubina TV, Skvortsov AP (1990) J Luminesc 46: 97
108. Rossman H, Schülzgen A, Henneberger F, Müller M (1990) Phys Stat Sol (b) 159: 287
109. Mei G, Carpenter S, Persans PD (1991) Solid State Comm 80: 557
110. Wang Y, Herron N, Moller K, Bein T (1991) Solid State Comm 77: 33
111. Liu X, Iu K-K, Thomas JK (1993) J Chem Soc Faraday Trans 89(11): 1861
112. Burdon J, Calvert P (1993) Mat Res Soc Symp Proc 286: 315
113. Choi KM, Shea KJ (1993) Chem Mater 5: 1067
114. Nosaka Y, Tohriiwa N, Kobayashi T, Fujii N (1993) Chem Mater 5: 930
115. Gacoin T, Boilot JP, Chaput F, Lecomte A (1992) Mat Res Soc Symp Proc 272: 21
116. Spanhel L, Schmidt H, Uhrig A, Klingshirn C (1992) Mat Res Soc Symp Proc 272: 53
117. Li C-Y, Wilson M, Haegel N, Mackenzie JD, Knobbe ET, Porter C, Reeves R (1992) Mat Res Soc Symp Proc 272: 41

118. Klein JD, Herrick II RD, Palmer D, Sailor MJ, Brumlik CJ, Martin CR (1993) Chem Mater 5: 902
119. Kanatzidis MG, Bessessur R, DeGroot DC, Schindler JL, Kannewurf CR (1993) Chem Mater 5: 595
120. Ruiz-Hitzky E (1993) Adv Mater 5: 334
121. Wang P, Metzger RM, Bandow S, Maruyama Y (1993) J Phys Chem 97: 2926
122. Xiao Y, Yao Z, Jin D (1993) Thin Solid Films 223:173
123. Sarkar D, Paul A, Misra TN (1993) Thin Solid Films 227: 105
124. Yang J, Peng X-G, Zhang Y, Wang H, Li T-J (1993) J Phys Chem 97: 4484
125. Shashidhar R, Ho YS, Baral S, Chow GM, Ratna BR, Chien CL (1993) private communication
126. Messersmith PB, Stupp SI (1992) J Mater Res 7: 2599
127. Novak BM (1993) Adv Mater 5: 422
128. Lee GR, Crayston JA (1993) Adv Mater 5: 434
129. Yaacob II, Bhandarkar S, Bose A (1993) J Mater Res 8: 573
130. Roger C, Hampden-Smith MJ (1992) J Mater Chem 2(10): 1111

Author Index Volumes 101-113

Subject Index

Springer-Verlag
and the Environment

We at Springer-Verlag firmly believe that an international science publisher has a special obligation to the environment, and our corporate policies consistently reflect this conviction.

We also expect our business partners – paper mills, printers, packaging manufacturers, etc. – to commit themselves to using environmentally friendly materials and production processes.

The paper in this book is made from low- or no-chlorine pulp and is acid free, in conformance with international standards for paper permanency.